Field-Effect Transistors Technology

The text provides a comprehensive exploration of the transitions occurring in the field-effect transistor technology, covering the historical evolution, current advancements, and future trends.

Features

- Highlights the significance of field-effect transistors in VLSI and post-complementary metal-oxide-semiconductor design strategies.
- Discusses design challenges, different modeling aspects of field-effect transistors, and emerging materials in semiconductor design.
- Showcases the importance of simulation in forecasting device behavior, enhancing performance, and investigating novel device designs.
- Covers topics such as quantum computing, device simulation process on technology computer-aided design, carbon nanotubes, and organic field-effect transistors.

The text is primarily written for senior undergraduates, graduate students, and academic researchers in the fields of electrical engineering, electrical and communications engineering, materials science, nanoscience, and nanotechnology.

Field-Effect Transistors Technology

From Sustainability to Next-Generation VLSI Design

Edited by
Ashish Raman, Prabhat Singh, Naveen Kumar, and Sarabdeep Singh

CRC Press
Taylor & Francis Group
Boca Raton London New York

CRC Press is an imprint of the
Taylor & Francis Group, an **informa** business

Cover Image: Shutterstock

First edition published 2026
by CRC Press
2385 NW Executive Center Drive, Suite 320, Boca Raton FL 33431

and by CRC Press
4 Park Square, Milton Park, Abingdon, Oxon, OX14 4RN

CRC Press is an imprint of Taylor & Francis Group, LLC

ISBN: 9781032876078 (hbk)
ISBN: 9781032876085 (pbk)
ISBN: 9781003533535 (ebk)

DOI: 10.1201/9781003533535

Typeset in Sabon
by Newgen Publishing UK

I dedicate this book to my beloved mother, Veena Saxena, my father, R. R. Saxena, my wife, Deepti Saxena, and my daughter, Arshika Saxena. Your unwavering love and support have been the cornerstone of my success. I am deeply grateful for everything you have done and continue to do for me. This book is a testament to your belief in me and a symbol of my heartfelt appreciation.

— Dr. Ashish Raman

I dedicate this book to my loving mother Shakuntala Singh, father Dinesh Singh, wife Sadhana Singh and brother Prasoon Singh, as a token of my appreciation for everything you have done and continue to do for me. Your love and support are the foundation of my success, and I am blessed to have you as a part of my life. Your love and belief in me mean everything. This book is dedicated to you, as a symbol of my gratitude for all you've done.

— Dr. Prabhat Singh

This book is dedicated to my beloved wife, Nisha Chaudhary, and our wonderful son, Zishaan Kumar, whose love and encouragement have been my greatest motivation. Special thanks to my parents, Smt. Indu Devi, and Shri. Rajpal Singh, for their unwavering support and sacrifices, and to my brother, Nitish Kumar, for always being there for me. I also extend my heartfelt appreciation to my supervisors, colleagues, and friends, whose support and encouragement have been instrumental in the completion of this work.

— Dr. Naveen Kumar

This book is lovingly dedicated to my dear wife, Manmeet Kour, and our amazing son, Gurshabad Singh, whose love and inspiration have been my greatest driving force. I am profoundly grateful to my parents, Ramnik Singh and Amrit Kaur, for their constant support, selfless sacrifices, and unwavering presence in my life. My deepest appreciation also goes to my mentors, colleagues, and friends, whose invaluable guidance and encouragement have played a crucial role in the successful completion of this work.

— Dr. Sarabdeep Singh

I dedicate this book to my beloved mother Veena Saxena, my father Rakesh Saxena, my wife Deepti Saxena, and my daughter Arshi Saxena. Your unwavering love and support have been the cornerstone of my success. I am deeply grateful for [illegible] you have done and continue to do for me. [illegible] belief [illegible] and a symbol of my [illegible].

— Dr. [illegible] Saxena

I dedicate this book to my loving mother [illegible] Singh, my father [illegible] Singh, wife [illegible] Singh and brother [illegible] Singh, as a token of my appreciation for [illegible] you have done [illegible] Thank you for [illegible] my [illegible] love and belief [illegible]

— Dr. [illegible]

This book is dedicated to my [illegible] and my [illegible] support [illegible]

— Dr. Naveen [illegible]

This book is lovingly dedicated to my [illegible] and [illegible] my parents [illegible] selfless [illegible] My deepest appreciation [illegible] whose invaluable guidance [illegible] have played a crucial role in the successful completion of this work.

— Dr. [illegible]

Contents

About the Editors

Ashish Raman earned his PhD from the Dr. B. R. Ambedkar National Institute of Technology Jalandhar, India, where he is an Associate Professor. Dr. Raman's main areas of interest are electronics devices and circuits, device modeling, device design and simulation, MEMS, high-power devices, etc. He works as a principal investigator and member of various projects funded by the Science and Engineering Research Board (SERB), Ministry of Electronics, IT (MeitY), FIST, ISRO, and many more. He has contributed research articles/papers in SCI, Scopus, and other reputed journals like IEEE Transactions of Electron Devices, IEEE Transactions on Nanotechnology, Elsevier and Springer journals, and at international conferences. Dr. Raman is a member of the IEEE Electron Devices Society, the IEEE Solid-State Circuits Society, and the Institution of Engineers Society, India.

Prabhat Singh earned his PhD from the National Institute of Technology, Hamirpur, Himachal Pradesh, India. Dr. Prabhat Singh is an Assistant Professor in the Indian Institute of Information Technology Senapati, Manipur, India. His research revolves around different semiconductor devices including cryogenic CMOS, ultra-scaled FETs, solar cells, Quantum Dots, etc. and their prospective applications. He is currently working on unravelling the interfacial behavior of different FET-based biological sensors using different classical and semi-quantum analytical models. His main areas of research interest include semiconductor device physics, solid-state devices, analog complementary metal oxide semiconductors (CMOS) integrated circuits, and nanoscale device design and simulation.

Naveen Kumar earned his PhD from the Dr. B. R. Ambedkar National Institute of Technology, Jalandhar, India. Dr. Naveen Kumar is a post-doctoral research associate in the Device Modelling Group in the electronic and nanoscale engineering division, at the University of Glasgow, United Kingdom. His research revolves around different semiconductor devices including ultra-scaled FETs, solar cells, photodiodes, HEMT, quantum dots, etc. and their prospective applications. He is currently working on unravelling the interfacial behavior of different FET-based chemical/biological sensors using different classical and semi-quantum analytical models. His main areas of research interest include semiconductor device physics, MEMS/NEMS, and spintronics.

Sarabdeep Singh earned his BE degree from the Mahant Bachittar Singh College of Engineering and Technology, University of Jammu, India, in 2007, and his MTech degree in VLSI design from the Centre for Development of Advanced Computing, (CDAC) Mohali India, in 2011. He earned his PhD from the Dr. B. R. Ambedkar National Institute of Technology (NIT) Jalandhar in 2021. He served at Chandigarh Group of Colleges (CGC) Landran, Mohali as an Assistant Professor for about 5 years in the department of electronics and communication engineering. He is currently working at the Model Institute of Engineering and Technology Jammu as Assistant Professor in the department of electronics and communication engineering. His current research interests are microelectronics, sensors, and semiconductor devices, focusing on nanowire FETs, IMOS FETs, and negative-capacitance FETs. He has, to his credit, over 20 research papers published in national and international journals, indexed in SCI/SCOPUS, and conference proceedings. He has also guided many students toward their MTech degrees. He has been a reviewer of multiple international peer-reviewed journals.

Contributors

Rachana Arya
ECE Department, Bipin Tripathi Kumaon Institute of Technology
Dwarahat, Uttrakhand, India

K. Baskaran
Alagappa Chettiar Government College of Engineering and Technology
Karaikudi, Tamil Nadu, India

Atanu Chowdhury
Calcutta Institute of Technology Uluberia
Howrah, West Bengal, India

R. S. Gamad
Shri G. S. Institute of Technology and Science
Indore, Madhya Pradesh, India

Abheek Gupta
University of Engineering and Management
Jaipur, Rajasthan, India

Angshuman **Khan**
University of Engineering and Management
Jaipur, Rajasthan, India

Mamta Khosla
National Institute of Technology
Jalandhar, Punjab, India

C.N.S. Vinoth Kumar
College of Engineering and Technology
SRM Institute of Science and Technology
Chennai, Tamil Nadu, India

Harshita Kushwah
Shri G.S. Institute of Technology and Science
Indore, Madhya Pradesh, India

Kaushik Mazumdar
Department of Electronics Engineering
IIT (ISM) Dhanbad, India

Subarna Mondal
National Institute of Technology Durgapur
West Bengal, India

Ayash Kanto Mukherjee
Indian Institute of Technology Patna
Bihta, Patna, Bihar, India

R. Naresh
College of Engineering and Technology
SRM Institute of Science and Technology
Chennai, Tamil Nadu, India

S. Ponmudi
Alagappa Chettiar Government College of Engineering and Technology
Karaikudi, Tamil Nadu, India

Delshi Howsalya Devi. R
Karpaga Vinayaga College of Engineering and Technology
Tamil Nadu, India.

Ashish Raman
National Institute of Technology
Jalandhar, Punjab, India

C. Uma Rani
Alagappa Chettiar Government College of Engineering and Technology
Karaikudi, Tamil Nadu, India

S. Maheswari
Saveetha Institute of Medical and Technical Sciences
Chennai, Tamil Nadu, India

Satyajeet Sahoo
Vignan's Foundation for Science, Technology and Research
Guntur, India

Swastik Kumar Sahu
Department of Electronics Engineering,
IIT (ISM), Dhanbad, India

Samayun Saikh
Indian Institute of Technology Patna
Bihta, Patna, Bihar, India

Aswini **Kumar Samantaray**
Manipal Institute of Technology Bengaluru Guntur
Manipal Academy of Higher Education
Manipal, India

Soumya Sen
National Institute of Technology
Jalandhar, Punjab, India

Sarita Sharma
Arihant College
Indore, Madhya Pradesh, India

S. Sharmila
Karpaga Vinayaga College of Engineering and Technology
Tamil Nadu, India

S. Agnes Shifani
Saveetha Institute of Medical and Technical Sciences
Chennai, Tamil Nadu, India

B.K. Singh
Director
Birla Institute of Applied Sciences
Uttarakhand, India

Mahesh Kumar Singh
National Institute of Technology
Delhi, India

Shriti Suman
National Institute of Technology
Delhi, India

Manisha Tanwer
Dr. D. Y. Patil College of Engineering
Akurdi, Pune, Maharashtra, India

Jasmeet Kaur Tandon
Poonamchand Gupta Vocational College
Indore, Madhya Pradesh, India

Kirti B. Zare
Dr. D. Y. Patil Institute of Engineering
Management and Research
and
D. Y. Patil International University
Akurdi, Pune, Maharashtra, India

Chapter 1

Semiconductors

The Backbone of Modern Technological Advancements

Jasmeet Kaur Tandon and Sarita Sharma

1.1 INTRODUCTION

1.1.1 Context and Importance of Semiconductors in Modern Technology

Semiconductors (Figure 1.1) are the cornerstone of modern electronics, enabling the functionality of virtually all electronic devices, from smartphones and computers to industrial machinery and medical equipment. At the heart of this technology lies the ability of semiconductors to control electrical current, making them essential components in transistors, integrated circuits (ICs), and other electronic devices. The miniaturization and enhancement of semiconductor devices have been key drivers in the evolution of technology, enabling the rapid advancement of computing power, data storage, and energy efficiency [1].

1.2 THE ROLE OF SILICON IN SEMICONDUCTOR DESIGN

For decades, silicon has been the material of choice for semiconductor manufacturing due to its abundant availability, well-understood properties, and excellent performance in a wide range of applications (Figure 1.2). Silicon's ability to form a stable oxide layer (silicon dioxide) is crucial for creating the insulating layers in transistors, which are the fundamental building blocks of most electronic devices. This has allowed silicon-based technologies to scale down transistor sizes following Moore's Law, which predicts the doubling of transistors on a microchip approximately every two years, leading to continuous improvements in processing power and reductions in cost per transistor [2, 3].

DOI: 10.1201/9781003533535-1

Figure 1.1 Semiconductor.

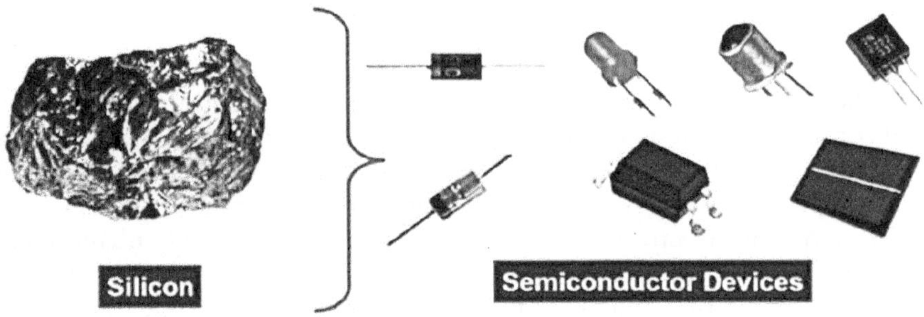

Figure 1.2 An illustration of the role of silicon in semi-conducting devices.

1.3 CHALLENGES FACED BY TRADITIONAL SILICON-BASED SEMICONDUCTORS

However, as the industry approaches the physical limits of silicon-based technology, several challenges have emerged (Figure 1.3):

- **Scaling Limits:** The ongoing miniaturization of silicon transistors is reaching a point where further reductions in size lead to issues like quantum tunneling, where electrons can pass through insulating layers, causing leakage currents and power inefficiencies [4].
- **Performance Limitations:** Silicon's electron mobility, while adequate for many applications, is not optimal for high-speed or high-frequency devices. As the demand for faster and more efficient electronics grows, silicon's limitations in these areas become more pronounced [5].

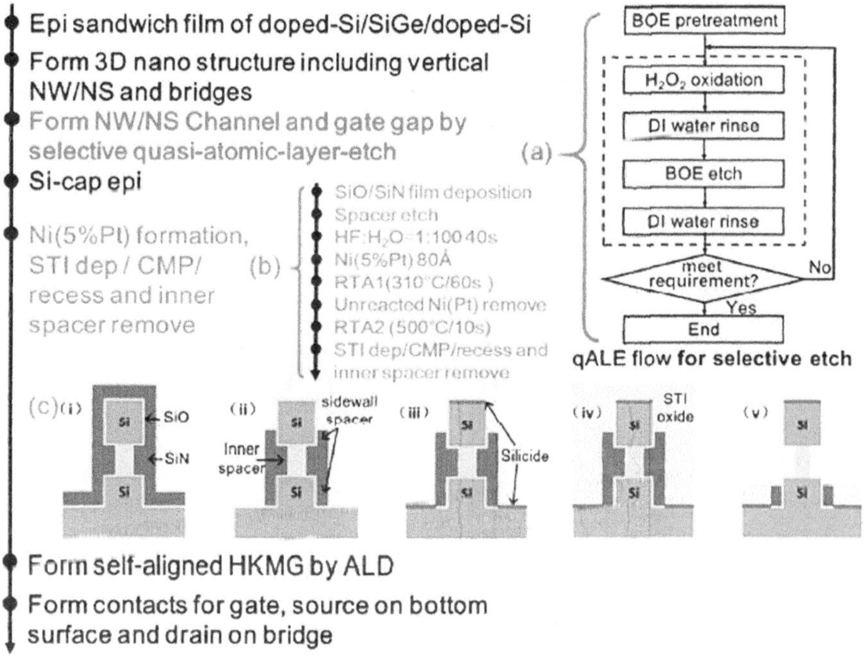

Figure 1.3 Illustration of the challenges faced by traditional silicon-based semiconductors.

- **Heat Dissipation:** As devices become smaller and more powerful, managing heat generation and dissipation becomes increasingly difficult (Figure 1.4). Silicon's thermal conductivity, while reasonable, is not sufficient to efficiently manage the heat produced in high-performance devices [5].
- **Power Consumption:** With the proliferation of portable and wearable electronics, there is a growing demand for materials that can operate at lower power levels without sacrificing performance, an area where silicon is facing increasing pressure (Figures 1.5 and 1.6).

These challenges have prompted extensive research into alternative materials that can overcome the limitations of silicon, leading to the exploration of emerging materials in semiconductor design. These new materials promise to extend the capabilities of electronic devices, enabling further advancements in technology and addressing the growing needs of the modern world [6].

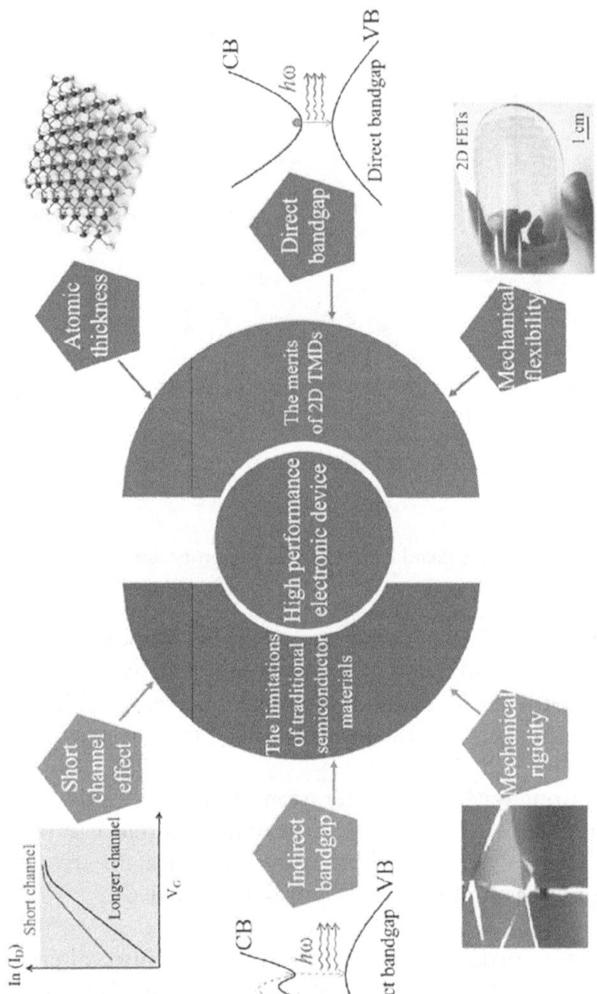

Figure 1.4 Heat consumption.

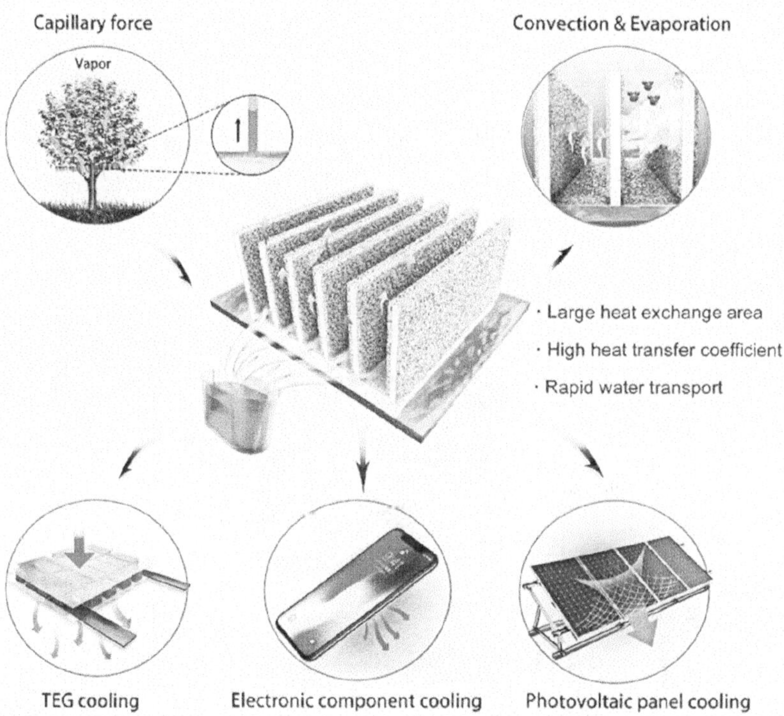

Figure 1.5 Power consumption.

1.4 EMERGING MATERIALS IN SEMICONDUCTOR DESIGN

As the limitations of silicon-based semiconductors become increasingly apparent, the search for alternative materials that can deliver enhanced performance, scalability, and energy efficiency has intensified. Emerging materials in semiconductor design (Figure 1.7) offer promising solutions to the challenges posed by traditional silicon technology, enabling the continued evolution of electronics in the face of these obstacles [7].

1.4.1 Graphene

- Properties: Graphene (Figure 1.8), a single layer of carbon atoms arranged in a hexagonal lattice, has garnered significant attention due to its exceptional electrical conductivity, mechanical strength, and flexibility. Its high electron mobility—up to 200 times that of silicon—makes it an attractive candidate for high-speed transistors and other electronic devices [5].

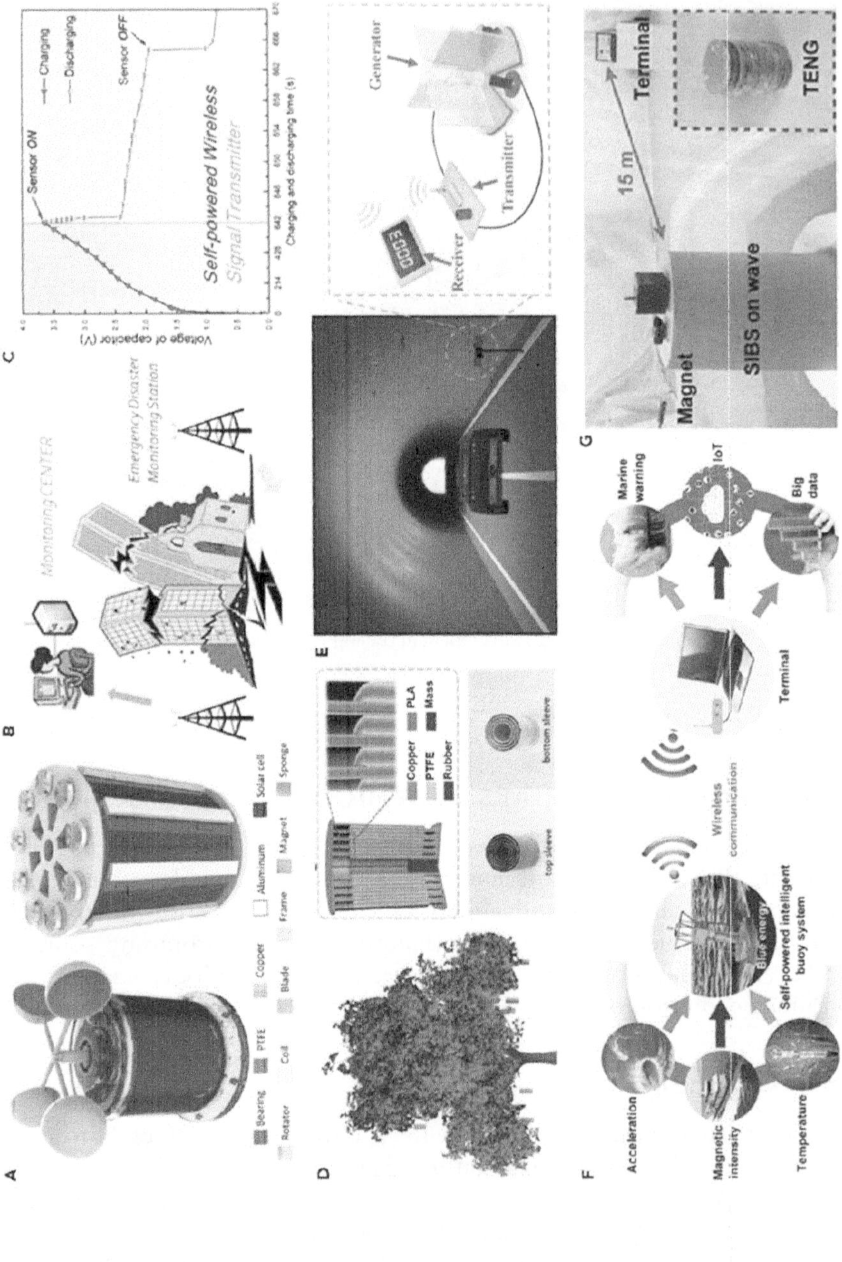

Figure 1.6 Power supply and consumption tools.

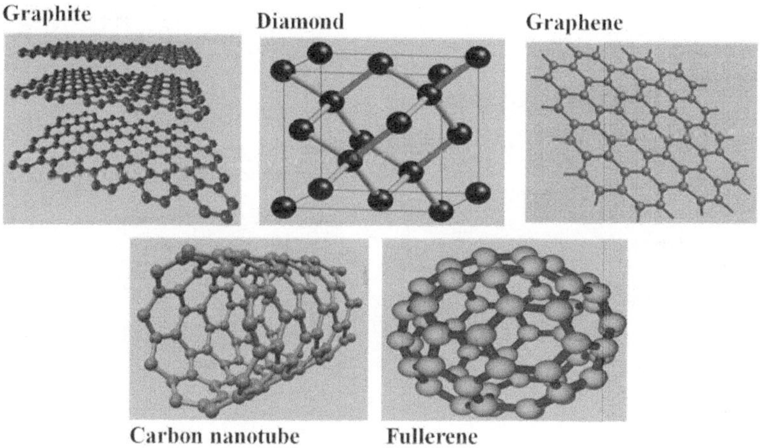

Figure 1.7 Emerging materials in semiconductor design.

- Applications: Graphene is being explored for use in ultra-fast transistors, flexible displays, and advanced sensors [4]. Its potential to enable faster and more efficient electronic components positions it as a leading material for next-generation devices (Figure 1.8).

1.4.2 Transition Metal Dichalcogenides (TDMs)

- Properties: TMDs, such as molybdenum disulfide (MoS_2) and tungsten disulfide (WS_2), are layered materials with unique electronic properties (Figure 1.9). Unlike graphene, TMDs have a natural bandgap, making them suitable for use in field-effect transistors (FETs) and other semiconductor devices [8].
- Applications: TMDs are being investigated for use in low-power, high-performance electronics, flexible and wearable devices, and optoelectronics. Their ability to function effectively in ultra-thin layers offers opportunities for continued device miniaturization.

1.4.3 Silicon Carbide (SiC) and Gallium Nitride (GaN)

- Properties: SiC (Figure 1.10) and GaN are wide-bandgap semiconductors known for their superior thermal conductivity and ability to handle high voltages and temperatures. These properties make them ideal for high-power and high-frequency applications where silicon falls short [9].

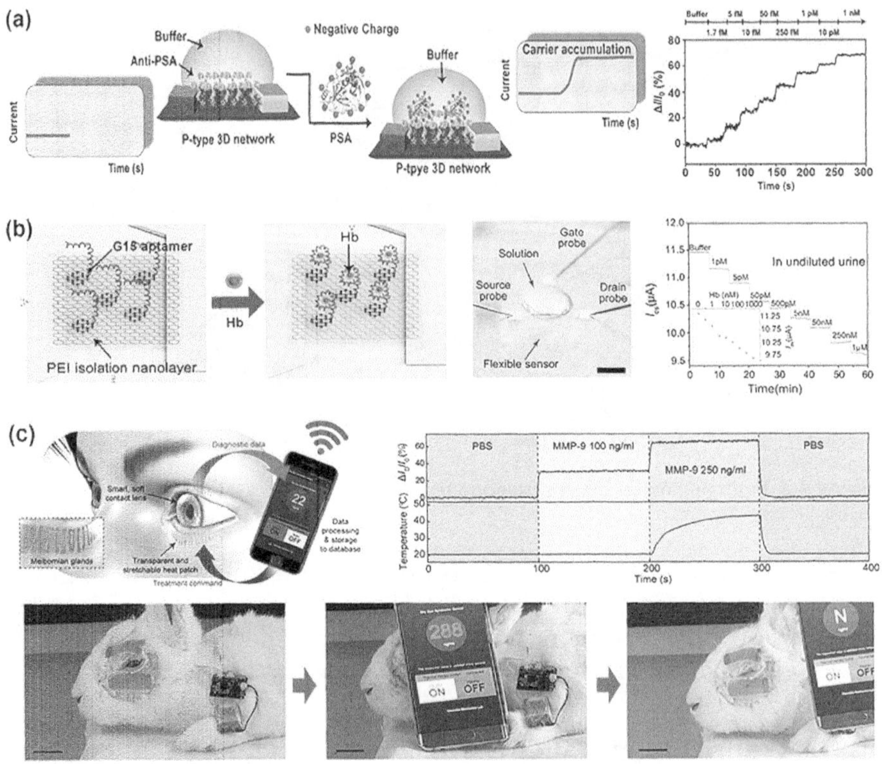

Figure 1.8 Application of graphene.

- Applications: SiC and GaN are increasingly used in power electronics, such as in electric vehicles, renewable energy systems, and 5G communication technologies. Their efficiency in converting and controlling electrical power contributes to significant energy savings and improved performance in demanding environments.

1.4.4 Perovskites

- Properties: Perovskites are a class of materials with a unique crystal structure that has shown promise in optoelectronic applications (Figure 1.11). They exhibit strong light absorption, efficient charge transport, and can be processed at low temperatures, making them cost-effective compared to traditional semiconductors [10].
- Applications: Perovskites are being explored for use in next-generation solar cells, light-emitting diodes (LEDs), and photodetectors. Their

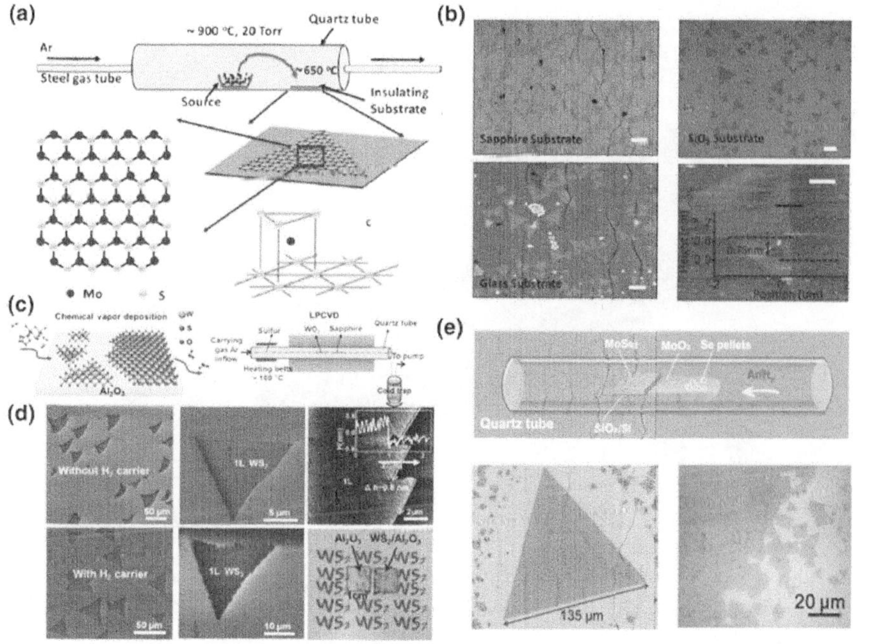

Figure 1.9 Transition metal dichalcogenides.

potential to revolutionize the field of optoelectronics lies in their ability to achieve high performance at a lower cost.

1.4.5 Topological Insulators

- Properties: Topological insulators are materials that behave as insulators in their bulk but have conductive surface states (Figure 1.12). These surface states are protected by the material's topological order, making them resistant to scattering and defects, which is advantageous for electronic applications [11].
- Applications: Topological insulators are being studied for their potential in quantum computing, spintronics, and low-power electronics. Their unique properties offer opportunities to develop devices that leverage quantum effects for enhanced functionality and efficiency.

1.4.6 2D Materials Beyond Graphene

- Properties: In addition to graphene, other two-dimensional (2D) materials like phosphorene (a single layer of black phosphorus) and

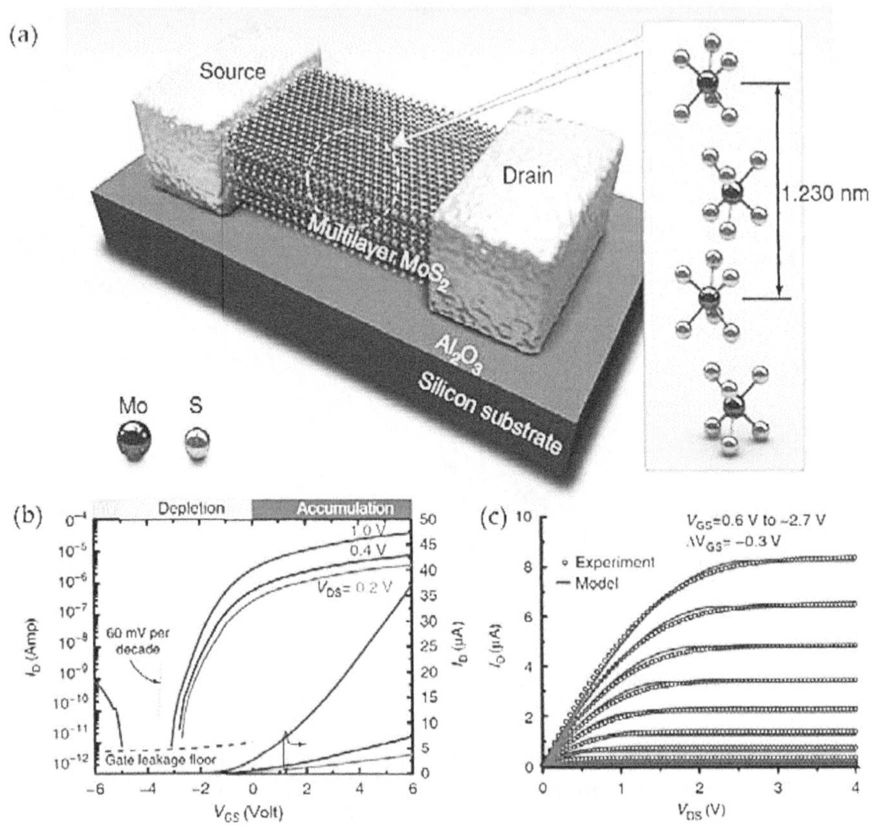

Figure 1.10 Silicon carbide.

borophene (a single layer of boron) are being researched for their distinctive electronic properties, such as tunable bandgaps and high anisotropic conductivity.

- Applications: These 2D materials are being considered for use in transistors, sensors, and flexible electronics. Their tunable properties allow for the design of customized electronic components tailored to specific applications [12] (Figure 1.13).

1.4.7 CONCLUSION OF THIS SECTION

These emerging materials represent a new frontier in semiconductor design, offering solutions to the challenges posed by silicon and enabling the continued advancement of electronic devices. As research progresses, these materials hold the potential to redefine the limits of what is possible

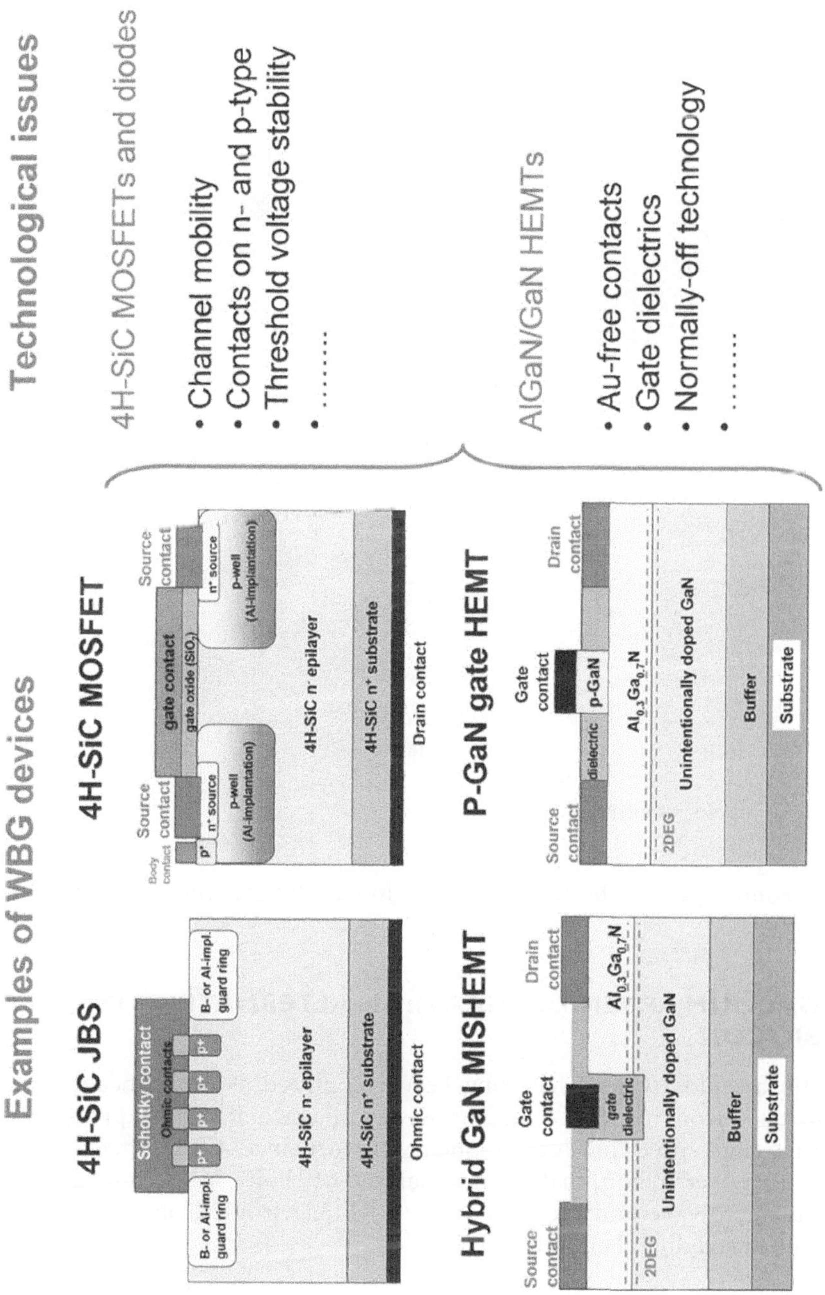

Figure 1.11 Perovskites.

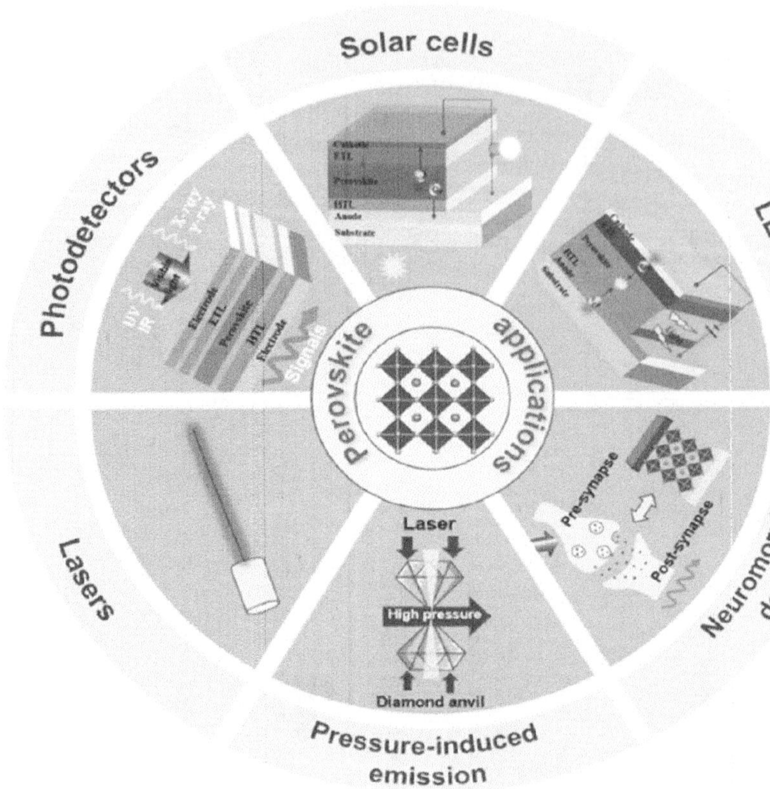

Figure 1.12 Topologies insulators.

in electronics, paving the way for faster, more efficient, and more versatile technologies.

1.5 GALLIUM ARSENIDE (GAAS): AN ALTERNATIVE TO SILICON

Gallium arsenide (GaAs) has long been recognized as a significant alternative to silicon in semiconductor design, particularly for applications requiring high-speed and high-frequency performance. GaAs offers several advantages over silicon, making it a material of choice in specialized fields such as microwave and radio frequency (RF) electronics, optoelectronics, and high-speed digital circuits [13, 14].

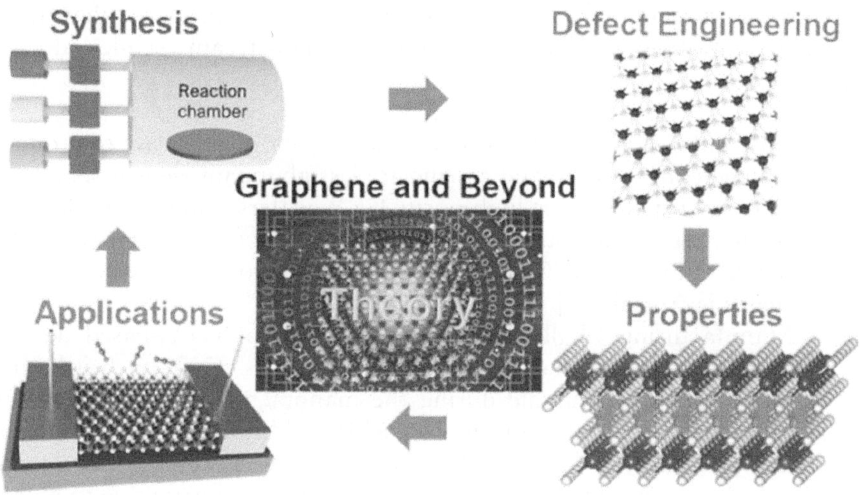

Figure 1.13 2D Materials beyond graphene.

Advantages of GaAs

- Higher Electron Mobility: One of the most notable advantages of GaAs over silicon is its higher electron mobility. GaAs electrons can move more than five times faster than those in silicon, allowing for faster signal processing and higher operating frequencies. This makes GaAs particularly suitable for high-speed devices such as RF amplifiers, microwave integrated circuits (MMICs), and ultra-fast digital circuits.
- Direct Bandgap: GaAs has a direct bandgap, meaning it can efficiently emit and absorb light. This property makes GaAs an ideal material for optoelectronic devices, including light-emitting diodes (LEDs), laser diodes, and solar cells. Silicon, on the other hand, has an indirect bandgap, which makes it less efficient for these applications [14].
- Thermal Stability: GaAs devices can operate at higher temperatures than their silicon counterparts without significant degradation in performance. This makes GaAs a good choice for applications in harsh environments, such as space or military electronics.
- Radiation Resistance: GaAs is less susceptible to radiation damage compared to silicon, which is a crucial advantage in space applications and nuclear environments where radiation can cause significant degradation of electronic components [13].

Limitations of GaAs

Despite its advantages, GaAs also has several limitations that have prevented it from replacing silicon in mainstream semiconductor applications:

- Cost and Availability: GaAs is significantly more expensive to produce than silicon. Gallium and arsenic are less abundant elements, and the process of synthesizing high-purity GaAs wafers is more complex and costly than producing silicon wafers. This cost factor limits the widespread adoption of GaAs in consumer electronics, where cost efficiency is crucial [12].
- Manufacturing Challenges: The fabrication of GaAs-based devices is more complex than that of silicon devices. GaAs is brittle and more difficult to handle during the manufacturing process, leading to higher defect rates and lower yields. Additionally, the equipment and processes used to manufacture GaAs devices are less mature and less standardized than those for silicon, further contributing to higher production costs [3].
- Integration with Existing Technologies: The semiconductor industry has heavily invested in silicon-based manufacturing infrastructure. Integrating GaAs with existing silicon technology requires significant changes in manufacturing processes, which can be prohibitively expensive. Moreover, the lack of compatibility with silicon-based CMOS (complementary metal-oxide-semiconductor) technology limits the ability to combine GaAs with other silicon components on the same chip [4].
- Toxicity: Arsenic, a key component of GaAs, is highly toxic and poses environmental and health risks during the manufacturing process and disposal of GaAs devices. This has led to stringent regulations and safety requirements, further increasing the cost and complexity of working with GaAs [5].

1.5.1 Conclusion of the Section

Gallium arsenide offers substantial performance benefits over silicon, particularly in high-speed, high-frequency, and optoelectronic applications. However, its high cost, manufacturing challenges, and limited compatibility with existing silicon technology have restricted its use to niche markets rather than mainstream semiconductor applications. As a result, while GaAs remains an important material in specific high-performance applications, it is unlikely to replace silicon as the dominant material in semiconductor design [12].

1.6 REVIEW OF RELATED LITERATURE

1.6.1 Fundamental Role of Semiconductors

Semiconductors are pivotal in modern technology, serving as the foundation for a vast array of electronic devices. They are crucial for the operation of transistors, diodes, and integrated circuits, which are integral to computing, communication, and consumer electronics [13].

Moore's Law and Progress

Historical Perspective: According to Moore's Law, the number of transistors on a microchip doubles approximately every two years, leading to exponential increases in computing power and reductions in cost per transistor [9]. This trend has been the driving force behind the rapid advancements in semiconductor technology, enabling more powerful and compact devices [14].

Economic Impact

Market Influence: The semiconductor industry significantly impacts the global economy. It supports a wide range of sectors, including consumer electronics, automotive, telecommunications, and healthcare. The industry's growth is closely tied to technological innovation and consumer demand.

1.6.2 Challenges with Traditional Semiconductor Materials

Silicon Dominance

Performance Limits: Silicon has been the cornerstone of semiconductor technology due to its well-understood properties and manufacturing processes. However, as silicon devices approach physical limits, challenges such as leakage currents, heat dissipation, and scaling issues become more pronounced.

Material Limitations

Physical Constraints: The miniaturization of silicon-based transistors is constrained by fundamental physical limits. Issues such as quantum tunneling and increased power consumption at smaller scales necessitate exploration of alternative materials [10, 14].

1.6.3 Emerging Materials in Semiconductor Design

Graphene

Properties and Applications: Graphene's exceptional electrical, thermal, and mechanical properties make it a candidate for next-generation transistors and flexible electronics [2]. Its high electron mobility and thinness offer potential improvements in performance and miniaturization [12].

Transition Metal Dichalcogenides (TMDs)

Thin-Film Semiconductors: TMDs like MoS_2 and WS_2 exhibit unique electronic properties in monolayer form, including direct bandgaps and high carrier mobility. These properties are beneficial for applications in flexible electronics and low-power transistors [8].

Silicon Carbide (SiC) and Gallium Nitride (GaN)

Wide Bandgap Materials: SiC and GaN are key materials for high-power and high-frequency applications due to their wide bandgaps and thermal stability. They are increasingly used in power electronics and RF applications where traditional silicon devices are inadequate [7, 8].

Perovskites

Optoelectronic Potential: Perovskites have shown promise in photovoltaic applications due to their high power conversion efficiency and tunable bandgap. They are also being explored for use in light-emitting diodes (LEDs) and other optoelectronic devices [9].

Topological Insulators

Unique Electronic Properties: Topological insulators possess conducting surface states with insulating bulk. This property enables novel electronic and spintronic devices with potential applications in quantum computing and advanced sensors [10].

1.6.4 Impact and Future Directions

Technological Advancements

Next-Generation Devices: Emerging materials are expected to drive advancements in semiconductor devices, including faster, more efficient transistors and novel optoelectronic components. These materials offer solutions to the limitations of traditional silicon technology and enable new functionalities [11].

Research and Innovation

Ongoing Research: Continued research is essential for overcoming the challenges associated with emerging materials, including issues related to synthesis, scalability, and integration into existing technologies [4]. Innovation in material science and manufacturing processes will shape the future of semiconductor technology.

Sustainability and Environmental Concerns

Environmental Impact: The production and disposal of semiconductor materials, particularly those with hazardous components, raise environmental and health concerns. Addressing these issues through sustainable practices and safer materials is critical for the long-term viability of semiconductor technologies [12].

1.7 EMERGING MATERIALS IN SEMICONDUCTOR DESIGN

1.7.1 Graphene: A Revolutionary Material for Semiconductors

Graphene, a single layer of carbon atoms arranged in a two-dimensional hexagonal lattice, has captured the imagination of scientists and engineers since its discovery. Its extraordinary properties have positioned it as one of the most promising emerging materials in semiconductor design. Graphene's unique combination of high electron mobility, mechanical flexibility, and strength offers significant advantages over traditional semiconductor materials like silicon, making it a strong candidate for next-generation electronic devices (Figure 1.14).

Key Properties of Graphene

- **High Electron Mobility**
  Graphene is renowned for its exceptionally high electron mobility, which can exceed 200,000 cm^2/Vs at room temperature. This is more than 100 times higher than that of silicon, enabling the possibility of ultra-fast transistors and high-frequency devices. High electron mobility allows for quicker charge carrier movement, resulting in

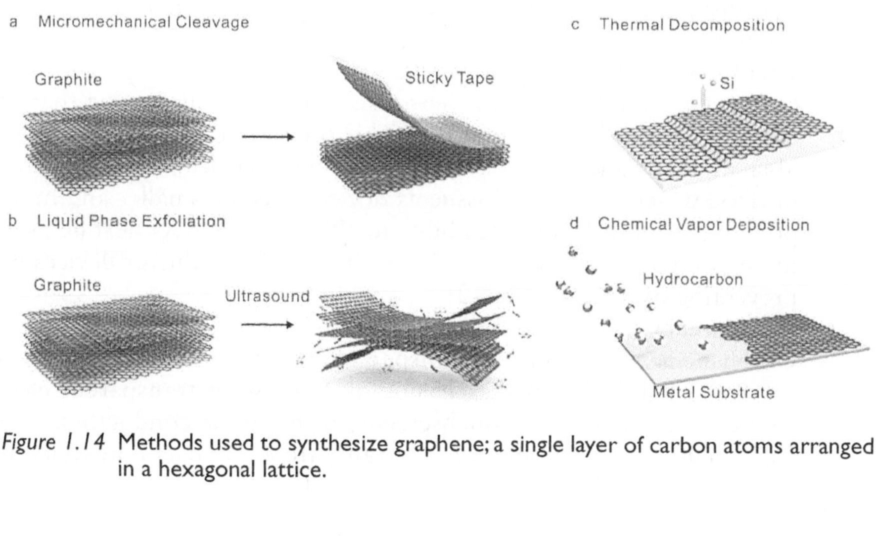

Figure 1.14 Methods used to synthesize graphene; a single layer of carbon atoms arranged in a hexagonal lattice.

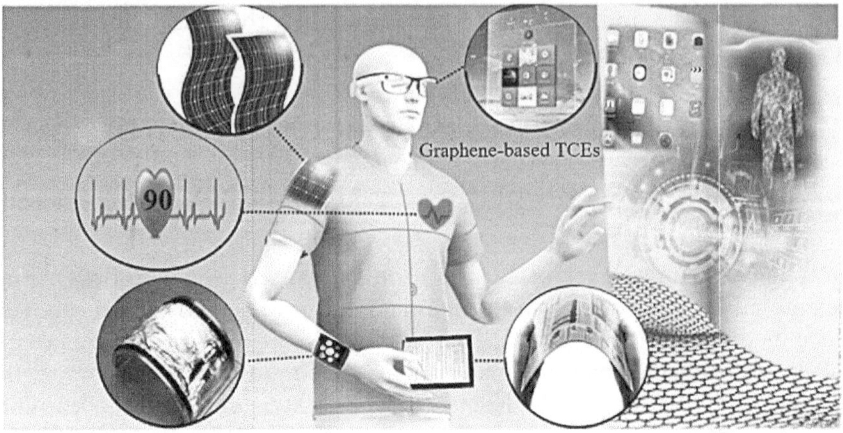

Figure 1.15 Potential applications of graphene-based technology in wearable electronics and healthcare.

faster switching speeds in transistors and reduced energy losses in electronic circuits [14].

- **Flexibility and Mechanical Strength**
  Graphene is not only extremely thin—just one atom thick—but also remarkably strong, with a tensile strength about 200 times greater than steel. Despite this strength, it is highly flexible, which opens up possibilities for flexible and wearable electronics. This combination of strength and flexibility means that graphene-based devices could be both durable and adaptable, ideal for applications in foldable displays, wearable sensors, and other flexible electronic systems (Figure 1.15).
- **Thermal Conductivity**
  Graphene exhibits excellent thermal conductivity, surpassing that of silicon by an order of magnitude. This property is particularly valuable in managing heat dissipation in high-performance electronic devices, which is a critical issue as devices become smaller and more powerful (Figure 1.16). The ability to efficiently conduct heat helps in maintaining the reliability and longevity of semiconductor devices by preventing overheating [8–12].
- **Transparency**
  Graphene is almost entirely transparent, absorbing only about 2.3% of visible light. This makes it suitable for use in transparent electronic devices, such as touchscreens, transparent conductors, and solar cells. The combination of transparency and conductivity is

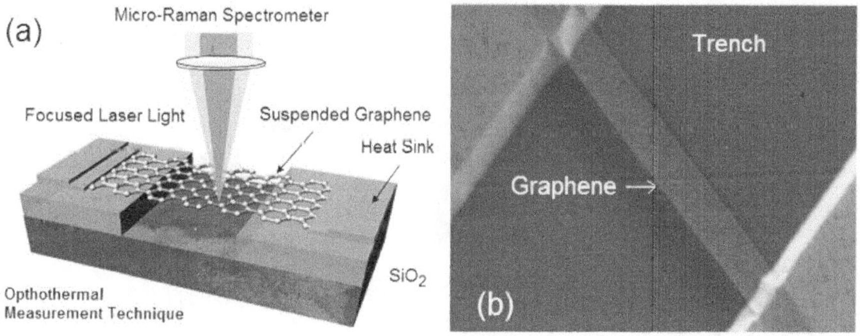

Figure 1.16 Use of graphene-based technologies in wearable electronics and healthcare applications.

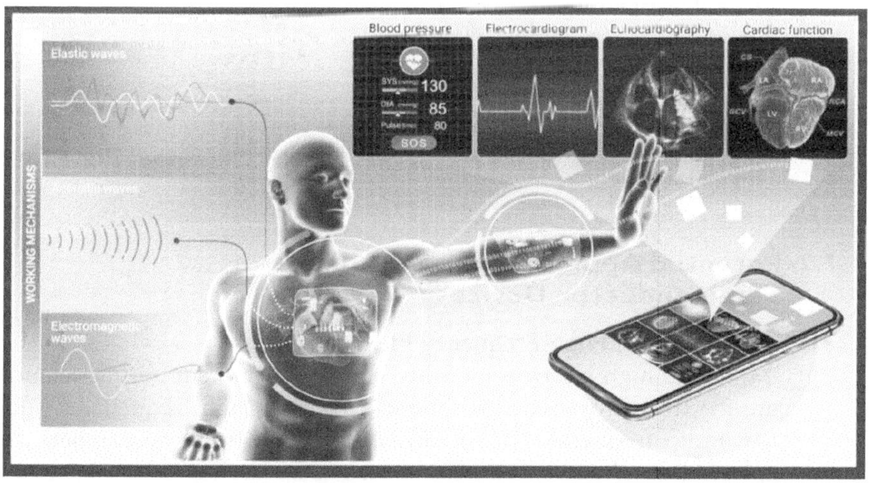

Figure 1.17 Potential applications of graphene-based technology in wearable electronics and healthcare.

rare and highly desirable for certain applications in optoelectronics (Figure 1.17).

- **Chemical Stability and Surface Area**
  Graphene is chemically stable, meaning it does not readily react with other substances, which is beneficial for the longevity of devices [12]. Additionally, it has a high surface area, which makes it an excellent material for sensors and energy storage devices, where surface interactions play a crucial role (Figure 1.18).

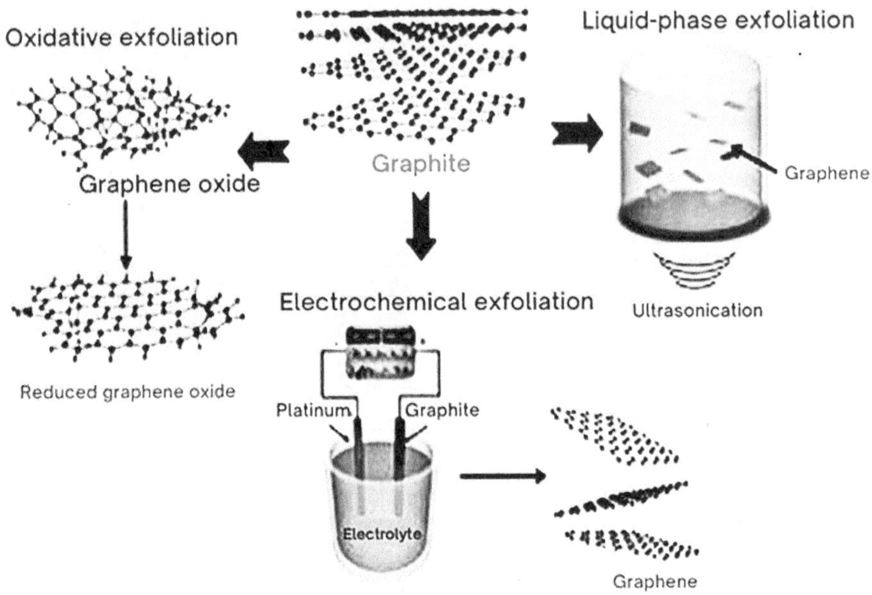

Figure 1.18 Applications of graphene-based technology in wearable electronics and healthcare.

1.7.2 Potential Applications of Graphene in Semiconductor Devices

- Transistors and High-Frequency Electronics
  Graphene's high electron mobility and speed make it an ideal candidate for transistors that operate at extremely high frequencies, such as in radio frequency (RF) and microwave devices. Researchers are exploring graphene-based field-effect transistors (GFETs) that could surpass the performance of traditional silicon-based transistors, especially in applications requiring high-speed data processing and communication [4] (Figure 1.19).
- Flexible and Wearable Electronics
  The flexibility of graphene enables the development of foldable, bendable, and stretchable electronic devices. Graphene can be integrated into flexible circuits, sensors, and displays, leading to innovations in wearable technology, smart clothing, and flexible medical devices. Its mechanical properties allow these devices to withstand deformation without losing functionality [2] (Figure 1.20).
- Transparent Conductors
  Graphene's transparency and conductivity make it a promising material for transparent conductors, which are essential components

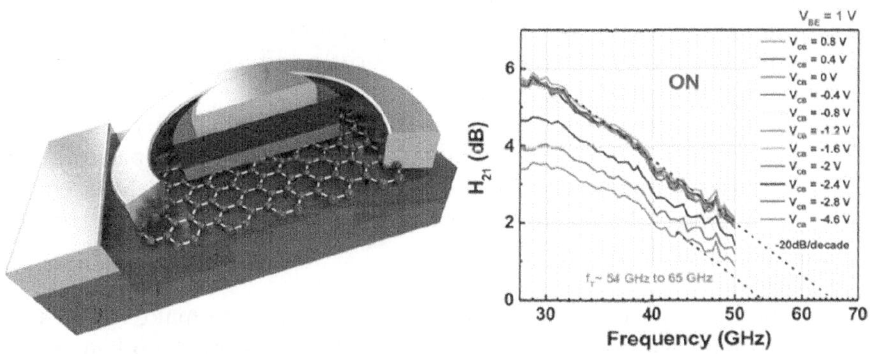

Figure 1.19 Use of graphene in wearable technologies and healthcare applications.

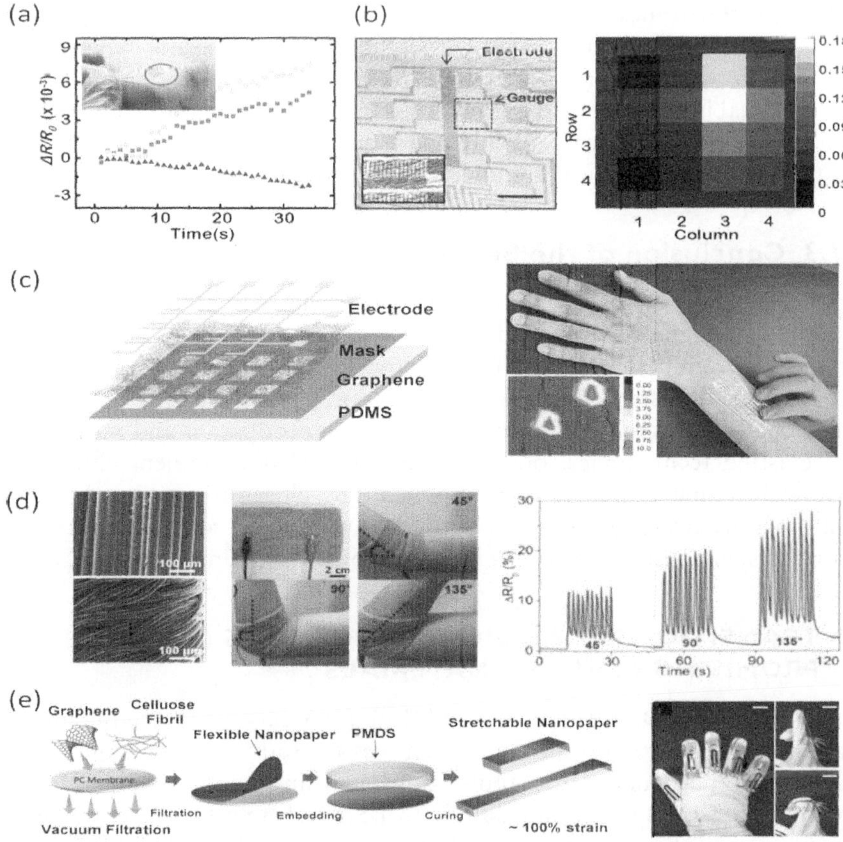

Figure 1.20 Application of graphene in flexible and wearable electronics.

in touchscreens, displays, and photovoltaic cells (Figure 1.21). Graphene could potentially replace indium tin oxide (ITO), the current industry standard, which is brittle and less abundant.

- **Energy Storage Devices**
  Graphene's large surface area and excellent conductivity make it an ideal material for energy storage devices like supercapacitors and batteries. Graphene-based electrodes can store and deliver energy more efficiently, leading to faster charging times and longer battery life (Figure 1.22).
- **Sensors**
  The high surface area and sensitivity of graphene make it an excellent material for chemical and biological sensors. Graphene-based sensors can detect single molecules, making them incredibly sensitive for applications in environmental monitoring, healthcare diagnostics, and security [12] (Figure 1.23).
- **Optoelectronics**
  Due to its direct interaction with light, graphene is being explored for use in optoelectronic devices such as photodetectors, modulators, and optical interconnects (Figure 1.24). Graphene's ability to convert light into electrical signals quickly and efficiently could lead to faster and more efficient optical communication systems.

1.7.3 Conclusion of the Section

Graphene stands out as a material with the potential to revolutionize semiconductor design. Its extraordinary properties, including high electron mobility, flexibility, and thermal conductivity, open up new avenues for the development of faster, more efficient, and more versatile electronic devices. While challenges remain in terms of large-scale production and integration into existing technologies, ongoing research and development efforts are steadily bringing graphene-based applications closer to reality. As these challenges are overcome, graphene could play a pivotal role in the future of electronics, enabling innovations that were previously unimaginable [13].

1.8 TRANSITION METAL DICHALCOGENIDES (TMDS): A PROMISING CLASS OF MATERIALS

Transition Metal Dichalcogenides (TMDs) are a class of layered materials that have emerged as promising alternatives to traditional semiconductors like silicon, particularly for applications in thin-film transistors (TFTs). TMDs such as molybdenum disulfide (MoS_2) and tungsten disulfide (WS_2) are composed of a transition metal atom (e.g., molybdenum or tungsten) sandwiched between two chalcogen atoms (e.g., sulfur). These materials exhibit unique properties that make them suitable for next-generation

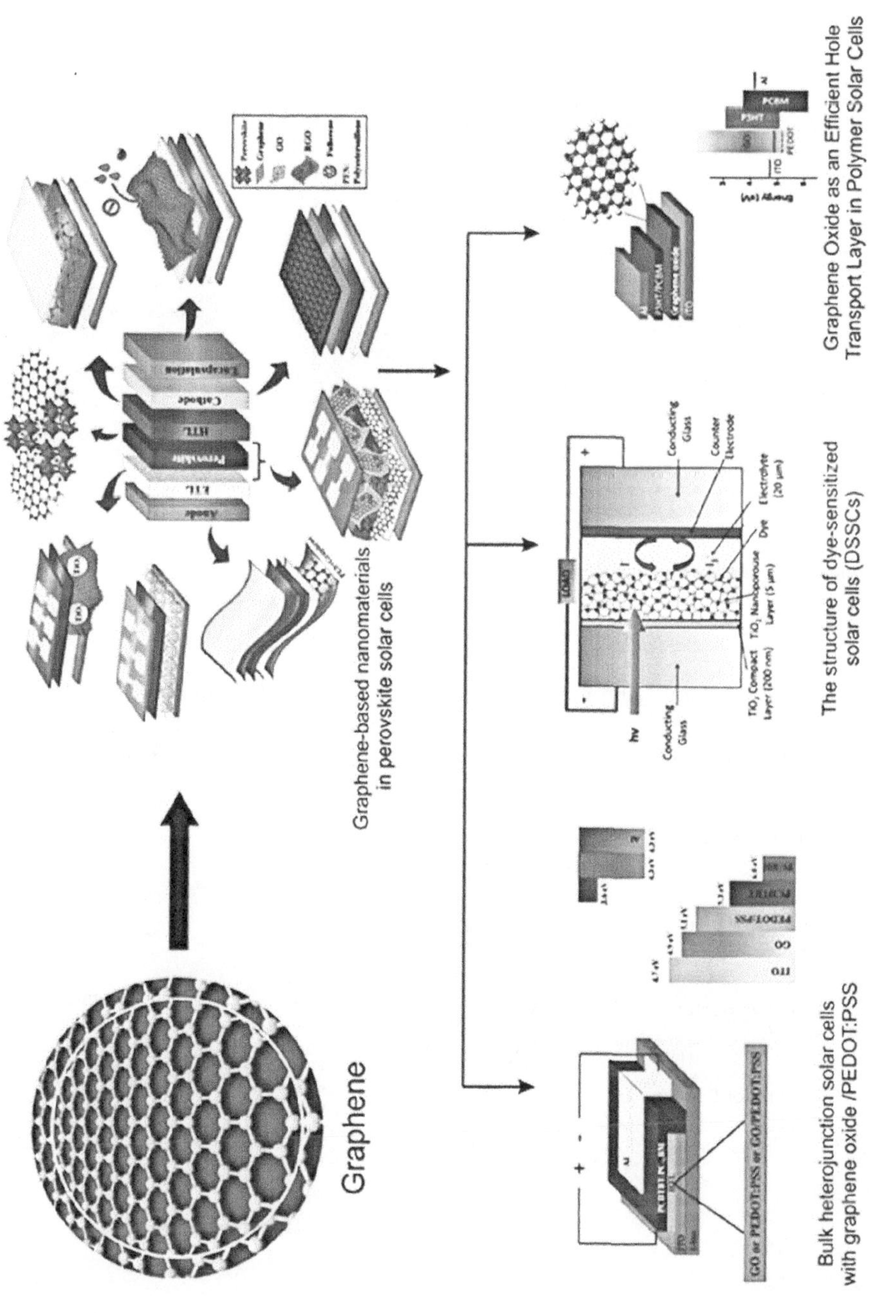

Figure 1.21 Transparent conductors.

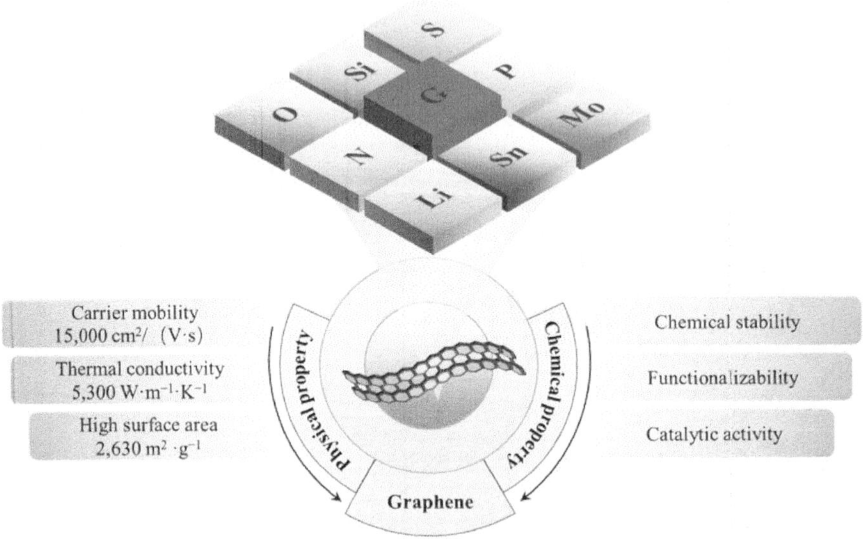

Figure 1.22 Energy storage devices.

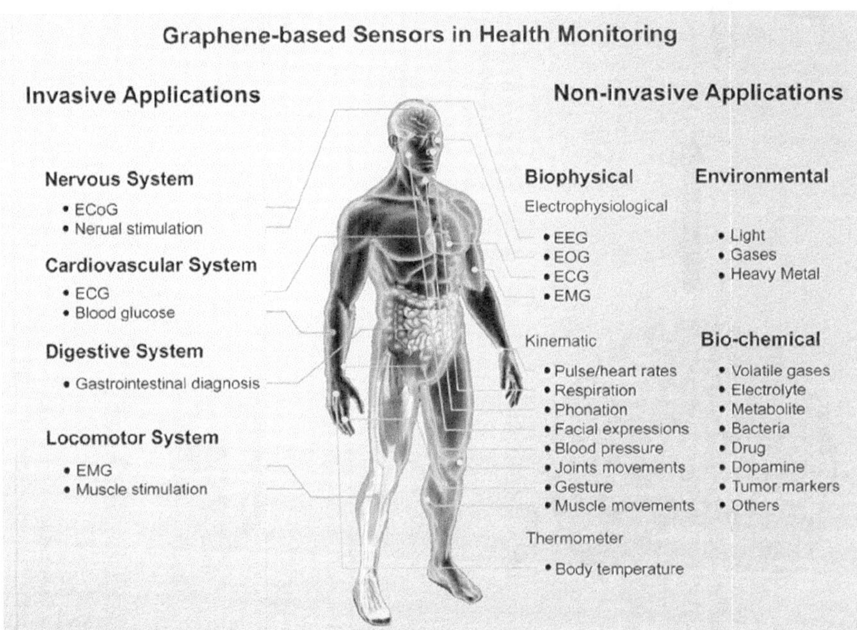

Figure 1.23 Sensors.

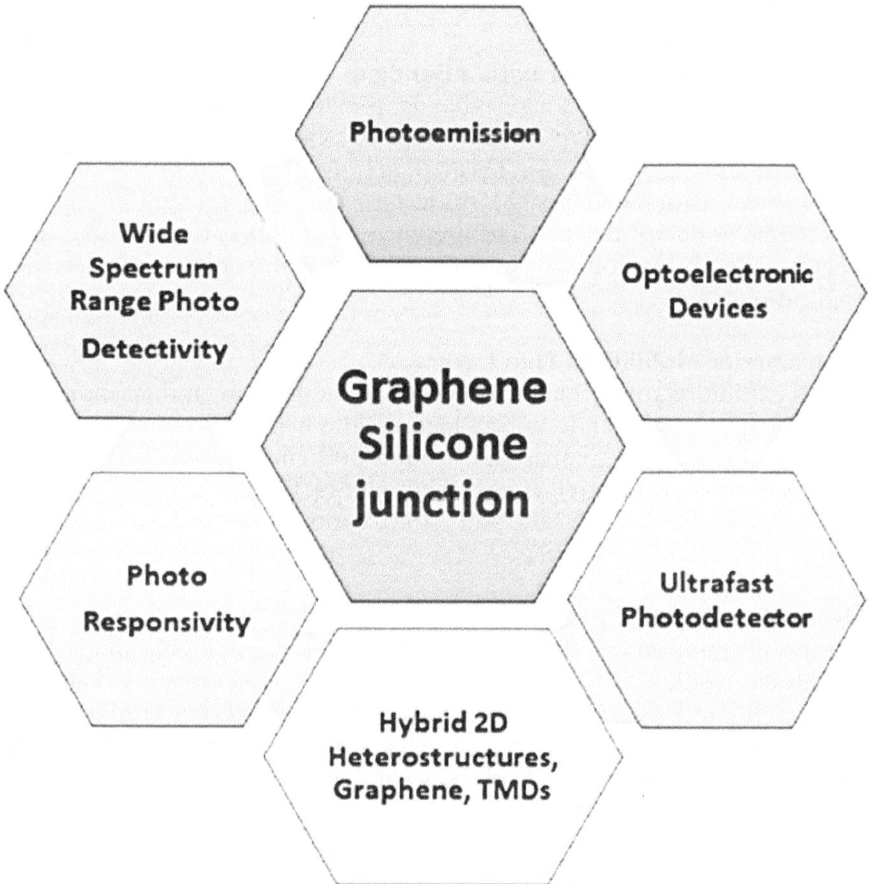

Figure 1.24 Optoelectronics.

electronic devices, especially as the demand for thinner, more flexible, and high-performance transistors grows [5].

1.8.1 Key Properties of TMDs

Layered Structure and Thickness

TMDs have a naturally layered structure, where individual layers are held together by weak van der Waals forces. This allows for the exfoliation of TMDs into atomically thin sheets, sometimes as thin as a single layer, akin to graphene. The ability to create ultra-thin films is crucial for developing transistors that are not only small but also flexible and transparent (Figure 1.25).

Semiconducting Behavior with a Bandgap

Unlike graphene, which is a zero-bandgap material, TMDs like MoS_2 and WS_2 have a direct bandgap when reduced to a monolayer, typically ranging from 1 to 2 eV (Figure 1.26). This bandgap is essential for creating on/off states in transistors, making TMDs ideal for use in digital logic circuits and other semiconductor devices. The presence of a direct bandgap also enables efficient light absorption and emission, which is beneficial for optoelectronic applications [4–8].

High Carrier Mobility in Thin Layers

TMDs exhibit relatively high carrier mobility, especially in their monolayer form (Figure 1.27). While not as high as graphene, the electron mobility of MoS_2, for example, can reach around 200 cm^2/Vs, which is sufficient for many electronic applications. This allows for the efficient transport of charge carriers in thin-film transistors, contributing to faster and more responsive electronic devices.

Mechanical Flexibility

The atomic thinness and flexibility of TMDs make them highly adaptable to flexible electronics. TMD-based transistors can be integrated into bendable, stretchable, and wearable electronic systems, maintaining their performance even under mechanical strain. This is particularly advantageous for the development of flexible displays, wearable sensors, and other emerging technologies (Figure 1.28).

Chemical Stability and Environmental Robustness

TMDs are generally chemically stable and resistant to oxidation, which enhances their durability in various environments (Figure 1.29). This stability is crucial for the reliability and longevity of TMD-based devices, particularly those exposed to harsh conditions or requiring long-term use [2–9].

1.8.2 Applications of TMDs in Thin-Film Transistors (TFTs)

Low-Power Electronics

The direct bandgap of TMDs like MoS_2 allows for efficient switching with low power consumption, making them ideal for low-power electronics. Thin-film transistors based on TMDs can operate at lower voltages compared to traditional silicon transistors, reducing energy consumption and heat generation, which is critical for battery-powered and portable devices.

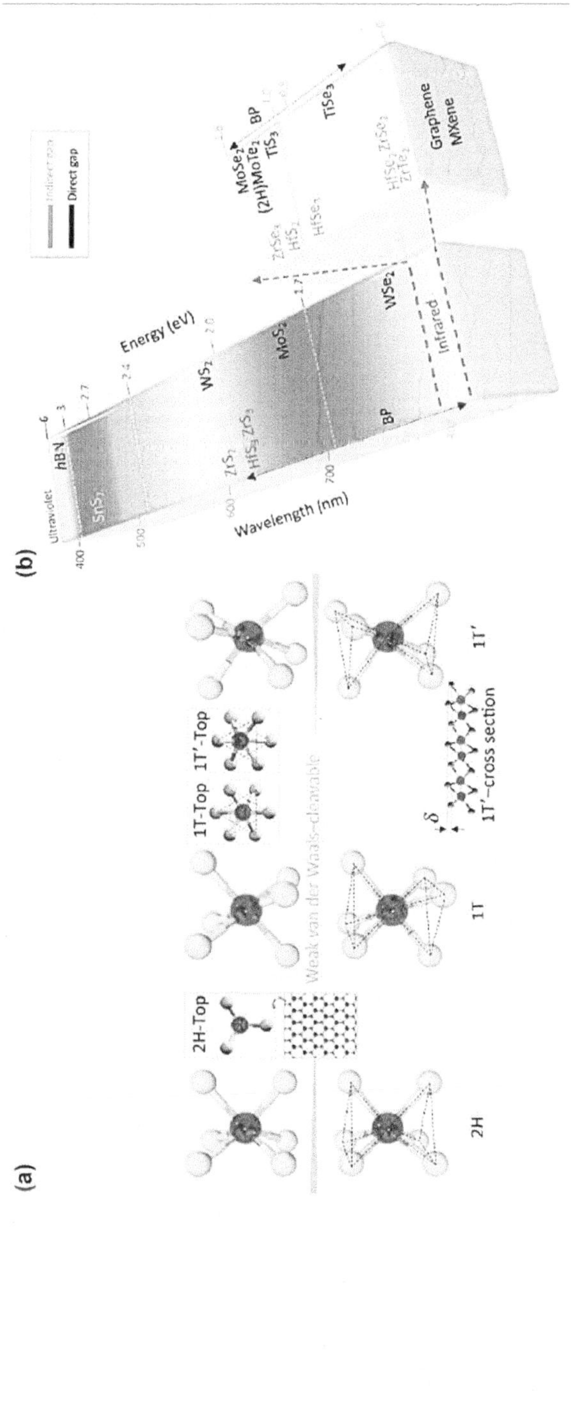

Figure 1.25 Layered structure and thickness.

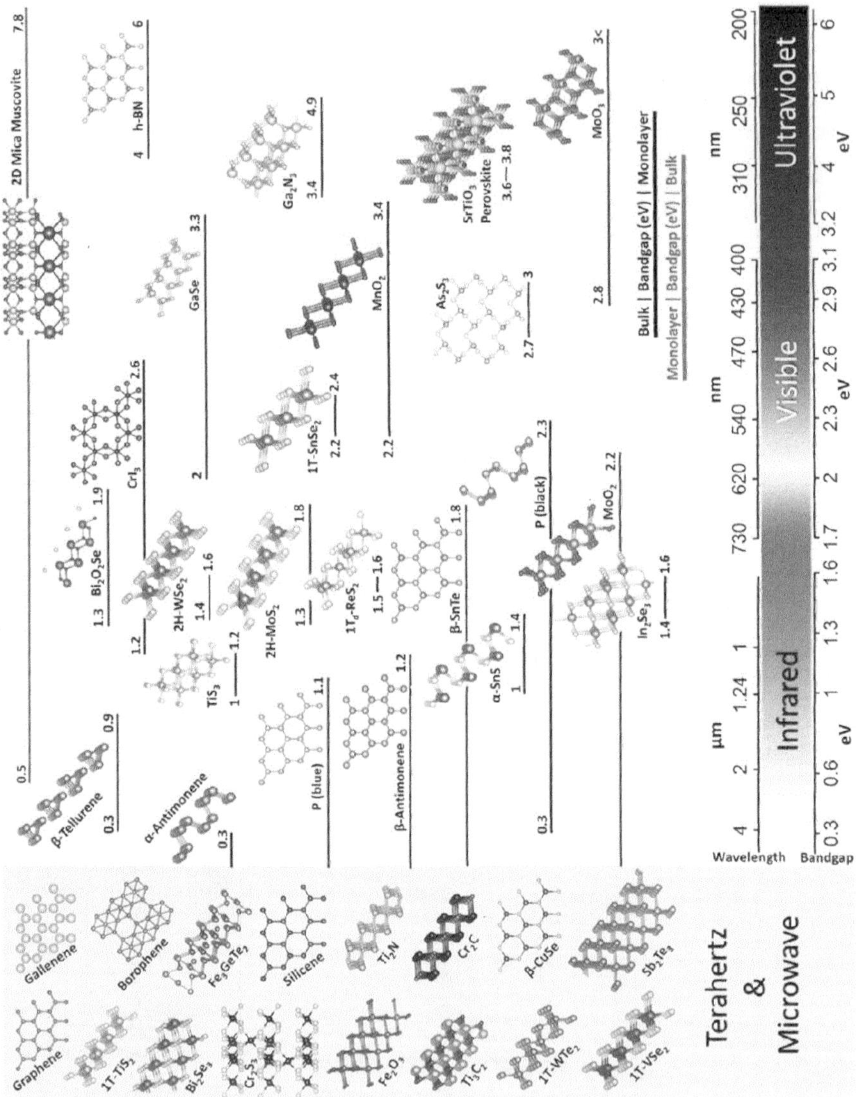

Figure 1.26 Semiconducting behavior with a bandgap.

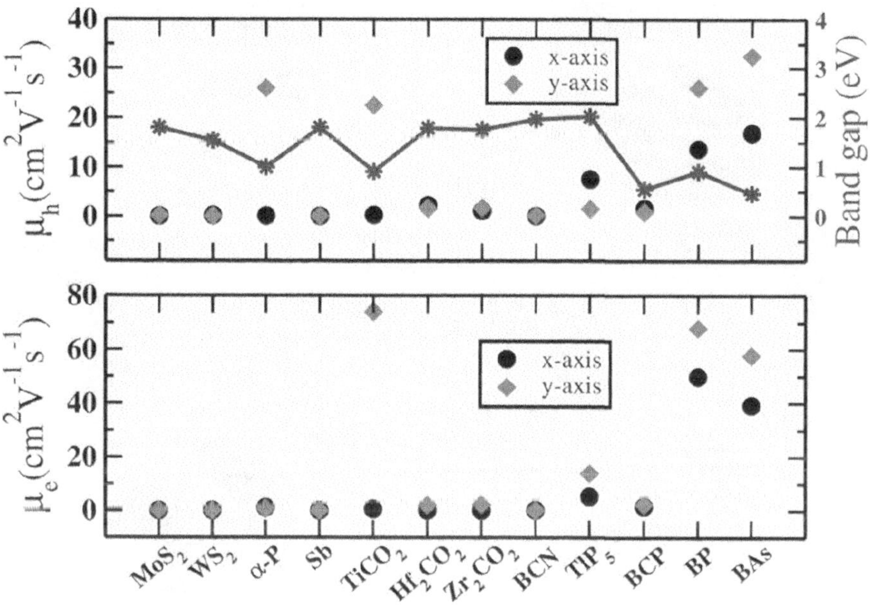

Figure 1.27 High carrier mobility in thin layers.

Flexible and Wearable Devices:

TMD-based TFTs can be fabricated on flexible substrates, enabling the development of bendable and wearable electronics. These transistors maintain their electronic properties even when subjected to mechanical deformation, making them suitable for applications in smart clothing, flexible displays, and wearable health monitors.

Transparent Electronics

The atomic thinness of TMDs allows them to be nearly transparent, especially when used in combination with transparent substrates like glass or plastic. This transparency is essential for developing transparent electronics, such as see-through displays, heads-up displays (HUDs), and transparent solar cells, where both performance and aesthetics are critical [7].

Next-Generation Displays

TMDs like MoS_2 and WS_2 are being explored for use in advanced display technologies, including organic light-emitting diode (OLED) displays and quantum dot displays. Their semiconducting properties and ability to be used as thin films make them suitable for driving pixels in high-resolution, energy-efficient displays.

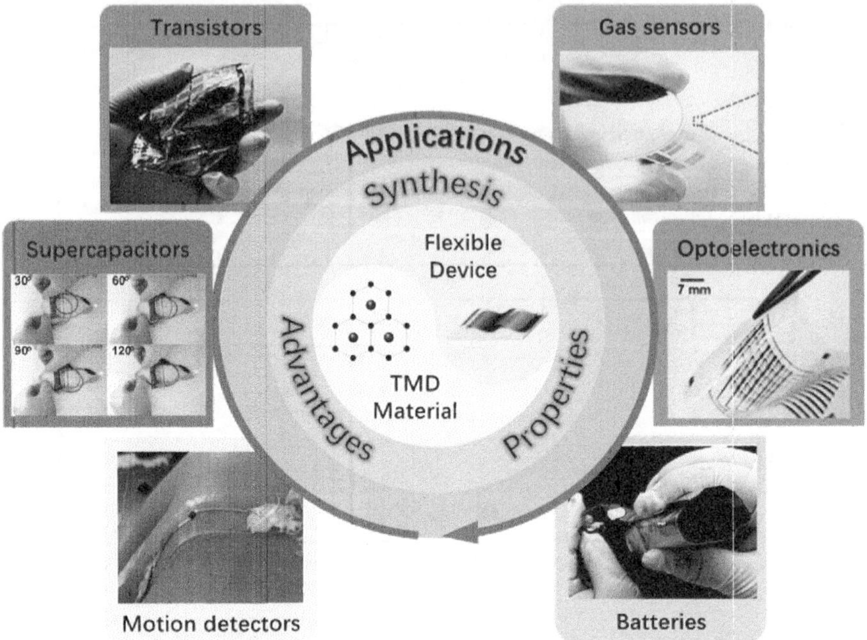

Figure 1.28 Mechanical flexibility.

Optoelectronics

Due to their direct bandgap, TMDs are also being investigated for optoelectronic applications, including photodetectors, light-emitting diodes (LEDs), and solar cells. The ability to efficiently absorb and emit light makes TMDs versatile materials for integrating electronic and optical functions on the same platform [8].

High-Performance Logic Circuits

TMD-based TFTs are being developed for use in logic circuits, where their ability to switch between on and off states with low power consumption is a significant advantage. These circuits could be used in various computing applications, from basic logic gates to more complex processing units, potentially enhancing the performance and efficiency of future computing devices.

1.8.3 Conclusion of the Section

Transition Metal Dichalcogenides (TMDs) like MoS_2 and WS_2 represent a significant advancement in semiconductor materials, offering unique properties that are well-suited for the next generation of thin-film transistors.

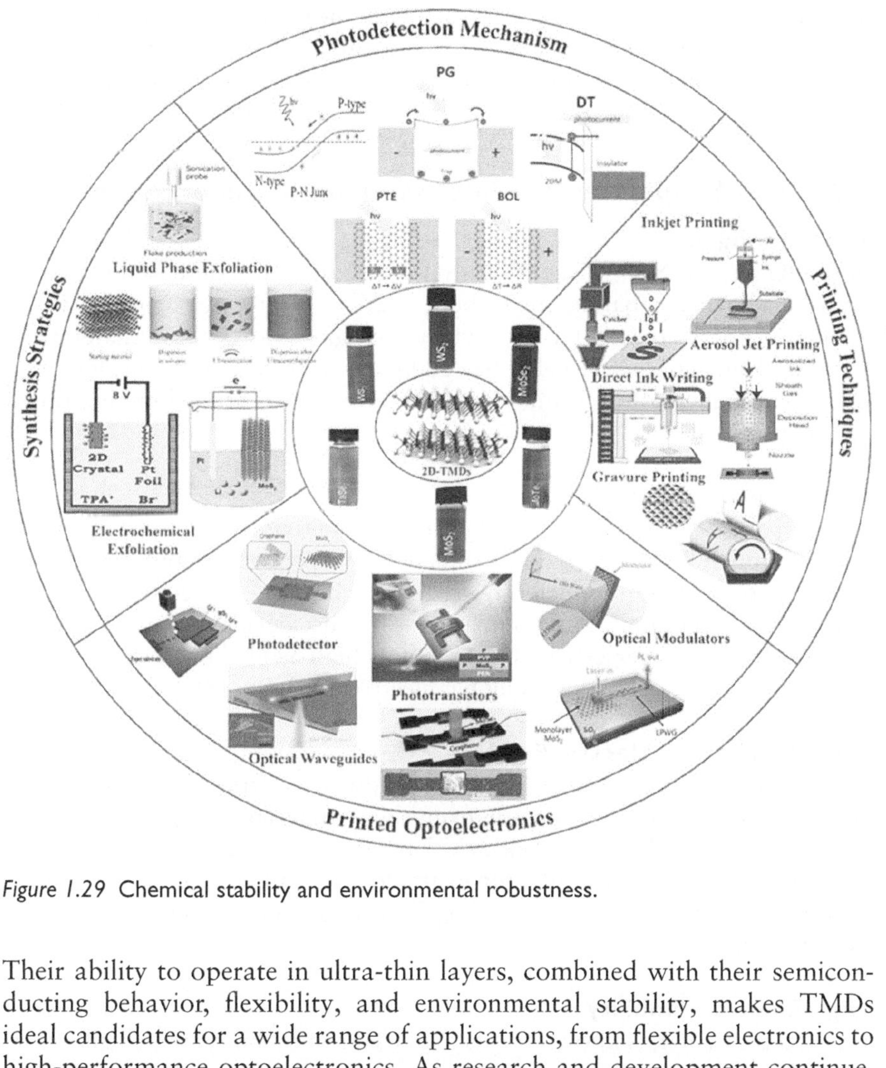

Figure 1.29 Chemical stability and environmental robustness.

Their ability to operate in ultra-thin layers, combined with their semiconducting behavior, flexibility, and environmental stability, makes TMDs ideal candidates for a wide range of applications, from flexible electronics to high-performance optoelectronics. As research and development continue, TMDs are likely to play an increasingly important role in the evolution of semiconductor technology, enabling more efficient, versatile, and innovative electronic devices [5–12].

1.9 SILICON CARBIDE (SIC) AND GALLIUM NITRIDE (GAN): WIDE-BANDGAP SEMICONDUCTORS FOR HIGH-POWER AND HIGH-FREQUENCY APPLICATIONS

As the demand for higher efficiency, power density, and performance in electronic devices grows, wide-bandgap (WBG) semiconductors like Silicon

Carbide (SiC) and Gallium Nitride (GaN) have emerged as crucial materials in the development of next-generation high-power and high-frequency devices. These materials offer significant advantages over traditional silicon, making them ideal for applications in power electronics, RF (radio frequency) systems, and various high-performance sectors.

1.9.1 Key Properties of Wide-Bandgap Semiconductors

Wide Bandgap

Both SiC and GaN have significantly wider bandgaps compared to silicon, with SiC having a bandgap of about 3.26 eV and GaN around 3.4 eV, compared to silicon's 1.1 eV. A wider bandgap allows these materials to withstand higher electric fields before breaking down, enabling devices to operate at higher voltages and temperatures, which is critical in high-power applications [6].

High Thermal Conductivity

SiC, in particular, has excellent thermal conductivity (about 3.7 W/cm·K), which is much higher than that of silicon. This property allows SiC devices to dissipate heat more efficiently, reducing the need for complex cooling systems and enabling more compact and robust device designs. GaN also has good thermal properties, though not as high as SiC, but it compensates with other advantages such as higher electron mobility [2].

High Breakdown Electric Field

SiC and GaN have a much higher breakdown electric field compared to silicon. This allows devices made from these materials to operate at higher voltages without the risk of electrical breakdown, making them ideal for high-power applications. SiC, for example, can handle electric fields up to 10 times higher than silicon, enabling the creation of smaller, more efficient power devices.

High Electron Mobility

GaN, in particular, is known for its high electron mobility, which allows for faster switching speeds and higher efficiency in high-frequency applications. This makes GaN an excellent material for RF devices, such as those used in wireless communication, radar systems, and satellite communication.

Temperature Resistance

Both SiC and GaN can operate at much higher temperatures than silicon, with SiC being able to function reliably at temperatures exceeding 600°C. This makes them ideal for harsh environments where cooling is difficult, such as in automotive, aerospace, and industrial applications [3].

1.9.2 Applications of Silicon Carbide (SiC)

Power Electronics
SiC is widely used in power electronics for applications that require high efficiency and compact designs. SiC-based power devices, such as MOSFETs, Schottky diodes, and JFETs, are commonly used in electric vehicles (EVs), renewable energy systems (like solar inverters), and industrial motor drives. These devices offer lower switching losses, higher efficiency, and can operate at higher temperatures and voltages than their silicon counterparts [2].

Electric Vehicles (EVs)
SiC devices are playing a crucial role in the advancement of electric vehicles. SiC-based inverters and power converters help increase the overall efficiency of EVs, leading to longer driving ranges and reduced charging times. The ability to operate at higher temperatures also reduces the cooling requirements, allowing for lighter and more compact power systems in vehicles [3].

Aerospace and Defense
The high thermal stability and radiation resistance of SiC make it an ideal material for aerospace and defense applications. SiC devices are used in power supplies for aircraft, satellites, and military equipment, where reliability in extreme conditions is paramount [4].

Renewable Energy
In the renewable energy sector, SiC power devices are used in solar inverters, wind turbine converters, and energy storage systems. SiC's efficiency and high-voltage operation capabilities help maximize energy conversion and reduce losses, contributing to more sustainable energy solutions [1].

1.9.3 Applications of Gallium Nitride (GaN)

High-Frequency RF Devices
GaN is particularly suited for high-frequency RF applications due to its high electron mobility and efficiency. GaN-based transistors, such as high-electron-mobility transistors (HEMTs), are widely used in RF amplifiers, radar systems, satellite communications, and wireless infrastructure. These devices offer superior performance in terms of speed and power density compared to traditional silicon-based RF devices.

5G and Beyond
The rollout of 5G networks and the development of future communication technologies rely heavily on GaN-based devices. GaN's ability to operate at high frequencies and power levels makes it ideal for the next generation

of wireless communication systems, where high efficiency and small form factors are critical.

Power Supply and Conversion

GaN is increasingly being used in power supply and power conversion applications, such as in high-efficiency chargers for consumer electronics, data centers, and industrial power supplies. GaN-based power devices can switch faster than silicon-based devices, resulting in higher efficiency and smaller, lighter power systems.

Electric Vehicles and Charging Infrastructure

GaN is also finding its way into the automotive industry, particularly in onboard chargers and fast-charging infrastructure for electric vehicles. GaN's high efficiency and compact size enable faster and more efficient charging solutions, contributing to the broader adoption of EVs [12].

Optoelectronics

GaN is the material of choice for blue and ultraviolet LEDs, as well as laser diodes. These devices are used in a wide range of applications, from general lighting to data storage and medical devices. GaN-based LEDs are also key components in solid-state lighting, offering high efficiency and long lifespans [5].

1.9.4 Conclusion of the Section

Silicon Carbide (SiC) and Gallium Nitride (GaN) are at the forefront of semiconductor technology, offering significant advantages over traditional silicon for high-power and high-frequency applications. SiC's high thermal conductivity and voltage capabilities make it ideal for power electronics, electric vehicles, and renewable energy systems. Meanwhile, GaN's high electron mobility and efficiency make it indispensable for RF devices, 5G communications, and fast-charging technologies. As these materials continue to be developed and integrated into more applications, they are set to play a critical role in the future of electronics, enabling devices that are faster, more efficient, and more reliable [4–13].

1.10 PEROVSKITES: A NEW FRONTIER IN OPTOELECTRONICS

Perovskites are a class of materials with a specific crystal structure that has garnered significant attention for their potential in optoelectronics. The perovskite structure, named after the mineral perovskite ($CaTiO_3$), can accommodate a wide variety of compositions, including organic-inorganic hybrid materials and purely inorganic compounds. These materials have shown remarkable properties that make them highly promising for applications in photovoltaics (solar cells) and light-emitting diodes (LEDs), among other optoelectronic devices.

1.10.1 Key Properties of Perovskites

High Absorption Efficiency
Perovskites exhibit excellent light absorption properties across a broad spectrum, including visible and near-infrared regions. This high absorption efficiency is a key factor in their effectiveness as photovoltaic materials, allowing for more efficient conversion of sunlight into electrical energy. The absorption coefficient of perovskite materials is significantly higher than that of silicon, enabling the use of thinner films without sacrificing performance [2].

Tunable Bandgap
The bandgap of perovskites can be easily tuned by altering their composition or structure. This flexibility allows for the optimization of light absorption and emission properties for specific applications. For example, adjusting the bandgap can improve the performance of perovskite solar cells or tailor the emission wavelength of perovskite LEDs to desired colors.

High Carrier Mobility
Perovskite materials exhibit high charge carrier mobility, which contributes to their efficient charge transport properties. This high mobility is crucial for reducing energy losses in solar cells and enhancing the brightness and efficiency of LEDs. The long carrier diffusion lengths in perovskites also help in minimizing recombination losses, which further improves device performance [13].

Low Processing Costs
Many perovskite materials can be processed using low-cost methods such as solution-based techniques or printing. This is in contrast to traditional semiconductor materials, which often require expensive and complex fabrication processes. The ability to manufacture perovskite devices using simple and scalable techniques makes them a cost-effective option for commercial applications.

Flexible and Lightweight
Perovskite films can be deposited on flexible substrates, making them suitable for applications in lightweight and flexible electronic devices. This characteristic is beneficial for developing innovative form factors in optoelectronic devices, including bendable solar panels and wearable LEDs [14].

1.10.2 Applications of Perovskites in Optoelectronics

Photovoltaics (Solar Cells)

High Efficiency: Perovskite solar cells have achieved remarkable efficiency improvements in a relatively short period. The power conversion efficiencies of perovskite solar cells have surpassed 25%, making them competitive with and potentially superior to traditional silicon solar cells. The high absorption efficiency, tunable bandgap, and excellent charge transport properties contribute to their impressive performance.

Cost-Effectiveness: The low-cost processing methods used for perovskite solar cells, such as solution processing or roll-to-roll printing, make them an attractive alternative to traditional photovoltaic technologies. This cost advantage could lead to more affordable solar energy solutions and wider adoption of renewable energy [11].

Flexibility and Lightweight: The ability to deposit perovskite layers on flexible and lightweight substrates opens up new possibilities for integrating solar cells into a variety of surfaces, including building facades, windows, and portable electronic devices.

Light-Emitting Diodes (LEDs)

Bright and Efficient Emission: Perovskite LEDs have shown high brightness and efficiency, with excellent color purity and tunability. The ability to precisely control the emission wavelength through composition adjustments allows for the production of LEDs in various colors, including red, green, and blue. This makes perovskite LEDs suitable for applications in display technologies, general lighting, and color displays.

Low Fabrication Costs: Similar to perovskite solar cells, perovskite LEDs can be manufactured using cost-effective techniques, such as solution processing or inkjet printing. This capability supports the development of inexpensive and high-performance LED lighting solutions.

Flexible and Transparent Displays: The flexible nature of perovskite films enables the creation of flexible and transparent LED displays. These displays can be used in innovative applications, including bendable screens, wearable electronics, and smart windows.

Photodetectors

Perovskites are also being explored for use in photodetectors due to their high absorption efficiency and charge transport properties. Perovskite-based photodetectors can be used in a variety of applications, including imaging systems, environmental monitoring, and optical communication [10].

Optoelectronic Devices
Beyond photovoltaics and LEDs, perovskites hold potential for other optoelectronic applications, such as lasers and optical sensors. Their tunable optical properties and efficient charge transport make them versatile materials for a range of optoelectronic devices.

1.10.3 Conclusion of the Section

Perovskites represent a revolutionary advancement in optoelectronics, offering outstanding properties that enable high-performance photovoltaics and light-emitting diodes. Their high absorption efficiency, tunable bandgap, and low processing costs make them competitive alternatives to traditional materials, with potential applications ranging from flexible solar panels to bright and efficient LEDs. As research and development continue, perovskites are expected to play a significant role in shaping the future of optoelectronic technologies, contributing to more efficient and cost-effective solutions in renewable energy and lighting [8].

1.11 TOPOLOGICAL INSULATORS: UNVEILING A NEW CLASS OF MATERIALS

Topological insulators (TIs) represent a novel class of materials that exhibit unique electronic properties due to their topological nature. Unlike conventional insulators, which block the flow of electric current throughout the material, topological insulators have a fascinating combination of insulating behavior in the bulk and conducting behavior on their surface. This unusual property arises from their underlying electronic structure, which is protected by topological invariants that define their quantum state.

1.11.1 Key Properties of Topological Insulators

Conducting Surface States
The hallmark feature of topological insulators is their robust surface states that conduct electricity while the bulk of the material remains insulating. These surface states are typically described by Dirac-like or Weyl-like fermions, which are massless or nearly massless particles with relativistic dispersion relations. These surface states are protected by time-reversal symmetry, meaning they are immune to scattering by non-magnetic impurities and defects, leading to highly efficient conduction along the surface [9].

Bulk Insulating Behavior
In contrast to their conductive surface states, the bulk of a topological insulator behaves as an insulator. This insulating behavior occurs because the bulk states are separated from the surface states by a large energy gap,

which prevents the flow of current through the interior of the material. The presence of this gap is a result of the strong spin-orbit coupling in the material, which influences the electronic band structure and leads to the formation of a topological bandgap [2–12].

Time-Reversal Symmetry Protection

The surface states of topological insulators are protected by time-reversal symmetry, which implies that any perturbations that do not break this symmetry will not destroy the topological nature of the surface states. This protection makes the surface states robust against disorder and impurity scattering, contributing to their high conductivity and stability.

Topological Order

Topological insulators exhibit a type of order known as topological order, which is characterized by nontrivial topological invariants. These invariants arise from the electronic band structure of the material and cannot be altered by smooth changes in the material's parameters. The topological invariants are related to the number of surface states crossing the Fermi level and are indicative of the material's topological classification [12].

Spin-Momentum Locking

In topological insulators, the spin of the surface electrons is locked perpendicular to their momentum, a phenomenon known as spin-momentum locking. This means that the direction of an electron's spin is correlated with its direction of motion, which can lead to novel spintronic applications and influences the material's electronic and magnetic properties [10, 12].

1.11.2 Applications and Implications of Topological Insulators

Quantum Computing

Topological insulators have potential applications in quantum computing due to their unique electronic properties. The robustness of their surface states against perturbations makes them suitable candidates for hosting anyons, exotic quasiparticles that are theoretically predicted to exhibit non-Abelian statistics. These anyons could be used to construct topologically protected quantum gates, which are less susceptible to decoherence and errors.

Spintronics

The spin-momentum locking in topological insulators offers opportunities for developing advanced spintronic devices. These devices leverage the spin of electrons for information processing and storage, potentially leading to faster and more energy-efficient electronics. Topological insulators could be used to create spin-based transistors and memory devices with improved performance [13].

Thermoelectric Material

The unique surface states of topological insulators can also be exploited in thermoelectric applications. The high conductivity of the surface states, combined with the insulating bulk, can enhance the efficiency of thermoelectric materials, which convert temperature gradients into electrical energy. This could lead to more effective thermoelectric generators and coolers [11].

Optoelectronics

Topological insulators are being investigated for use in optoelectronic devices, such as photodetectors and light-emitting devices. Their surface states can interact with light in novel ways, potentially leading to improved performance in these devices. Additionally, the topological nature of the surface states can affect the emission and absorption properties of light in these materials [12].

Fundamental Research

The study of topological insulators contributes to the fundamental understanding of quantum phenomena and materials science. Investigating their electronic properties, surface states, and interactions with other quantum systems can lead to new insights into condensed matter physics and the nature of topological phases of matter.

1.11.3 Conclusion of the Section

Topological insulators represent a groundbreaking development in materials science, characterized by their unique combination of insulating bulk and conducting surface states. Their ability to host robust surface states protected by time-reversal symmetry opens up exciting possibilities for advanced technological applications, including quantum computing, spintronics, thermoelectrics, and optoelectronics. As research continues to explore and harness the properties of topological insulators, they are poised to play a transformative role in the development of new technologies and the advancement of fundamental science.

1.12 APPLICATIONS OF EMERGING MATERIALS IN HIGH-PERFORMANCE COMPUTING

Emerging materials are poised to revolutionize high-performance computing (HPC) by offering significant improvements in speed, efficiency, and functionality compared to traditional semiconductor materials. These advancements are driven by the unique properties of materials such as graphene, transition metal dichalcogenides (TMDs), silicon carbide (SiC), gallium nitride (GaN), and topological insulators. Here's how these materials are contributing to the evolution of HPC:

Graphene

- **High Electron Mobility**: Graphene's exceptional electron mobility allows for faster electron transport and reduced resistance in electronic components. This leads to the potential for faster transistors and logic gates, which can significantly enhance processor speeds and overall computational performance [2].
- **Scalability**: The thin, two-dimensional nature of graphene enables the creation of extremely small transistors, potentially extending Moore's Law beyond its traditional limits. This scalability can lead to more powerful processors that are smaller and more energy-efficient [3].
- **Thermal Management**: Graphene's high thermal conductivity helps dissipate heat more effectively, addressing one of the major challenges in high-performance computing related to heat generation and management. This could result in more reliable and stable processors, especially as clock speeds continue to increase [5].

Transition Metal Dichalcogenides (TMDs)

- **High Carrier Mobility**: TMDs like MoS_2 and WS_2 exhibit high carrier mobility, which can enhance the performance of transistors used in processors. This allows for faster switching speeds and more efficient data processing [6].
- **Flexible Electronics**: The ability to fabricate TMDs on flexible substrates offers opportunities for creating novel computing architectures, such as flexible and wearable computing devices. This flexibility can lead to innovative forms of high-performance computing that are more adaptable to various applications [7].
- **Low Power Consumption**: TMD-based transistors can operate at lower voltages, which reduces power consumption and heat generation. This is particularly beneficial for mobile and portable high-performance computing devices, where energy efficiency is critical [8].

Silicon Carbide (SiC)

- **High-Voltage Operation**: SiC's ability to operate at high voltages and temperatures makes it suitable for power electronics that support high-performance computing systems. Efficient power management is essential for maintaining the stability and performance of HPC systems.
- **High Thermal Conductivity**: The excellent thermal conductivity of SiC helps manage heat in high-performance processors, allowing for higher power densities and reduced cooling requirements. This can lead to more compact and efficient computing systems [9].

Gallium Nitride (GaN)

- **Fast Switching Speeds**: GaN's high electron mobility and high breakdown voltage enable faster switching speeds and improved efficiency in power electronics. This can enhance the performance of power management circuits in HPC systems, contributing to overall speed and efficiency.
- **Efficiency in Power Conversion**: GaN-based power devices offer high efficiency in power conversion, which is crucial for optimizing the power supply systems of high-performance computing clusters. This results in reduced energy consumption and improved operational efficiency [10].

Topological Insulators

- **Robust Surface States**: The conducting surface states of topological insulators are immune to certain types of scattering, leading to more reliable and stable performance in electronic devices. This property can improve the performance of processors by reducing error rates and increasing operational stability.
- **Quantum Computing Potential**: Topological insulators are also being explored for their potential in quantum computing. The robustness of their surface states could lead to the development of topologically protected qubits, which may advance the field of quantum computing and offer exponential increases in computational power.

1.13 POTENTIAL IMPACTS ON HIGH-PERFORMANCE COMPUTING

Increased Processing Speed: Emerging materials such as graphene and TMDs enable faster switching speeds and higher electron mobilities, which directly translate to increased processing speeds in HPC systems. This allows for more complex computations and faster data processing, crucial for applications ranging from scientific simulations to artificial intelligence.

Enhanced Energy Efficiency: The reduced power consumption and heat generation associated with emerging materials like TMDs and GaN contribute to improved energy efficiency in HPC systems. This not only lowers operational costs but also supports sustainable computing practices by reducing the environmental impact of large-scale computing infrastructure [12].

Improved Thermal Management: The high thermal conductivity of materials like graphene and SiC helps manage heat more effectively, allowing HPC systems to operate at higher performance levels without overheating. This can lead to more compact and powerful computing architectures.

Innovative Computing Architectures: The unique properties of emerging materials enable the development of novel computing architectures and devices. Flexible electronics, quantum computing potentials, and advanced power management systems are just a few examples of how emerging materials can drive innovation in HPC.

Scalability and Miniaturization: Materials like graphene and TMDs offer scalability and miniaturization beyond traditional silicon technologies. This enables the creation of smaller, more powerful processors and memory devices, contributing to the continued advancement of HPC technology [14].

1.13.1 Conclusion of the section

Emerging materials are set to play a transformative role in high-performance computing by enhancing processing speeds, improving energy efficiency, and enabling innovative computing architectures. As research and development in these materials progress, they will likely lead to more powerful, efficient, and sustainable HPC systems, driving advancements across various scientific, industrial, and technological fields.

1.14 POWER ELECTRONICS: ENHANCING EFFICIENCY WITH SILICON CARBIDE (SIC) AND GALLIUM NITRIDE (GAN)

In the field of power electronics, materials like Silicon Carbide (SiC) and Gallium Nitride (GaN) are revolutionizing the efficiency and performance of electronic devices. Power electronics involve the conversion, control, and management of electrical power, and optimizing these processes is crucial for a wide range of applications, from consumer electronics to industrial systems and renewable energy. SiC and GaN, as wide-bandgap semiconductors, offer significant advantages over traditional silicon-based devices, leading to improved efficiency, reduced energy losses, and enhanced performance [13].

Silicon Carbide (SiC)

1. High-Voltage Operation
   Wide Bandgap: SiC has a wide bandgap of about 3.26 eV, which allows it to withstand higher voltages compared to silicon. This property enables SiC-based devices to operate efficiently at high voltages, making them ideal for power conversion applications that require handling large amounts of power.

High Breakdown Voltage: SiC devices can handle electric fields up to 10 times higher than silicon devices. This high breakdown voltage capability reduces the need for complex and bulky insulation in power electronic systems, leading to more compact and reliable designs.

2. **High Thermal Conductivity**
   Efficient Heat Dissipation: SiC exhibits excellent thermal conductivity (approximately 3.7 W/cm·K), which helps in efficient heat dissipation. This is critical in high-power applications where heat management is a significant challenge. The ability to dissipate heat effectively allows SiC devices to operate at higher power densities and temperatures without overheating.

   Reduced Cooling Requirements: Improved thermal management reduces the need for extensive cooling systems, which can lower the overall cost and complexity of power electronic systems.

3. **Low Switching Losses**
   Fast Switching Speeds: SiC devices, such as MOSFETs and Schottky diodes, have low switching losses due to their high electron mobility and low on-resistance. This leads to higher efficiency in power conversion processes, as less energy is wasted during switching events.

   Higher Efficiency: Reduced switching losses contribute to overall energy savings and improved efficiency in power electronic systems, including inverters, motor drives, and power supplies.

4. **High-Temperature Operation**
   Robust Performance: SiC can operate reliably at higher temperatures (up to 600°C), which is beneficial in harsh environments such as automotive and aerospace applications. This capability reduces the need for complex cooling solutions and enhances the reliability and lifespan of power electronic systems.

Gallium Nitride (GaN)

1. **High Electron Mobility**
   Fast Switching Speeds: GaN has high electron mobility and a wide bandgap of around 3.4 eV, enabling fast switching speeds and high-frequency operation. GaN-based transistors, such as High Electron Mobility Transistors (HEMTs), offer superior performance in switching and amplification applications.

   Efficiency Improvements: The ability to switch quickly and efficiently reduces power losses and enhances the overall efficiency of power electronic devices, leading to more compact and effective power converters.

2. High Breakdown Voltage
   High Voltage Handling: GaN devices can handle high voltages with minimal energy loss due to their wide bandgap and high breakdown voltage capabilities. This makes them suitable for high-power applications, such as power supplies, electric vehicles, and RF amplifiers.

   Reduced Size and Weight: GaN's high voltage capability allows for smaller and lighter components, contributing to more compact and portable power electronic systems.

3. Low Conduction Losses
   Efficient Power Conversion: GaN transistors exhibit low on-resistance and minimal conduction losses, which improves the efficiency of power conversion processes. This results in reduced energy wastage and lower operational costs for power electronic systems.

   Enhanced Performance: The high efficiency of GaN devices enhances the performance of power electronics in applications such as data centers, telecommunications, and industrial automation.

4. High Thermal Conductivity
   Effective Heat Management: GaN has good thermal conductivity, which helps in managing heat effectively in high-power applications. This characteristic supports high-performance operation and contributes to the longevity and reliability of power electronic devices.

   Improved Reliability: Efficient heat management reduces thermal stress and enhances the reliability of GaN-based power electronics, making them suitable for demanding environments and applications.

Applications and Impact

1. Power Supplies
   Data Centers: SiC and GaN devices are used in power supplies for data centers to improve efficiency and reduce energy consumption. Their high switching speeds and low losses contribute to more efficient power conversion and cooling.

   Consumer Electronics: GaN-based chargers and adapters for consumer electronics offer faster charging times and higher efficiency compared to traditional silicon-based chargers.

2. Electric Vehicles (EVs)
   Inverters and Converters: SiC and GaN devices are employed in EV inverters and converters to enhance power conversion efficiency and

reduce energy losses. This leads to improved driving range and performance of electric vehicles.

Fast Charging: GaN technology enables high-speed charging solutions for EVs, reducing charging times and improving the convenience for users.

3. **Renewable Energy Systems**
   Solar Inverters: SiC and GaN devices are used in solar inverters to maximize energy conversion efficiency and reduce losses in photovoltaic systems. This contributes to more effective utilization of solar energy.

   Wind Turbine Converters: High-efficiency power converters based on SiC and GaN help optimize energy conversion in wind turbines, leading to more reliable and efficient renewable energy generation.

4. **Industrial Applications**
   Motor Drives: SiC and GaN devices are used in industrial motor drives to enhance performance, efficiency, and reliability. Their ability to handle high voltages and temperatures makes them suitable for demanding industrial environments.

1.14.1 Conclusion of the Section

Silicon Carbide (SiC) and Gallium Nitride (GaN) are transforming power electronics by offering significant improvements in efficiency, performance, and thermal management. SiC's high-voltage operation, thermal conductivity, and low switching losses, combined with GaN's fast switching speeds, high breakdown voltage, and low conduction losses, are driving advancements in a wide range of applications. These materials contribute to more efficient power supplies, faster charging solutions, and reliable renewable energy systems, ultimately leading to lower energy consumption, reduced operational costs, and enhanced performance in electronic devices.

1.15 FLEXIBLE ELECTRONICS: THE ROLE OF GRAPHENE IN ADVANCING WEARABLE AND FLEXIBLE TECHNOLOGIES

Flexible electronics represent a rapidly evolving field that seeks to create electronic devices that are not only flexible and lightweight but also capable of conforming to various surfaces and being integrated into diverse applications. Graphene, a single layer of carbon atoms arranged in a two-dimensional honeycomb lattice, has emerged as a key material in advancing flexible and wearable electronics due to its exceptional properties. Here's how graphene contributes to the development of these technologies:

Key Properties of Graphene for Flexible Electronics

1. Exceptional Mechanical Flexibility
   High Flexibility: Graphene is incredibly flexible, with a high strain tolerance of up to 20% strain without breaking. This flexibility allows it to be used in applications requiring significant bending, stretching, or conforming to irregular surfaces.

   Durability: The mechanical robustness of graphene means it can withstand repeated flexing and deformation, which is crucial for the longevity and reliability of flexible electronic devices.

2. High Electrical Conductivity
   Low Electrical Resistance: Graphene exhibits extraordinary electrical conductivity, with minimal electrical resistance. This property ensures efficient charge transport and low power consumption in electronic devices.

   High Carrier Mobility: The high electron mobility of graphene contributes to fast electronic switching and high-speed performance, making it suitable for high-frequency applications.

3. Thermal Conductivity
   Efficient Heat Dissipation: Graphene has one of the highest thermal conductivities among known materials, which helps in managing heat in flexible electronic devices. This efficient heat dissipation prevents overheating and ensures stable performance.

4. Transparency
   Optical Transparency: Graphene is highly transparent to visible light, with over 97% transparency in a single layer. This property allows for the development of transparent conductive films and touchscreens that are both flexible and visually clear.

5. Lightweight
   Low Mass: The two-dimensional nature of graphene makes it extremely lightweight, which is advantageous for applications in wearable electronics where minimizing weight is crucial for user comfort and practicality.

1.15.1 Applications of Graphene in Flexible Electronics

1. Flexible Displays
   OLED Screens: Graphene-based electrodes can be used in organic light-emitting diodes (OLEDs) to create flexible and bendable displays.

These displays can be integrated into various surfaces, including curved or foldable screens for smartphones and other devices.

E-Paper: Graphene can be used in e-paper technologies, providing flexible and lightweight displays for electronic books and other portable reading devices.

2. **Wearable Electronics**

   Health Monitoring: Flexible graphene-based sensors can be integrated into wearable health monitoring devices, such as smartwatches or fitness trackers. These sensors can monitor physiological parameters like heart rate, body temperature, and sweat composition.

   Smart Textiles: Graphene can be incorporated into fabrics to create smart textiles that have electronic functionalities, such as temperature regulation, health monitoring, or interactive features. These textiles can be used in clothing, sportswear, and medical garments.

3. **Flexible Touchscreens**

   Touch Sensors: Graphene-based touchscreens offer flexibility and durability while maintaining high sensitivity and accuracy. These touchscreens can be used in foldable phones, flexible tablets, and other devices with unconventional form factors.

   Interactive Surfaces: Graphene's transparency and flexibility allow for the creation of interactive surfaces that can be integrated into various objects, including furniture, walls, and vehicles.

4. **Flexible Energy Storage**

   Supercapacitors: Graphene-based supercapacitors offer high energy and power density in a flexible form factor. These energy storage devices can be used in wearable electronics and other flexible devices where traditional batteries may not be practical.

   Flexible Batteries: Graphene-enhanced batteries can be made flexible and lightweight, making them suitable for use in flexible electronics and wearable devices.

5. **Flexible Sensors**

   Environmental Sensors: Graphene-based sensors can be used to detect environmental changes, such as gas concentrations, humidity, or temperature. These sensors can be integrated into flexible platforms for applications in environmental monitoring and smart packaging.

   Biological Sensors: Flexible graphene sensors can be used for detecting biological markers in medical diagnostics. These sensors can be incorporated into wearable health monitors for continuous health monitoring.

1.15.2 Challenges and Future Directions

1. **Scalability and Cost**
   Manufacturing Challenges: Producing high-quality graphene at scale and integrating it into electronic devices can be challenging and costly. Research is ongoing to develop cost-effective and scalable manufacturing methods for graphene-based materials.

2. **Integration with Existing Technologies**
   Compatibility: Integrating graphene with existing electronic technologies and materials can pose technical challenges. Researchers are working on optimizing the compatibility of graphene with various substrates and electronic components.

3. **Long-Term Stability**
   Degradation: Ensuring the long-term stability and durability of graphene-based devices under various environmental conditions is crucial. Further research is needed to address issues related to the stability and longevity of graphene in practical applications.

1.15.3 Conclusion of the Section

Graphene is transforming the field of flexible electronics by offering unparalleled mechanical flexibility, electrical conductivity, thermal management, and transparency. Its unique properties enable the development of a wide range of flexible and wearable devices, including displays, health monitors, smart textiles, and energy storage systems. While there are challenges to overcome in terms of scalability, cost, and integration, ongoing research and technological advancements are likely to expand the applications of graphene and drive further innovation in flexible electronics.

1.16 OPTOELECTRONICS: THE ROLE OF PEROVSKITES IN SOLAR CELLS AND LEDS

Optoelectronics involves the study and application of electronic devices that source, detect, and control light. Materials like perovskites are making significant contributions to this field, particularly in the development of advanced solar cells and light-emitting diodes (LEDs). Perovskites are a class of materials with a specific crystal structure that has shown remarkable potential in optoelectronic applications due to their unique optical and electronic properties.

1.16.1 Perovskites in Solar Cells

1. **High Power Conversion Efficiency**
   Efficient Light Absorption: Perovskite solar cells have demonstrated high light absorption coefficients across a broad range of the solar spectrum. This allows them to absorb more light and convert it into electricity more efficiently compared to traditional silicon-based solar cells.

   Rapid Efficiency Gains: Since their introduction, perovskite solar cells have achieved rapid improvements in power conversion efficiency (PCE). Laboratory-scale devices have reached efficiencies exceeding 25%, comparable to or even surpassing those of silicon solar cells.

2. **Low-Cost Manufacturing**
   Simple Processing: Perovskite materials can be processed using low-cost techniques such as spin-coating, spray-coating, and inkjet printing. This simplicity and versatility in manufacturing methods contribute to reduced production costs compared to conventional silicon solar cells, which require complex and expensive fabrication processes.

   Flexible Substrates: Perovskite solar cells can be deposited on flexible substrates, enabling the development of lightweight, bendable solar panels. This opens up new possibilities for integrating solar energy harvesting into a wide range of surfaces, including textiles and portable devices.

3. **Tunable Bandgap**
   Customizable Properties: The bandgap of perovskite materials can be easily tuned by adjusting their composition. This tunability allows for the optimization of light absorption properties to match the solar spectrum and improve the overall efficiency of solar cells.

   Multijunction Cells: By combining perovskites with other materials in multijunction solar cells, researchers can further enhance the efficiency of solar energy conversion. This approach leverages the complementary absorption properties of different materials to capture a wider range of the solar spectrum.

4. **Stability and Durability**
   Challenges: Despite their high efficiency, perovskite solar cells face challenges related to long-term stability and environmental degradation. Research is ongoing to address issues such as moisture sensitivity and the stability of perovskite materials under prolonged exposure to sunlight and harsh conditions.

Improvement Strategies: Advances in material engineering and encapsulation techniques are being developed to enhance the stability and longevity of perovskite solar cells. These improvements are crucial for making perovskite technology viable for commercial applications.

1.16.2 Perovskites in Light-Emitting Diodes (LEDs)

1. **High Luminous Efficiency**
   Bright Emission: Perovskite LEDs (PeLEDs) exhibit high luminous efficiency and brightness. The efficiency of light emission from perovskite materials is comparable to or better than that of traditional LED materials, such as organic LEDs (OLEDs).

   Wide Color Gamut: Perovskites offer a broad range of emission wavelengths, allowing for the production of LEDs with vibrant colors and high color purity. This property is beneficial for applications requiring accurate color rendering and high-quality displays.

2. **Solution-Processed Fabrication**
   Cost-Effective Production: Similar to solar cells, perovskite LEDs can be fabricated using solution-processing techniques, which are less expensive and less complex than traditional methods used for other types of LEDs. This enables scalable production and the potential for low-cost manufacturing of high-performance LEDs.

   Flexible and Transparent Devices: Perovskite materials can be used to create flexible and transparent LEDs, which can be integrated into various applications such as flexible displays, lighting panels, and wearable electronics.

3. **High Quantum Efficiency**
   Efficient Light Generation: Perovskites exhibit high photoluminescence quantum efficiency, which translates to efficient light generation in LEDs. This high quantum efficiency enhances the performance and brightness of perovskite-based light-emitting devices.

4. **Stability and Reliability**
   Material Stability: While perovskites offer significant advantages in terms of efficiency and cost, there are ongoing challenges related to the stability and long-term reliability of perovskite LEDs. Research is focused on improving the material stability and encapsulation techniques to ensure the durability and performance of these devices over time.

1.16.3 Future Directions and Challenges

1. **Scalability and Commercialization**
   Manufacturing Scale-Up: Scaling up the production of perovskite-based solar cells and LEDs from laboratory prototypes to commercial products involves addressing challenges related to consistency, reproducibility, and quality control. Efforts are being made to develop scalable and reliable manufacturing processes.

2. **Environmental and Health Concerns**
   Lead Contamination: Many perovskite materials contain lead, raising concerns about environmental and health impacts. Research is focused on developing lead-free or lead-reduced perovskite materials and improving the safety and sustainability of perovskite technologies.

3. **Long-Term Stability**
   Durability Improvements: Enhancing the stability and durability of perovskite solar cells and LEDs is critical for their commercial viability. Advances in material composition, encapsulation methods, and environmental protection are being explored to address these issues.

1.16.4 Conclusion of the Section

Perovskites are driving significant advancements in optoelectronics, particularly in the fields of solar cells and light-emitting diodes. Their high power conversion efficiency, low-cost manufacturing, and tunable properties make them highly attractive for solar energy applications, while their high luminous efficiency and wide color gamut enhance the performance of LEDs. Despite challenges related to stability, scalability, and environmental concerns, ongoing research and development efforts are likely to overcome these hurdles and pave the way for widespread adoption of perovskite-based technologies in the optoelectronics industry.

1.17 CHALLENGES AND FUTURE DIRECTIONS: MATERIAL SYNTHESIS AND SCALABILITY

As emerging materials like graphene, perovskites, silicon carbide (SiC), gallium nitride (GaN), and transition metal dichalcogenides (TMDs) make their way into semiconductor technologies, several challenges arise related to material synthesis and scalability. Addressing these challenges is crucial for integrating these advanced materials into existing manufacturing processes and realizing their full potential.

1.17.1 Material Synthesis Challenges

1. Quality and Purity
   Controlling Defects: Achieving high-quality material with minimal defects is crucial for maintaining the desired properties and performance of advanced semiconductors. For materials like graphene and perovskites, controlling defects at the atomic or molecular level during synthesis can be challenging, impacting the consistency and reliability of the final product.

   Purity Requirements: Ensuring the purity of materials such as SiC and GaN is essential for their performance in high-power applications. Contaminants or impurities can degrade the electrical properties and reduce the efficiency of these materials.

2. Cost of Raw Materials
   Expensive Precursors: Some advanced materials, such as certain perovskites and TMDs, require expensive raw materials or precursors. This can drive up the cost of production and impact the commercial viability of these technologies.

   Economic Viability: Reducing the cost of raw materials and finding more affordable alternatives are crucial for making these technologies economically viable for widespread adoption.

3. Scalable Fabrication Techniques
   Transitioning from Lab to Production: Many advanced materials are synthesized using lab-scale methods that are not directly transferable to large-scale manufacturing. Developing scalable fabrication techniques that maintain material quality and performance is a key challenge.

   Process Optimization: Optimizing synthesis processes to scale up production while maintaining high quality and uniformity is critical. For example, methods like chemical vapor deposition (CVD) for graphene or solution processing for perovskites must be adapted for industrial-scale production.

1.17.2 Integration into Existing Manufacturing Processes

1. Compatibility with Current Technologies
   Material Integration: Integrating new materials into existing semiconductor manufacturing processes can be complex. Compatibility issues with established processes, such as photolithography or etching, need to be addressed to ensure that new materials can be seamlessly incorporated into current device architectures.

Interfacing with Traditional Materials: Developing methods to interface new materials with traditional semiconductor materials, such as silicon, requires innovation in fabrication techniques and material engineering.

2. **Manufacturing Infrastructure**
   Modifying Equipment: Existing semiconductor manufacturing equipment may need modifications or upgrades to accommodate new materials. This includes changes to deposition systems, etching tools, and characterization instruments.

   Investment in New Facilities: Significant investment may be required to build or upgrade facilities for the production of advanced materials. This includes setting up clean rooms, specialized processing lines, and quality control systems.

3. **Uniformity and Reproducibility**
   Consistent Quality: Achieving uniform material quality across large areas and batches is essential for reliable performance in semiconductor devices. Variability in material properties can lead to inconsistent device performance and reduced yields.

   Standardization: Developing standardized procedures for material synthesis and integration helps ensure reproducibility and consistency across different production runs and manufacturing sites.

1.17.3 Future Directions

1. **Advanced Synthesis Techniques**
   Innovation in Methods: Research into novel synthesis methods, such as chemical vapor deposition (CVD) for large-area graphene or inkjet printing for perovskites, aims to improve scalability and reduce costs. Advances in these techniques can lead to more efficient and cost-effective production.

   Automation and Control: Implementing advanced automation and process control systems can enhance the precision and reproducibility of material synthesis, addressing challenges related to quality and uniformity.

2. **Hybrid Approaches**
   Combining Materials: Hybrid approaches that combine new materials with traditional ones can leverage the strengths of both. For instance, integrating graphene with silicon-based electronics or combining

perovskites with existing solar cell technologies can create hybrid devices with enhanced performance and functionality.

Multi-Material Integration: Developing techniques for integrating multiple materials into a single device can enable new functionalities and improve performance. This includes combining different 2D materials or integrating new semiconductors with established substrates.

3. **Environmental and Sustainability Considerations**
   Sustainable Practices: Developing environmentally friendly synthesis methods and recycling processes for advanced materials can address concerns about resource use and waste management. This includes finding ways to reduce the environmental impact of raw material extraction and processing.

   Lead-Free Alternatives: For materials like perovskites that contain toxic elements such as lead, research into lead-free or less harmful alternatives is essential for improving the sustainability and safety of these technologies.

4. **Collaborative Research and Development**
   Industry-Academia Partnerships: Collaboration between industry and academic institutions can accelerate the development and commercialization of advanced materials. Joint research efforts can address technical challenges, share knowledge, and foster innovation.

   Public-Private Initiatives: Government and industry initiatives aimed at supporting research, funding, and infrastructure development can facilitate the advancement of new materials and their integration into semiconductor manufacturing.

1.17.4 Conclusion of the Section

Material synthesis and scalability present significant challenges for the integration of advanced materials into semiconductor technologies. Addressing issues related to material quality, cost, and compatibility with existing manufacturing processes is crucial for realizing the potential of materials like graphene, perovskites, SiC, GaN, and TMDs. Future directions include advancing synthesis techniques, exploring hybrid approaches, and focusing on sustainability. Collaborative research and development efforts will play a key role in overcoming these challenges and enabling the widespread adoption of these innovative materials in semiconductor applications.

1.18. PERFORMANCE VS. COST: TRADE-OFFS IN EMERGING SEMICONDUCTOR MATERIALS

The integration of emerging materials into semiconductor technologies often involves a complex balance between performance gains and associated costs. While these materials offer significant performance improvements, they also present challenges that can impact their overall cost-effectiveness. Here's a detailed discussion on the trade-offs between performance and cost in emerging semiconductor materials:

1.18.1 Performance Gains

1. **Enhanced Electrical Properties**
   High Electron Mobility: Materials like graphene and gallium nitride (GaN) exhibit exceptionally high electron mobility, leading to faster electronic switching and improved performance in high-frequency applications. This enables the development of high-speed transistors and efficient power amplifiers.

   Wide Bandgap: Wide-bandgap materials such as silicon carbide (SiC) and GaN can operate at higher voltages and temperatures, making them ideal for high-power and high-frequency applications. They offer superior performance in power electronics and radio frequency applications compared to traditional silicon.

2. **Improved Energy Efficiency**
   Low Power Consumption: Advanced materials can significantly reduce power consumption in electronic devices. For example, perovskite solar cells demonstrate high power conversion efficiency, which enhances the energy harvesting capabilities of photovoltaic systems.

   High Luminous Efficiency: Perovskite LEDs exhibit high luminous efficiency and color purity, improving the performance of lighting and display technologies.

3. **Flexibility and Integration**
   Flexible Electronics: Materials like graphene and transition metal dichalcogenides (TMDs) enable the development of flexible and wearable electronics. Their flexibility, combined with high performance, allows for new applications in smart textiles, flexible displays, and sensors.

1.18.2 Cost Considerations

1. Material Costs:
   Raw Materials and Precursor Costs: The cost of raw materials and precursors for emerging materials can be high. For example, certain high-purity chemicals used in the synthesis of perovskites or TMDs can drive up production costs.

   Economic Viability: The economic feasibility of using emerging materials is often affected by their cost relative to the performance benefits they offer. High material costs can impact the overall affordability of devices and technologies.

2. Manufacturing and Fabrication Costs
   Complex Processing: Many advanced materials require complex and precise fabrication techniques. For instance, the production of high-quality graphene often involves costly processes such as chemical vapor deposition (CVD) or epitaxial growth, which can increase manufacturing costs.

   Scalability Challenges: Scaling up production from laboratory to industrial levels presents challenges that can affect cost. Issues related to process optimization, uniformity, and yield must be addressed to make large-scale production economically viable.

3. Integration Costs
   Compatibility with Existing Technologies: Integrating new materials with existing semiconductor technologies may require modifications to current manufacturing processes and equipment. These adjustments can lead to additional costs in terms of equipment upgrades and process adaptation.

   Infrastructure Investment: Developing new manufacturing infrastructure or modifying existing facilities to accommodate emerging materials can be capital-intensive. Investments in clean rooms, specialized processing lines, and quality control systems are often necessary.

4. Balancing Performance and Cost
   Performance Gains vs. Costs: When evaluating emerging materials, it is essential to perform a cost-benefit analysis to determine whether the performance gains justify the additional costs. For example, while SiC and GaN offer superior performance in high-power applications, their higher material costs must be weighed against the potential efficiency improvements and longevity of the devices.

   Long-Term Savings: In some cases, the higher initial costs of advanced materials can be offset by long-term savings due to improved efficiency and reduced maintenance costs. For instance, the high efficiency and durability of perovskite solar cells may result in lower overall costs for solar energy systems over their lifetime.

5. **Market and Application-Specific Considerations**
   Target Applications: The suitability of emerging materials often depends on the specific application and market. For high-end applications where performance is critical, such as aerospace or telecommunications, the cost may be justified. In contrast, for consumer electronics, cost considerations may play a more significant role.

   Competitive Advantage: Emerging materials can provide a competitive edge in terms of performance and functionality, which may justify higher costs. For instance, advanced materials can enable the development of innovative products with unique features that set them apart in the market.

6. **Research and Development**
   Cost Reduction Through Innovation: Ongoing research and development efforts aim to reduce the costs associated with emerging materials. Innovations in synthesis methods, processing techniques, and material formulations can lead to more cost-effective solutions without compromising performance.

   Scaling Economies: As production volumes increase and technologies mature, economies of scale can help reduce costs. Advances in manufacturing processes and increased market adoption can lead to lower costs for emerging materials over time.

1.18.3 Conclusion of this section

The trade-off between performance and cost in emerging semiconductor materials involves balancing the significant performance gains offered by materials such as graphene, perovskites, SiC, GaN, and TMDs against their associated costs. While these materials offer enhanced electrical properties, energy efficiency, and flexibility, their higher raw material, manufacturing, and integration costs present challenges. A thorough cost-benefit analysis, market-specific considerations, and ongoing research and development efforts are essential to optimizing the use of these materials and achieving a favorable balance between performance and cost.

1.19 ENVIRONMENTAL AND HEALTH CONCERNS: PRODUCTION AND DISPOSAL OF EMERGING SEMICONDUCTOR MATERIALS

The production and disposal of emerging semiconductor materials, such as graphene, perovskites, silicon carbide (SiC), gallium nitride (GaN), and transition metal dichalcogenides (TMDs), raise important environmental and health considerations. Addressing these concerns is crucial for ensuring the sustainability and safety of semiconductor technologies [8].

1.19.1 Environmental Impact

1. Resource Extraction and Use
   Raw Material Mining: The extraction of raw materials required for advanced semiconductors, such as rare earth elements and metals, can lead to significant environmental impacts, including habitat destruction, soil erosion, and water pollution. For example, mining for elements used in perovskites or TMDs can be environmentally invasive.

   Energy Consumption: The synthesis of advanced materials often requires substantial energy input. High-temperature processes and complex fabrication techniques can result in considerable energy consumption, contributing to greenhouse gas emissions if fossil fuels are used.

2. Waste Generation
   Manufacturing Waste: The production of advanced semiconductor materials can generate hazardous waste, including chemical byproducts and solvents. Proper management and disposal of these wastes are essential to minimize environmental contamination.

   End-of-Life Disposal: The disposal of electronic devices containing emerging materials can pose environmental challenges. Materials like perovskites may contain toxic elements, such as lead, which require careful handling and recycling to prevent environmental contamination.

3. Recycling and Circular Economy
   Material Recovery: Developing efficient recycling methods for advanced materials is crucial for reducing environmental impact. Recycling can recover valuable materials from end-of-life devices and reduce the need for raw material extraction.

   Circular Economy Practices: Emphasizing circular economy principles, such as designing for disassembly and reuse, can minimize waste and enhance the sustainability of semiconductor technologies.

1.19.2 Health Risks

1. Exposure to Hazardous Materials
   Chemical Hazards: The production of materials like perovskites involves the use of hazardous chemicals, such as lead and solvents. Exposure to these chemicals during manufacturing or handling can pose health risks to workers and the surrounding community.

Nanomaterial Risks: For materials like graphene, which are used in nanomaterial form, there may be concerns about potential health effects from inhalation or skin contact. Nanoparticles can exhibit different properties compared to their bulk counterparts and may pose unique health risks.

2. Toxicity of Materials
   Lead and Heavy Metals: Perovskites often contain lead, which is a toxic element. Long-term exposure to lead can cause health problems, including neurological damage and other serious health conditions. Research into lead-free perovskites and safe handling procedures is ongoing to mitigate these risks [7].

   Gallium and Silicon Compounds: The production and disposal of GaN and SiC devices may involve compounds containing gallium or silicon, which can be hazardous if not properly managed.

3. Worker Safety
   Protective Measures: Ensuring worker safety in facilities that manufacture advanced materials requires implementing stringent health and safety protocols. This includes providing appropriate personal protective equipment (PPE), controlling exposure to hazardous substances, and conducting regular health monitoring.

   Training and Awareness: Workers should be trained on the safe handling of chemicals and materials, as well as emergency procedures in case of exposure or accidents.

4. Mitigation Strategies
   Developing Safer Materials, such as Lead-Free Alternatives: Research into lead-free or less toxic alternatives to traditional semiconductor materials can reduce environmental and health risks. For example, developing lead-free perovskite formulations or exploring other nontoxic materials can mitigate concerns associated with lead.

   Green Chemistry: Adopting green chemistry principles in the synthesis of advanced materials can minimize the use of hazardous substances and reduce environmental impact. This includes using less toxic solvents and developing more sustainable processing methods.

5. Improving Waste Management
   Efficient Recycling Systems: Developing efficient recycling systems for electronic waste, including devices containing emerging materials, can help recover valuable materials and reduce environmental contamination. Advances in recycling technologies and infrastructure are crucial for effective waste management [6].

Waste Reduction: Implementing waste reduction strategies in manufacturing processes, such as minimizing the use of hazardous chemicals and optimizing material use, can lower the environmental footprint of semiconductor production.

6. **Regulations and Standards**
   Environmental Regulations: Compliance with environmental regulations and standards, such as those related to hazardous waste management and emissions, is essential for minimizing environmental impact. Adhering to these regulations helps ensure that semiconductor production and disposal are conducted in an environmentally responsible manner.

   Health and Safety Standards: Establishing and enforcing health and safety standards for the handling and production of advanced materials protects workers and communities from potential health risks. Regular audits and inspections can ensure compliance with safety standards.

7. **Research and Innovation:**
   Sustainable Practices: Ongoing research into more sustainable practices, including energy-efficient manufacturing techniques and safer material alternatives, can address environmental and health concerns. Innovations in material science and manufacturing technology are key to reducing the overall impact of semiconductor technologies [12].

1.19.3 Conclusion of this section

The production and disposal of emerging semiconductor materials present significant environmental and health challenges, including resource extraction impacts, hazardous waste generation, and potential toxicity. Addressing these concerns involves developing safer materials, improving waste management, and adhering to environmental and health regulations. Research and innovation play a crucial role in advancing sustainable practices and reducing the negative impact of semiconductor technologies on the environment and human health.

1.20 SUMMARY OF KEY POINTS, LOOKING TO THE FUTURE, AND FINAL CONCLUSION

1.20.1 Emerging Materials Discussed

1. **Graphene**
   Properties: High electron mobility, flexibility, and mechanical strength.

   Impact: Potential for faster transistors, high-performance electronic devices, and flexible electronics.

2. **Transition Metal Dichalcogenides (TMDs)**
   Examples: MoS_2, WS_2.

   Properties: Thin-film semiconductors with direct bandgaps.

   Impact: Used in thin-film transistors, potential for flexible and low-power electronic applications.

3. **Silicon Carbide (SiC)**
   Properties: Wide bandgap, high thermal conductivity.

   Impact: Suitable for high-power, high-frequency, and high-temperature applications, such as power electronics and RF devices.

4. **Gallium Nitride (GaN)**
   Properties: Wide bandgap, high electron mobility.

   Impact: Improves efficiency in power electronics and RF devices, capable of high-speed and high-power operation.

5. **Perovskites**
   Properties: High power conversion efficiency, tunable bandgap.

   Impact: Promises significant advancements in solar cells and LEDs, with potential for cost-effective and high-performance optoelectronic devices.

6. **Topological Insulators**
   Properties: Conducting surface states with insulating bulk.

   Impact: Potential for new electronic and spintronic devices due to unique electronic properties.

1.20.2 Looking to the Future

1. **Continued Development**
   Materials Innovation: Ongoing research will focus on improving the performance and scalability of emerging materials. This includes optimizing synthesis methods, enhancing material properties, and developing cost-effective production techniques [1].

   Hybrid Technologies: Combining emerging materials with traditional semiconductor technologies could lead to innovative solutions that leverage the strengths of both. Hybrid devices could offer enhanced performance and functionality.

2. **Integration and Standardization**
   Manufacturing Integration: Efforts will continue to integrate emerging materials into existing semiconductor manufacturing processes.

This involves addressing compatibility issues, optimizing fabrication techniques, and modifying manufacturing infrastructure.

Standards Development: Establishing industry standards for emerging materials will be crucial for ensuring consistency, reliability, and widespread adoption in semiconductor technologies.

3. **Sustainability and Environmental Impact**
   Eco-Friendly Practices: Research will also focus on developing more sustainable practices for the production and disposal of emerging materials. This includes improving recycling methods, reducing hazardous waste, and finding safer material alternatives.

   Regulatory Compliance: Adhering to environmental regulations and health standards will be essential for minimizing the impact of new materials on the environment and human health.

4. **Commercialization and Market Adoption**
   Cost Reduction: As production techniques improve and economies of scale are realized, the costs associated with emerging materials are expected to decrease, making them more accessible for commercial applications [10].

   Diverse Applications: Emerging materials will likely find applications across various sectors, including consumer electronics, renewable energy, automotive, and telecommunications, driving innovation and new product development.

5. **Long-Term Prospects**
   Advancement in Performance: The continued development of emerging materials is expected to lead to significant advancements in semiconductor performance, including faster, more efficient, and more versatile electronic devices.

   Technological Breakthroughs: Future breakthroughs in material science and engineering could enable new technologies and applications that we cannot yet envision, shaping the future of semiconductor design and its impact on technology [12].

1.20.3 Final Conclusion

The exploration and integration of emerging materials in semiconductor design hold the promise of revolutionizing the industry with advancements in performance, efficiency, and functionality. As research progresses, these materials are expected to overcome current challenges related to cost, scalability, and environmental impact. The future of semiconductor design will likely be characterized by a continued emphasis on innovation, sustainability, and the development of new technologies that harness the unique properties of these advanced materials.

REFERENCES

1. Berkley, S. A., et al. (2020). "Emerging Semiconductor Materials for Advanced Technologies." *Journal of Materials Science*, 55(12), 4936–4953.
2. Geim, A. K., & Novoselov, K. S. (2007). "The Rise of Graphene." *Nature Materials*, 6(3), 183–191.
3. Asan, M. Z., & Kane, C. L. (2010). "Colloquium: Topological Insulators." *Reviews of Modern Physics*, 82(4), 3045–3067.
4. Hsu, W.-T., et al. (2019). "Challenges in the Synthesis and Application of Emerging Semiconductor Materials." *Advanced Materials*, 31(16), 1807007.
5. Huang, M., et al. (2017). "Challenges in Scaling Down Silicon Technology." *IEEE Transactions on Electron Devices*, 64(10), 4320–4330.
6. Jansen, P., et al. (2019). "Silicon Carbide and Gallium Nitride for Power Electronics." *Materials Science and Engineering R: Reports*, 137, 1–39.
7. Kang, J., et al. (2015). "High-Performance Thin-Film Transistors Using Transition Metal Dichalcogenides." *Nature Communications*, 6, 7816.
8. Kojima, A., et al. (2009). "Organometal Halide Perovskites as Visible-Light Sensitizers for Photovoltaic Cells." *Journal of the American Chemical Society*, 131(17), 6050–6051.
9. Moore, G. E. (1965). "Cramming More Components onto Integrated Circuits." *Electronics*, 38(8), 114–117.
10. Novoselov, K. S., et al. (2012). "A Roadmap for Graphene." *Nature*, 490(7419), 192–200.
11. SIA (2021). "Global Semiconductor Industry Outlook." *Semiconductor Industry Association.*
12. Sze, S. M. (2021). *Semiconductor Devices: Physics and Technology*. Wiley.
13. Wright, A. (2013). "The Evolution of Moore's Law." *IEEE Spectrum*, 50(1), 26–35.
14. Yuan, W., et al. (2021). "Environmental and Health Concerns in Semiconductor Production." *Journal of Cleaner Production*, 276, 124251.

Chapter 2

Materials Frontiers

Pioneering Semiconductor Design

Soumya Sen, Mamta Khosla, and Ashish Raman

2.1 INTRODUCTION

The semiconductor industry is the cornerstone of modern electronics, enabling the development of everything from microprocessors to solar cells. Materials with electrical conductivity involving insulators and conductors that can be altered by temperature changes, electric fields, and doping are classified as semiconductors. They are essential to many electronic devices, such as integrated circuits (ICs), sensors, diodes, and transistors, due to their capacity to manipulate electrical attributes.

The progression of semiconductor technology has followed a trajectory of continuous miniaturization and performance enhancement, driven by the industry's adherence to Moore's Law, which anticipated that transistor compactness would double on integrated circuits approximately every two years. This has led to exponential growth in computing power, data processing speeds, and energy efficiency. Conventional silicon (Si) is used as the primary material in the devices owing to its surplus, steady qualities, and solidified manufacturing procedures. Silicon oriented metal-oxide-semiconductor field-effect transistors (MOSFETs) are the mainstay of the semiconductor sector, powering the daily usage electronics gadgets.

However, silicon reaches its physical and performance limits as device dimensions shrink to the nanoscale. Challenges such as short-channel effects, increased leakage currents, and heat dissipation issues are increasingly complex to manage with silicon alone. This has spurred research into alternative materials that can complement or replace silicon in semiconductor devices. Pursuing these new materials is essential to sustaining innovation and meeting the demands for higher performance, energy efficiency, and integration in modern electronics [1,2].

2.2 IMPORTANCE OF EMERGING MATERIALS

The imperative drives the pursuit of emerging materials in semiconductor design to triumph over the inherent constraints of orthodox silicon-based

 DOI: 10.1201/9781003533535-2

technologies and meet modern electronic applications' escalating demands. They offer distinctive electrical, optical, thermal, and mechanical functionalities that can significantly enhance device performance, enable new functionalities, and open avenues for innovation across various technological domains. One of the primary motivations for exploring new materials is to achieve higher electron mobility, which directly translates to faster device operation and improved efficiency. Gallium arsenide (GaAs) and indium gallium arsenide (InGaAs) demonstrate greater electron mobility than silicon, making them well-suited for rapid and ultra-frequency deployments such as pioneering communication systems and radar technologies. Relative to silicon, InGaAs have substantially greater electron mobility, which makes them desirable for high-speed and frequency tasks like radar technology and sophisticated communication networks.

Functioning in restrictive circumstances is a further crucial aspect. Wide-bandgap semiconductors, like silicon carbide (SiC) and gallium nitride (GaN), are ideal for power electronics, electric cars, and aerospace purposes because they can operate efficiently at elevated temperatures, voltages, and frequencies. These materials enhance energy efficiency and system reliability, making them well-suited for sustainable and robust technological solutions, particularly in power electronics, electric vehicles, and aerospace applications. Robust and durable technical alternatives rely on deeper system dependability and energy efficacy, both of which are facilitated by these materials.

Emerging materials also facilitate the manufacturing of wearable and pliable electronics, which require mechanically flexible and electronically functional substrates. Organic semiconductors and two-dimensional (2D) materials like graphene and transition metal dichalcogenides (TMDs) offer the necessary flexibility. They may be integrated into an assortment of materials, such as fabrics and polymers, given that they can be handled at modest temperatures. This capability is pivotal for applications in healthcare monitoring, bright clothing, and flexible displays. In optoelectronics, materials with direct bandgaps and light-matter solid interactions are crucial. TMDs and perovskites have shown exceptional light emission and detection properties, smoothing the approach for efficient LEDs, lasers, and photodetectors. Effective LEDs, lasers, and photodetectors are rendered accessible by these materials, which additionally boost optical communication systems, imaging technologies, and solar energy harvesting, discharge, and recognition.

Furthermore, the advent of quantum computing and spintronics necessitates materials with novel quantum properties. Topological insulators, 2D materials, and certain heterostructures exhibit the phenomena essential for developing quantum bits (qubits) and spin-based electronic devices, which can revolutionize computing by offering unprecedented processing speeds and data storage capabilities. Integrating emerging materials into semiconductor devices also encourages the exploration of new device architectures and fabrication techniques. Recent developments in material synthesis techniques, specifically molecular beam epitaxy (MBE), chemical vapour deposition (CVD), and atomic layer deposition (ALD), have facilitated meticulous regulation of material composition and architecture at the atomic level. This precision facilitates the engineering of material properties and the realization of complex structures like superlattices and quantum wells, crucially required for next-generation electronic and photonic devices. Incorporating potentially synthesized materials into semiconductor devices also encourages the exploration of new device architectures and fabrication techniques.

Despite the promising advantages, incorporating emerging materials into mainstream semiconductor technology presents challenges, including material stability, scalability, compatibility with existing fabrication processes, and economic viability. Addressing these challenges requires concerted efforts in material science research, process development, and industry collaboration to ensure that the benefits of these materials can be fully realized in practical applications [3–7, 51,52].

2.3 III–V COMPOUND SEMICONDUCTORS

As illustrated in Figure 2.1, III–V compound semiconductors are created by blending elements from Group V (nitrogen, phosphorus, arsenic, and antimony) and Group III (aluminium, gallium, and indium) of the periodic table. In recognition of their exceptional electrical, optical, and thermal attributes, which frequently outperform those of silicon, the traditional semiconductor material, these compounds have garnered considerable attention in the semiconductor sector.

Their crystal formations and bonding features contribute to their distinct qualities. These materials are perfect for optoelectronic applications encompassing light-emitting diodes (LEDs), laser diodes, and photodetectors as they usually feature a straight bandgap, which enables effective electron–hole pair production and recombination. High electron mobility is another feature of many III–V compounds, which is helpful for enhanced featured devices.

The exploration and utilization have expanded significantly in recent decades, spurred on by a desire for advanced communication technologies, streamlined power electronics, and innovative optoelectronic devices. This section delves into III–V compound semiconductors' fundamental properties, applications, and challenges, highlighting their growing importance in the semiconductor industry [8–11].

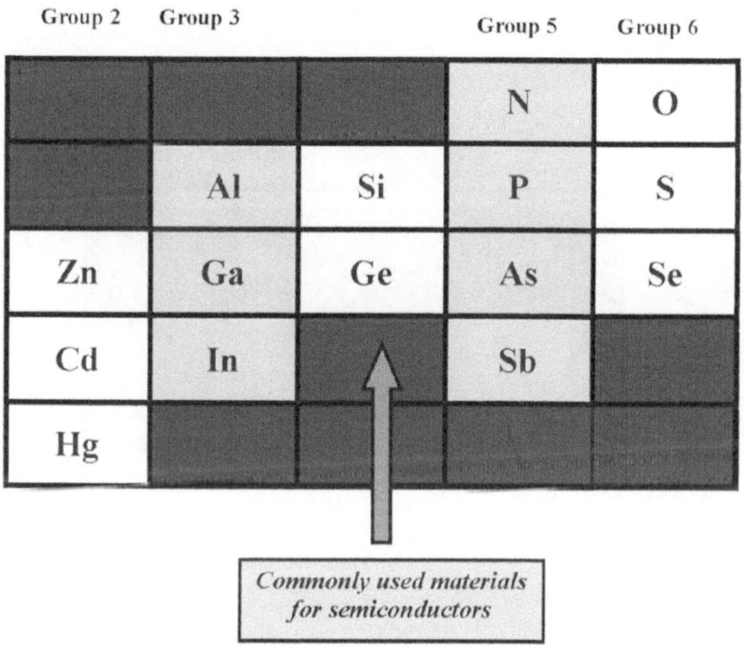

Figure 2.1 Group III-V elements represent compound semiconductors, whereas Group IV elements, such as silicon and germanium, represent conventional semiconductors.

2.3.1 Properties of III–V Compound Semiconductors

The critical properties have been summed up below:

1. **Direct Bandgap:** One of the most critical attributes is their direct bandgap. Unlike silicon, which has an indirect bandgap, materials like GaAs, InP, and GaN have direct bandgaps, allowing them to emit and absorb light efficiently. This property is particularly advantageous for optoelectronic devices, where efficient light emission or detection is crucial.

   For instance, GaAs have a direct bandgap of approximately 1.41 eV, making it suitable for infrared and visible light applications. In contrast, GaN has a broader bandgap of around 3.39 eV, which allows it to emit light in the ultraviolet (UV) spectrum and is extensively used in UV LEDs and robust electronic devices, illustrated in Figure 2.2.

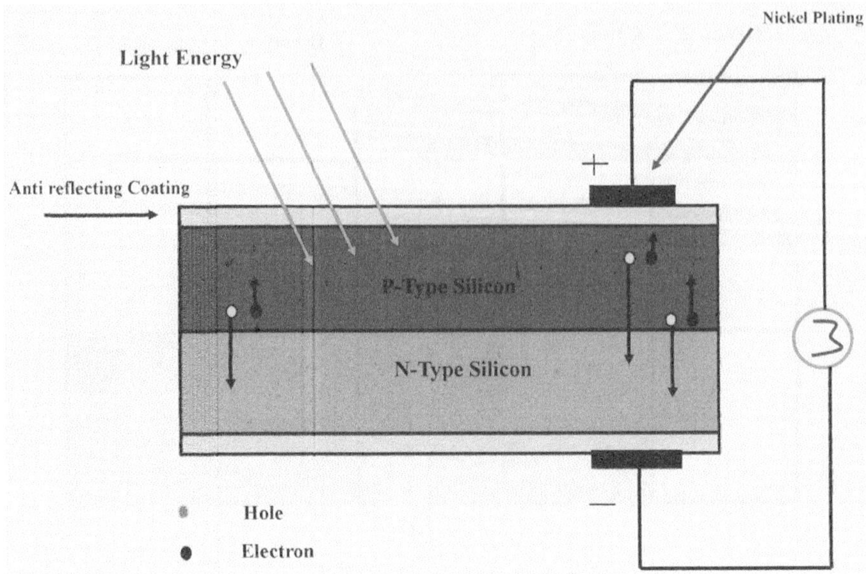

Figure 2.2 Light emitting diode.

2. **High Electron Mobility:** III–V semiconductors, especially GaAs and InP, are known for their superior electron mobility, which enables faster electron transport compared to silicon. High electron mobility is essential for devices that operate at high frequencies, such as radio frequency (RF) amplifiers and fast paced transistors. For example, the electron mobility in GaAs is about five times higher than that in silicon, making GaAs-based devices ideal for microwave and millimetre-wave applications.

   The high mobility also reduces power consumption and heat generation in electronic devices, making III–V semiconductors attractive for low-power, high-performance applications.
3. **Thermal Conductivity and Stability:** Certain III–V compounds, such as GaN and SiC (silicon carbide, often grouped with III–Vs for similar properties), exhibit excellent thermal conductivity and stability, making them suitable for optimized featured applications. GaN, for instance, can operate at much higher temperatures than silicon without degrading performance, which is critical for power electronics and devices exposed to harsh environments.

   The combination of high thermal conductivity and wide bandgap allows III–V semiconductors to handle large amounts of power while maintaining efficiency and reliability.

4. **Lattice Matching:** Another significant advantage is the ability to engineer heterostructures by layering different III–V compounds with closely matched lattice constants. This property enables the creation of quantum wells, superlattices, and other advanced structures with tailored electronic and optical properties. For example, combining GaAs and AlGaAs allows for creating high-electron-mobility transistors (HEMTs) and quantum cascade lasers (QCLs) with superior performance. Lattice matching also reduces the occurrence of defects at the interface of different materials, leading to higher-quality devices with improved performance and longevity.

2.3.2 Applications in High-Frequency Electronics

III–V compound semiconductors have revolutionized high-frequency electronics, where their superior electron mobility and high saturation velocity offer significant advantages over silicon-based devices. These materials enable the design and fabrication of transistors and other electronic components that operate at frequencies well beyond those achievable with silicon.

One of the most essential applications of III–V semiconductors in high-frequency electronics is in the field of radio frequency (RF) and microwave communication systems. GaAs and InP-based devices are commonly used in RF amplifiers, mixers, and oscillators, critical components in wireless communication systems, radar, and satellite communications. The high electron mobility in these materials allows for the rapid switching of transistors, leading to faster signal processing and higher data transmission rates.

GaAs, in particular, have become a material of choice for many RF utilities due to their proficiency at frequencies in the gigahertz (GHz) range. GaAs-based RF transistors, such as metal-semiconductor field-effect transistors (MESFETs) and HEMTs, offer low noise, high gain, and excellent thermal stability, making them ideal for use in low-noise amplifiers (LNAs) including power amplifiers in mobile phones, satellite receivers, and other communication devices.

InP is another III–V compound semiconductor that has found widespread use in high-frequency electronics, particularly in applications requiring even higher speeds and lower noise figures than GaAs. InP-based HEMTs and heterojunction bipolar transistors (HBTs) employed in millimetre-wave usages including advanced radar systems, high-speed fibre-optic communication, and imaging systems. InP's higher electron mobility and velocity saturation than GaAs allow superior performance in these demanding applications [12–15].

The development of GaN-based electronics has further expanded the capabilities of III–V semiconductors in enhanced applications. GaN's wide bandgap, elevated breakdown voltage, and electron mobility make it an

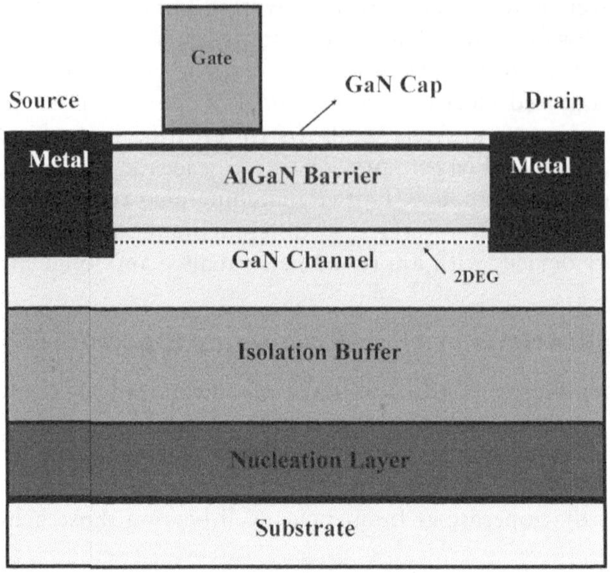

Figure 2.3 Gallium nitride-based high electron mobility transistors.

ideal material for power amplifiers, especially in applications requiring high efficiency and power density. GaN-based high-electron-mobility transistors (GaN HEMTs), as depicted in Figure 2.3, are now widely used in 5G communication systems, satellite communications, and radar systems, where they provide high power output, efficiency, and linearity.

In addition to RF and microwave applications, III–V semiconductors are also being explored in terahertz (THz) devices, which operate between microwave and infrared frequencies. Terahertz technology holds promise for imaging, spectroscopy, and communication applications, with potential uses in security screening, medical diagnostics, and high-speed wireless networks. III–V materials like InP and GaAs are considered promising candidates for THz devices due to their resilient electron mobility and proficiency to generate and detect THz radiation.

2.3.3 Heterostructure Integration and Benefits

Heterostructure integration refers to combining semiconductor materials with varying bandgaps, lattice constants, and other properties into a single device structure. With this approach, inventive semiconductor devices can be generated that operate superior to their counterparts, made of a homogeneous material by harnessing the inherent attributes of each material layer.

Layering III–V compound semiconductors is a prevalent way for developing heterostructures like GaAs/AlGaAs or InP/GaInAs. These heterostructures enable the engineering of band structures, carrier confinement, and other electronic properties, leading to significant enhancements in device performance. These heterostructures enable the engineering of band structures, carrier confinement, and other electronic properties, leading to significant enhancements in device performance.

1. **Bandgap Engineering:** One of the primary advantages of heterostructure integration is the ability to engineer the bandgap of a device. It is possible to create regions within a device with specific electronic properties by carefully selecting and combining materials with different bandgaps. For example, a GaAs/AlGaAs heterostructure can create quantum wells, where carriers are confined to a thin layer of GaAs with a lower bandgap, surrounded by AlGaAs with a higher bandgap. This confinement leads to discrete energy levels and enhanced optical and electronic properties, essential for devices like laser diodes and high-electron-mobility transistors (HEMTs).
2. **Carrier Confinement:** The use of heterostructures allows for the localization of charge carriers, such as electrons and holes, which in turn enhances the device's effectiveness. As an example, a two-dimensional electron gas (2DEG) is created at the interface of two materials with differing bandgaps, like AlGaAs and GaAs, in a high-efficiency metal-oxide semiconductor transistor (HEMT). The absence of impurity scattering causes the 2DEG to have high electron mobility, which in turn causes the device to operate quicker and consume minimal power [16–18].

   This carrier confinement is also crucial in optoelectronic devices like quantum well lasers, where the confinement of carriers in quantum wells enhances light emission efficiency and allows for generating coherent light with specific wavelengths.
3. **Strain Engineering:** Heterostructure integration also allows strain engineering, where the lattice mismatch between different materials induces strain in the crystal lattice. This strain can modify the electronic band structure, leading to changes in carrier mobility, bandgap, and other properties. Strained heterostructures are widely used in high-performance transistors, where the strain enhances carrier mobility and increases the device's speed.

   For example, strained silicon (Si) can be integrated with III–V materials to create high-speed transistors that combine the benefits of silicon's scalability with the superior electronic properties of III–V compounds.
4. **Thermal Management:** Using materials with different thermal conductivities in a heterostructure can improve device thermal

management. By incorporating materials with high thermal conductivity, such as GaN, into a device structure, heat generated during operation can be more effectively dissipated, improving device reliability and performance. This is particularly important in high-power and high-frequency devices, where thermal management is a critical factor in determining the overall efficiency and longevity of the device [19–21].

2.4 TRANSITION METAL DICHALCOGENIDES (TMDS)

Transition Metal Dichalcogenides (TMDs) represent a class of layered materials with the general formula MX_2, where M is a transition metal (e.g., Mo, W, Ti) and X is a chalcogen (e.g., S, Se, Te). These materials have recently gained significant attention due to their unique properties, which emerge when thinned to a single or few layers. TMDs are part of the broader family of two-dimensional (2D) materials, including graphene, but they offer distinct advantages, particularly in electronics and optoelectronics.

TMDs can exist in different phases, such as the semiconducting 2H phase, the metallic 1T phase, and the less common 1T' phase, often associated with topological properties. The ability to transition between these phases through external stimuli like strain, electric fields, or chemical doping adds another layer of functionality to these materials, making them highly versatile for various applications.

The most studied TMDs include molybdenum disulfide (MoS_2), tungsten diselenide (WSe_2), and molybdenum diselenide ($MoSe_2$). When reduced to a monolayer, these materials exhibit a transition from an indirect to a direct bandgap, which significantly enhances their optical properties. This feature, combined with their inherent mechanical flexibility and chemical stability, positions TMDs as promising candidates for next-generation electronic devices, sensors, and energy applications.

2.4.1 Electronic and Optical Properties

The electronic and optical properties of TMDs are profoundly influenced by their reduced dimensionality. In bulk form, TMDs generally possess an indirect bandgap; however, when thinned to a monolayer, they exhibit a direct bandgap. For instance, bulk MoS_2 has an indirect bandgap of about 1.2 eV, transforming into a direct bandgap of approximately 1.8 eV at the monolayer level. This direct bandgap makes TMDs highly efficient at absorbing and emitting light, crucial for optoelectronic devices like photodetectors, light-emitting diodes (LEDs), and lasers.

1. **Band Structure and Carrier Mobility:** The band structure of TMDs is characterized by a strong spin-orbit coupling, especially in heavier

transition metals like tungsten. This results in spin-split valence bands at the K and K' points of the Brillouin zone, giving rise to interesting spintronic and valleytronic phenomena. However, the carrier mobility in TMDs is generally lower compared to traditional semiconductors like silicon, primarily due to the solid electron-phonon interactions in these materials.

2. **Optical Absorption and Emission:** TMDs exhibit strong optical absorption across a wide range of the electromagnetic spectrum, from visible to near-infrared. This is complemented by their ability to emit light efficiently at specific wavelengths corresponding to their bandgap. The excitonic effects, where electron–hole pairs are firmly bound due to reduced dielectric screening in 2D materials, are prominent in TMDs, leading to sharp and intense photoluminescence peaks. These excitonic features can be tuned by changing the layer number, applying strain, or altering the dielectric environment, making TMDs highly versatile for tunable optoelectronic applications.
3. **Quantum Confinement and Tunability:** The quantum confinement in TMDs at the monolayer level enhances their electronic and optical properties, making them highly sensitive to external perturbations. The tunability of these properties through gating, chemical functionalization, or heterostructure engineering allows for designing custom devices tailored to specific needs, such as tunable photodetectors or flexible transistors0 [22–27].

2.4.2 Van der Waals Heterostructures and Applications

Van der Waals (vdW) heterostructures are created by layering various 2D materials, such as TMDs, on top of one another lacking the need for lattice matching. The weak vdW forces that hold the layers together allow for combining materials with vastly different electronic properties, creating heterostructures with novel functionalities. TMDs are crucial in these heterostructures, often used as the semiconducting component in conjunction with other 2D materials like graphene or hexagonal boron nitride (hBN).).

1. **Band Alignment and Charge Transfer:** In vdW heterostructures, the band alignment between the layers can be engineered to create type-I, type-II, or type-III heterojunctions. This flexibility in band alignment is critical for applications in photodetectors, where type-II heterojunctions can facilitate efficient charge separation, or in tunnelling transistors, where type-III heterojunctions can achieve steep subthreshold swings.
2. **Tunnelling Devices:** Another notable application is tunnelling field-effect transistors (TFETs). By stacking TMD layers with different

bandgaps and using vdW contacts, it is possible to create devices with ultra-low power consumption and high on/off ratios. These devices are up-and-coming for next-generation low-power electronics, where diminishing energy intake is critical.

3. **Optoelectronics and Photonics:** TMD-based vdW heterostructures are also being explored for advanced optoelectronic devices, like flexible photodetectors, LEDs, and solar cells. Combining materials with complementary attributes, namely a high carrier mobility layer with a high optical absorption layer, facilitates the creation of effective and adaptable devices. Additionally, the solid light-matter interaction in TMDs makes them ideal for photonic applications, including modulators and detectors in integrated photonic circuits.
4. **Sensors and Biosensors:** The enhanced surface-to-volume ratio and sensitivity to environmental changes establish TMDs and their heterostructures as excellent candidates for sensing applications. When integrated into vdW heterostructures, these materials can create highly sensitive biosensors capable of detecting low concentrations of biomolecules or gases. The ability to tailor the electronic properties through heterostructure engineering further enhances the selectivity and sensitivity of these sensors.
5. **Quantum Technologies:** The quantum confinement effects and the solid spin-orbit interaction in TMDs open up possibilities for quantum technologies. For example, TMD-based heterostructures can create single-photon emitters essential for quantum communication and computing. The ability to control the emission properties through heterostructure engineering offers a pathway to develop on-demand quantum light sources integrated into scalable quantum circuits.

Transition metal dichalcogenides' unique electronic and optical properties are central to developing next-generation semiconductors. Integrating TMDs into van der Waals heterostructures unlocks new possibilities for designing devices with customized properties and enhanced performance. As research continues to explore the potential of these materials, TMDs are likely to play a pivotal role in multiple possibilities, from flexible electronics to quantum technologies, driving innovation in semiconductor design [6, 28–33,54].

2.5 TWO-DIMENSIONAL MATERIALS

In material science and technology, the investigation of two-dimensional materials notably graphene and boron nitride has opened up novel possibilities for use. Contemporary electronic, photonic, and spintronic devices are becoming possible thanks to these materials' distinct mechanical, electrical, thermal, and optical characteristics. The promise of heterostructures is further enhanced by integrating diverse 2D materials into them, enabling the fabrication of devices with customized properties that surpass standard materials in numerous important areas. These materials are expected to be important as 2D material and heterostructure research develops.

2.5.1 Graphene: Properties and Applications

Graphene, a single layer of carbon atoms arranged in a hexagonal lattice, is one of the most revolutionary discoveries in materials science, as illustrated in Figure 2.4. Graphene has generated immense interest since its isolation in 2004 due to its extraordinary properties, positioning it as a foundational material for numerous technological advancements.

1. **Electronic Properties:** Graphene exhibits remarkable electronic properties, notably its high electron mobility, which can exceed 200,000 $cm^2/V{\cdot}s$ at room temperature. This high mobility is attributed to the linear dispersion relation of electrons near the Dirac points in graphene's band structure, which behaves as massless Dirac fermions. This unique characteristic results in graphene's potential to conduct electricity with minimal resistance and negligible scattering.

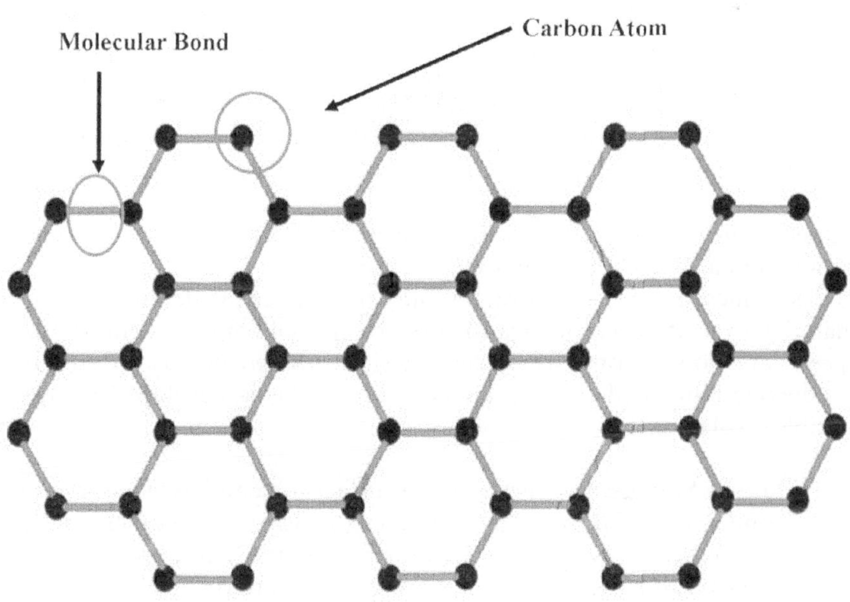

Figure 2.4 Graphene sheet showing the hexagonal structure and the individual atoms.

The absence of a bandgap, while a limitation for traditional semiconductor applications, also enables unique quantum effects such as the quantum Hall effect, possessing minimal resistance and negligible scattering.

2. **Mechanical Properties:** Graphene is renowned for its mechanical strength. It has a Young's modulus of approximately 1 TPa and an intrinsic tensile strength of around 130 GPa, making it one of the most robust materials ever measured. Despite its strength, graphene is also incredibly flexible and lightweight. These properties make graphene an ideal candidate for applications requiring durability and flexibility, such as flexible electronics, wearable devices, and even advanced composite materials used in aerospace and automotive industries.
   Despite its strength, graphene is also incredibly flexible and lightweight.
3. **Thermal Properties:** The thermal conductivity of graphene is exceptionally high, ranging between 3000 and 5000 W/m·K. This outstanding thermal property makes graphene highly effective in dissipating heat, which is particularly useful in managing thermal loads in electronic devices. Efficient heat dissipation prevents overheating in high-performance computing systems, LEDs, and other electronic components.
4. **Optical Properties:** Despite being only one atom thick, graphene absorbs about 2.3% of visible light, which is significant for a material of such minimal thickness. This, combined with its electrical conductivity, makes graphene suitable for transparent conductive films, critical components in displays, touchscreens, and solar cells. Modifying graphene's optical properties by chemical doping or structural alteration broadens its potential in photonic and optoelectronic applications.

Applications: Graphene's exceptional properties have led to its use being explored in a diverse range of applications:

Electronics: The high electron mobility of graphene makes it a promising material for next-generation transistors, which could operate at much higher speeds and lower power than current silicon-based devices. Transparent conductive films made from graphene are also being developed for use in flexible displays, OLEDs, and touchscreens.

1. **Energy Storage:** Graphene's large surface area and excellent electrical conductivity make it an ideal material for electrodes in supercapacitors and batteries. Graphene can improve charge rates, energy density, and overall battery lifespan in lithium-ion batteries.
2. **Sensors:** Graphene's sensitivity to changes in its environment makes it a robust material for sensors. Graphene-based sensors possess the

capability to identify individual molecules of gases, chemicals, and biomolecules, rendering them valuable for environmental monitoring, medical diagnostics, and safeguarding usages.

3. **Composite Materials:** When incorporated into polymers, metals, or ceramics, graphene significantly boosts the mechanical, thermal, and electrical properties of the host material. These composites are being applied in lightweight and robust materials-based areas of aerospace, automotive, and construction industries, where they are required [34–38,53].

2.5.2 Boron Nitride: Properties and Applications

Hexagonal boron nitride (hBN), often called "white graphene", is a two-dimensional material with a structure similar to graphene but composed of boron and nitrogen atoms. They exhibit a set of properties that complement those of graphene, making it an essential material in 2D materials.

1. **Electronic Properties:** Unlike graphene, hBN is an electrical insulator with a wide bandgap of approximately 5.9 eV. This insulating property makes it an ideal dielectric material for use in electronics, particularly as a substrate for other 2D materials like graphene. Its insulating nature helps maintain graphene's intrinsic electronic properties when used in electronic devices, minimizing interference and charge carrier scattering.
2. **Mechanical Properties:** hBN shares graphene's mechanical strength and flexibility, with a high thermal stability of up to 1000°C in air. This makes it suitable for applications in environments where high temperature and chemical stability are required. Additionally, hBN is known for its lubricating properties, which have been exploited in various industrial applications.
3. **Thermal Properties:** hBN exhibits excellent thermal conductivity, similar to graphene, which makes it valuable for thermal management in electronic devices. Its combination of high thermal conductivity and electrical insulation is particularly advantageous in heat spreaders and thermal interface materials, where efficient heat dissipation without electrical conduction is critical.
4. **Optical Properties:** hBN is transparent in the ultraviolet to the near-infrared range and has ultraviolet solid absorption capabilities. This makes it useful in applications like deep UV photonics, protective coatings, and transparent windows for optical devices.

Applications: Its unique properties enable its use in various applications across multiple industries:

1. **Substrates and Dielectrics:** hBN is widely used as a substrate for graphene and other 2D materials in electronic devices, providing an atomically flat, insulating surface. This helps to preserve the electronic properties of the overlying material, making the material an essential component in high-performance 2D material-based electronics.
2. **Thermal Management:** The high thermal conductivity and electrical insulation of hBN make it ideal for use in heat spreaders and thermal interface materials, which are essential in high-power electronic devices, LEDs, and other applications requiring efficient heat dissipation.
3. **Lubrication:** hBN is used as a lubricant in high-temperature environments where traditional lubricants fail. Its applications range from aerospace components to industrial machinery and high-performance engines, where reliable lubrication under extreme conditions is critical.
4. **Photonics and Optoelectronics:** hBN's wide bandgap and transparency suit deep UV photonics and optoelectronic applications. It is also being explored for UV lasers and detectors, whose properties offer significant advantages over conventional materials [39–42]

2.5.3 2D Heterostructures in Nanoelectronics and Spintronics

To form layered structures with novel properties, two-dimensional (2D) heterostructures are created by stacking different 2D materials, such as graphene, hBN, and transition metal dichalcogenides (TMDs). These heterostructures are held together by van der Waals forces, allowing the integration of materials with differing lattice constants and electronic properties.

1. **Nanoelectronics:** 2D heterostructures offer several advantages over traditional semiconductor materials in nanoelectronics. Devices with specific electronic characteristics can be built by meticulously specifying and stacking 2D substances. For example, graphene's high conductivity can be combined with hBN's insulating properties to create high-speed, low-power transistors with minimal charge carrier scattering. Additionally, the ability to control the band alignment between different layers enables the design of tunnelling transistors with steep subthreshold swings, which are essential for reducing power consumption in future electronic devices.
2. **Transistors:** 2D heterostructures can be used to develop transistors with high on/off ratios, low power consumption, and high speed. These transistors are being explored for use in next-generation computing and communication devices.

1. Sensors: The unique properties of 2D heterostructures make them ideal for developing susceptible sensors. These sensors can detect gases, chemicals, and biological agents at deficient concentrations, making them useful in environmental monitoring, medical diagnostics, and security applications.
2. Memory Devices: 2D materials can create non-volatile memory devices with high density and fast switching speeds, offering potential improvements over existing memory technologies.
3. **Spintronics:** Spintronics, or spin-based electronics, utilizes the spin of electrons and their charge to store and process information. 2D heterostructures are particularly well-suited for the applications due to the spin-orbit coupling and spin-filtering effects observed in certain 2D materials, such as TMDs and graphene.

Applications:

1. **Magnetic Tunnel Junctions:** Stacking layers of ferromagnetic and non-magnetic 2D materials can create magnetic tunnel junctions with high magnetoresistance. These are essential components in spintronic memory and logic devices, where manipulating electron spin is crucial for device operation.
2. **Spin Valves and Filters:** 2D heterostructures can be engineered to act as spin valves or filters, selectively allowing electrons with a particular spin orientation to pass through. This selective filtering enables the development of spintronic devices with low power consumption and high data processing speeds.
3. **Quantum Computing:** The spin properties of 2D materials and their ability to form heterostructures with precisely controlled interfaces offer potential pathways for developing qubits in quantum computing. In these systems, spin states can be used to encode and process quantum information, potentially developing highly efficient and scalable quantum computers [43–46].

2.6 CHALLENGES IN HETEROSTRUCTURE INTEGRATION

Integrating heterostructures, particularly those involving different semiconductor materials, presents significant challenges that must be addressed to exploit their potential in advanced electronic and photonic devices fully. These challenges arise from the complexities of combining materials with differing physical and chemical properties and the difficulties associated with large-scale manufacturing. This section elaborates on the critical challenges in heterostructure integration, including lattice mismatch and interface quality, epitaxial growth techniques, and scalability and manufacturing challenges. The integration of heterostructures, especially those comprising various semiconductor materials, poses considerable challenges that need to be resolved to maximize their usage in sophisticated electronic and photonic devices.

These challenges arise from the complexities of combining materials with differing physical and chemical properties, as well as the difficulties associated with large-scale manufacturing. This section elaborates on the key adversities in heterostructure integration, including lattice mismatch and interface quality, epitaxial growth techniques, and scalability and manufacturing challenges.

2.6.1 Lattice Mismatch and Interface Quality

Lattice mismatch occurs when two materials with different lattice constants are integrated into a heterostructure. The lattice constant is the distance between atoms in a crystal lattice, and when two materials with varied lattice constants are combined, strain and dislocations can occur at the interface. These dislocations can create defects in the crystal structure, which degrade the electrical and optical properties of the heterostructure. For example, dislocations can act as scattering centres for charge carriers in a heterostructure intended for high-speed electronics, reducing mobility and overall device performance.

Managing lattice mismatch is a critical challenge in heterostructure integration. To mitigate the mismatch effects, engineers often use buffer layers—intermediate layers of material with lattice constants that gradually transition between the two primary materials. While this technique can reduce strain and dislocations, it introduces additional complexity and potential defects into the heterostructure.

The interface quality between different materials in a heterostructure is crucial for device performance. An ideal interface should be defects-free and allow seamless charge transport across the junction. However, achieving such high-quality interfaces is difficult, especially when dealing with materials with different chemical bonding characteristics.

Interface defects, such as dangling bonds or impurities, can trap charge carriers, increasing recombination rates and reducing carrier lifetimes. These issues are particularly problematic in optoelectronic devices, where the efficiency of light emission or detection depends on the quality of the interface. In addition to electronic defects, differences in thermal expansion coefficients between materials can lead to mechanical stress and potential delamination during thermal cycling, further degrading interface quality.

Improving interface quality requires precise control over the growth process and surface passivation techniques to minimize defects. Advanced characterization methods, such as atomic resolution transmission electron microscopy (TEM) and scanning tunnelling microscopy (STM), are often employed to assess and improve interface quality.

2.6.2 Epitaxial Growth Techniques (MBE, CVD)

The practice of growing a crystalline layer of material on a substrate in a way that it aligns with the substrate's crystallographic alignment is known as epitaxial growth. This process is essential to producing heterostructures of excellent integrity with distinct interfaces and minimal deficiencies. Chemical Vapour Deposition (CVD) and Molecular Beam Epitaxy (MBE) are two of the most popular epitaxial growth techniques.

2.6.2.1 Molecular Beam Epitaxy (MBE)

MBE is a highly controlled epitaxial growth technique that involves the deposition of atomic or molecular beams of materials onto a heated substrate in an ultra-high vacuum environment, as shown in Figure 2.5. The critical advantage of MBE is its ability to precisely control the deposited layers' thickness, composition, and doping levels.

This precision makes MBE ideal for creating complex heterostructures, such as quantum wells and superlattices, where atomic-level control is essential. However, MBE has limitations, particularly in terms of scalability and cost. The ultra-high vacuum conditions and slow growth rates make

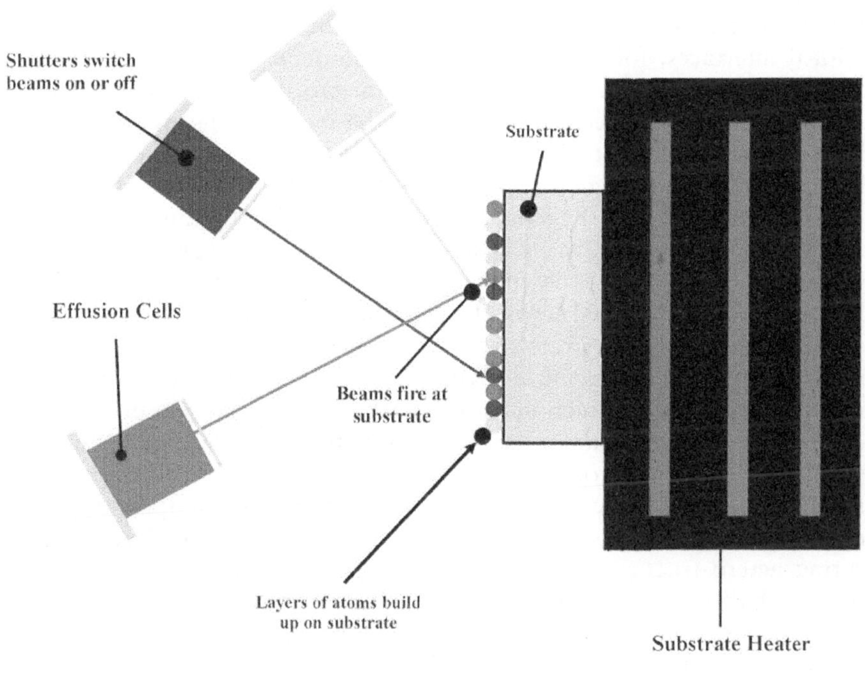

Figure 2.5 Molecular beam epitaxial growth process.

MBE relatively expensive and time-consuming. Additionally, the requirement for a high vacuum environment limits the size of the substrates that can be processed, making MBE less suitable for large-scale manufacturing.

2.6.2.2 Chemical Vapour Deposition (CVD)

CVD is another popular epitaxial growth technique that involves material deposition from a vapour phase onto a substrate. In CVD, precursor gases are introduced into a reaction chamber, decomposing or reacting on the heated substrate surface to form a solid film, depicted in Figure 2.6.

CVD is widely used for depositing various materials, including semiconductors, dielectrics, and metals. CVD offers several advantages over MBE, including higher growth rates and the ability to process larger substrates. This makes CVD more suitable for industrial-scale production of heterostructures. However, CVD also presents challenges, particularly in controlling the uniformity and purity of the deposited layers. The chemical reactions involved in CVD can produce unwanted by-products, leading to the growing film's contamination. Additionally, achieving sharp interfaces in CVD-grown heterostructures can be more challenging than in MBE, as the growth process is less controlled at the atomic level. To address the limitations of individual epitaxial growth techniques, researchers are exploring hybrid approaches that combine the strengths of MBE and CVD. For example, hybrid techniques may use MBE for the initial growth of critical interfaces, followed by CVD for the deposition of thicker layers. Such approaches aim to balance the need for precision with the demands of scalability and cost-effectiveness. One of the most significant challenges in heterostructure integration is scalability, the expertise to fabricate superior heterostructures in extensive proportions at an affordable expense.

While epitaxial growth techniques like MBE and CVD can produce high-quality heterostructures on small substrates, scaling these processes to larger substrates or wafer sizes is challenging. The uniformity of the deposited layers must be maintained across the entire substrate, which becomes increasingly difficult as the substrate size increases. In addition to uniformity, scalability involves reproducing the same high-quality heterostructures across multiple wafers or production runs. Variability in the growth process, such as temperature or precursor gas flow fluctuations, can lead to inconsistencies in the material properties and device performance. Addressing these scalability challenges requires advancements in growth equipment, process control, and in-situ monitoring techniques. Beyond scalability, manufacturing heterostructures for commercial applications presents several additional challenges. One major issue is the integration of heterostructures with existing semiconductor manufacturing processes. Traditional semiconductor fabrication relies heavily on silicon, and integrating new materials into these established processes can be complex and costly.

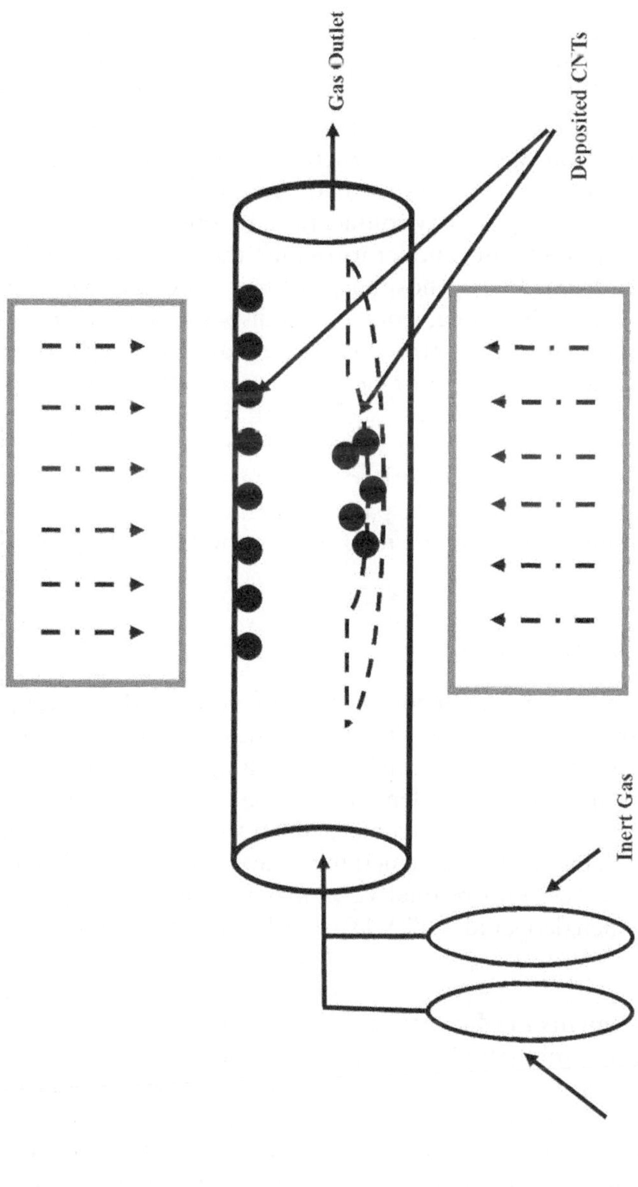

Figure 2.6 Chemical vapour deposition process.

For example, the introduction of III–V semiconductors or 2D materials into silicon-based fabrication lines may require new equipment, altered processing steps, or the development of entirely new fabrication techniques. Another manufacturing challenge is the need for high yield and reliability in heterostructure-based devices. Even small defects or variations in the heterostructure can significantly impact device performance, leading to lower yields and higher costs. Ensuring high yield and reliability requires rigorous process control, defect reduction strategies, and thorough testing and characterization of the heterostructures at multiple stages of production. The use of certain precursor materials and chemicals in epitaxial growth processes can pose environmental and safety risks. For example, some III–V compound semiconductors require the use of toxic or hazardous materials, such as arsine or phosphine, in their synthesis. Managing these risks requires careful handling, storage, and disposal of materials, as well as the implementation of safety protocols in the manufacturing environment.

Overcoming the challenges of heterostructure integration will require continued innovation in materials science, epitaxial growth techniques, and semiconductor manufacturing processes. Research efforts are focused on developing new materials with better lattice matching, improving interface quality through advanced surface passivation methods, and exploring novel growth techniques that combine the precision of MBE with the scalability of CVD.

2.6.3 Future Prospects

The ongoing research and incorporation of cutting-edge materials, notably in heterostructures, is expected to significantly impact semiconductor technology ahead. The constraints of existing silicon-based technologies becomes increasingly evident as the desire for increasingly potent, compact, and economical electronic gadgets rises. Prospective responses to these problems comprise emerging materials like III–V compound semiconductors, two-dimensional (2D) materials like graphene and transition metal dichalcogenides (TMDs), and sophisticated heterostructure topologies.

1. **Advancements in Material Science:** The exploration of new materials and their combinations are expected to yield breakthroughs in device performance. For instance, ongoing research into III–V semiconductors will likely result in new alloy compositions offering superior electron mobility, thermal conductivity, and bandgap tunability. Similarly, further studies on 2D materials are expected to unlock their potential in applications ranging from flexible electronics to quantum computing. Developing new synthesis techniques, such as atomic layer deposition (ALD) and chemical vapour deposition (CVD) at lower temperatures,

will be crucial in integrating these materials into existing semiconductor processes.

2. **Heterostructure Engineering:** The design and fabrication of heterostructures will become increasingly sophisticated, focusing on optimizing the interface quality, reducing defects, and achieving seamless integration of dissimilar materials. Advances in epitaxial growth techniques, such as hybrid molecular beam epitaxy (MBE)-CVD methods, will allow for the precise control of material deposition at the atomic scale while maintaining scalability for industrial applications. Additionally, the ability to engineer heterostructures with tailored electronic and optical properties will open new avenues for device innovation, particularly in optoelectronics, high-frequency communication, and sustainable computing.
3. **Quantum and Neuromorphic Computing:** Heterostructures based on III–V compounds and 2D materials are anticipated to assume a pivotal role in developing quantum and neuromorphic computing technologies. The distinctive electronic attributes of these materials, such as high carrier mobility, strong spin-orbit coupling, and tunable bandgaps, make them ideal candidates for quantum bits (qubits) and neuromorphic devices that mimic the human brain's neural networks. Integrating these advanced materials into quantum dots, topological insulators, and other quantum devices will be a primary focus of future research.
4. **Integration with Conventional Semiconductor Technology:** As emerging materials mature, their integration with conventional silicon-based semiconductor technology will become more critical. Hybrid approaches that combine the strengths of traditional silicon with the superior properties of III–V compounds and 2D materials will likely dominate future device architectures. This hybrid integration will enable the advancement of superior execution, energy-efficient devices that leverage the best attributes of each material system. For example, III–V compounds could be integrated with silicon in multi-junction solar cells or high-speed transistors. At the same time, 2D materials could be used for low-power, high-density memory and logic applications.
5. **Environmental and Sustainability Considerations:** As the semiconductor industry evolves, there will be an increasing emphasis on sustainability and environmentally friendly manufacturing processes. The use of toxic or rare materials in synthesizing advanced semiconductors will need to be carefully managed, focusing on recycling, waste reduction, and the development of alternative, eco-friendly materials. Additionally, the energy consumption of semiconductor fabrication processes will be a crucial consideration, driving the adoption of more energy-efficient techniques and materials [47–50].

2.7 CONCLUSION

Integrating emerging materials into semiconductor technology represents a paradigm shift that will redefine the capabilities and applications of electronic devices in the coming decades. The generation of more rapid, lessened, and additionally, energy-effective devices are made possible by the novel options presented by sophisticated heterostructure designs, 2D materials, and III–V compound semiconductors. These advancements provide a means for transcending the constraints of existing silicon-oriented technology. The challenges associated with heterostructure integration, such as lattice mismatch, interface quality, and scalability, are significant but not insurmountable. Continued advancements in material science, epitaxial growth techniques, and semiconductor manufacturing processes will be critical in addressing these obstacles and unlocking the utmost capacity of heterostructures. In the near term semiconductor technology will be shaped by the successful integration of these emerging materials, leading to innovations in an array of disciplines, from quantum computing and neuromorphic devices to flexible electronics and high-efficiency photovoltaics. The upcoming wave of semiconductor devices will deliver abilities that were previously only seen in science fiction, as engineers and researchers strive to push the envelope of whatever is feasible. From silicon to heterostructures, there is going to be a revolution that will change both the technological environment and our interactions with the digital world. This is not merely an evolution.

REFERENCES

[1] Shalf, John. "The future of computing beyond Moore's Law." *Philosophical Transactions of the Royal Society A* 378, no. 2166 (2020): 20190061. https://doi.org/10.1098/rsta.2019.0061

[2] Gupta, R.K. (Ed.). (2024). *Handbook of Semiconductors: Fundamentals to Emerging Applications* (1st ed.). CRC Press. https://doi.org/10.1201/9781003450146

[3] Terna, Augustine D., Elias E. Elemike, Justina I. Mbonu, Omosede E. Osafile, and Rachael O. Ezeani. "The future of semiconductors nanoparticles: Synthesis, properties and applications." *Materials Science and Engineering: B* 272 (2021): 115363. https://doi.org/10.1016/j.mseb.2021.115363

[4] Soumya Sen, Ashish Raman, Mamta Khosla, "A Literature Survey on Tunnel Field Effect Transistors", *Proceedings of International Conference on Women Researchers in Electronics and Computing, WREC2021*, AIJR (September, 2021), 506–512. https://doi.org/10.21467/proceedings.114.65

[5] Li, Yu, Jiawei Zhang, Qingguo Chen, Xinhui Xia, and Minghua Chen. "Emerging of heterostructure materials in energy storage: A review." *Advanced Materials* 33, no. 27 (2021): 2100855. https://doi.org/10.1002/adma.202100855

[6] Castellanos-Gomez, Andres, Xiangfeng Duan, Zhe Fei, Humberto Rodriguez Gutierrez, Yuan Huang, Xinyu Huang, Jorge Quereda, Qi Qian, Eli Sutter, and Peter Sutter. "Van der Waals heterostructures." *Nature Reviews Methods Primers* 2, no. 1 (2022): 58. https://doi.org/10.1038/nature12385

[7] Chen, Jiaxin, Weixin Ouyang, Wei Yang, Jr-Hau He, and Xiaosheng Fang. "Recent progress of heterojunction ultraviolet photodetectors: materials, integrations, and applications." *Advanced Functional Materials* 30, no. 16 (2020): 1909909. https://doi.org/10.1002/adfm.201909909

[8] Saurabh, J.S., and Kumar, M.J. (2016). *Tunnel Field-Effect Transistors* (1st ed.). pp. 39–77, TheTunnelFET, 9781119246299, CRC Press. https://doi.org/10.1002/9781119246312.ch3

[9] Cheng, Keh Yung, and Keh Yung Cheng. (2020). *III–V Compound Semiconductors and Devices: An Introduction to Fundamentals*. pp. 203–242. https://link.springer.com/book/10.1007/978-3-030-51903-2

[10] Mikhailova, Maiya Pavlovna, Konstantin Dmitrievich Moiseev, and Yu P. Yakovlev. "Discovery of III–V semiconductors: Physical properties and application." *Semiconductors* 53 (2019): 273–290. https://doi.org/10.1134/S1063782619030126

[11] Harbison, J. P., T. Sands, C. J. Palmstrom, T. L. Cheeks, L. T. Florez, and V. G. Keramidas. "New directions for III–V structures: metal/semiconductor heteroepitaxy." In *Gallium Arsenide and Related Compounds 1991, Proceedings of the Eighteenth INT Symposium, 9-12 September 1991, Seattle, USA*, pp. 1–17. CRC Press, 2020. https://doi.org/10.1201/9781003069638

[12] Zhao, Chao, Zhaonan Li, Tianyi Tang, Jiaqian Sun, Wenkang Zhan, Bo Xu, Huajun Sun et al. "Novel III-V semiconductor epitaxy for optoelectronic devices through two-dimensional materials." *Progress in Quantum Electronics* 76 (2021): 100313. https://doi.org/10.1016/j.pquantelec.2020.100313

[13] Chen, Yingjie, Baonan Jia, Xiaoning Guan, Lihong Han, Liyuan Wu, Pengfei Guan, and Pengfei Lu. "Design and analysis of III-V two-dimensional van der Waals heterostructures for ultra-thin solar cells." *Applied Surface Science* 586 (2022): 152799. DOI: https://doi.org/10.1016/j.apsusc.2022.152799

[14] Park, Jae-Seong, Mingchu Tang, Siming Chen, and Huiyun Liu. "Heteroepitaxial growth of III-V semiconductors on silicon." *Crystals* 10, no. 12 (2020): 1163. https://doi.org/10.3390/cryst10121163

[15] McNicholas, Kyle Marshall. "Emerging epitaxial materials for coherent III-V (opto) electronic heterostructure devices." PhD diss., 2019. http://dx.doi.org/10.26153/tsw/42287

[16] Paulauskas, Tadas, Vaidas Pačebutas, Andrejus Geižutis, Mindaugas Kamarauskas, Mantas Drazdys, Matas Rudzikas, Rokas Kondrotas et al. "Performance analysis of GaAsBi/InGaAs heterostructure for III-V multi-junction solar cells." *Solar Energy Materials and Solar Cells* 248 (2022): 112013. https://doi.org/10.1016/j.solmat.2022.112013

[17] Khalili, Adrien, Claire Abadie, Tung Huu Dang, Audrey Chu, Eva Izquierdo, Corentin Dabard, Charlie Gréboval et al. "Colloidal II–VI—Epitaxial III–V

heterostructure: A strategy to expand InGaAs spectral response." *Applied Physics Letters* 120, no. 5 (2022). https://doi.org/10.1063/5.0076708

[18] Sushkov, A. A., D. A. Pavlov, A. I. Andrianov, V. G. Shengurov, S. A. Denisov, V. Yu Chalkov, R. N. Kriukov, N. V. Baidus, D. V. Yurasov, and A. V. Rykov. "Comparison of III–V heterostructures grown on Ge/Si, Ge/SOI, and GaAs." *Semiconductors* 56, no. 2 (2022): 122–133. https://doi.org/10.1134/S1063782622010171

[19] Zhang, Xiangyi. "Heterostructures: New opportunities for functional materials." *Materials Research Letters* 8, no. 2 (2020): 49–59. https://doi.org/10.1080/21663831.2019.1691668

[20] Xie, Zhongjian, Bin Zhang, Yanqi Ge, Yao Zhu, Guohui Nie, YuFeng Song, Chang-Keun Lim, Han Zhang, and Paras N. Prasad. "Chemistry, functionalization, and applications of recent monoelemental two-dimensional materials and their heterostructures." *Chemical Reviews* 122, no. 1 (2021): 1127–1207. https://doi.org/10.1021/acs.chemrev.1c00165

[21] Zhu, Yuntian, Kei Ameyama, Peter M. Anderson, Irene J. Beyerlein, Huajian Gao, Hyoung Seop Kim, Enrique Lavernia et al. "Heterostructured materials: Superior properties from hetero-zone interaction." *Materials Research Letters* 9, no. 1 (2021): 1–31. https://doi.org/10.1080/21663831.2020.1796836

[22] Fu, Qiang, Jiecai Han, Xianjie Wang, Ping Xu, Tai Yao, Jun Zhong, Wenwu Zhong et al. "2D transition metal dichalcogenides: Design, modulation, and challenges in electrocatalysis." *Advanced Materials* 33, no. 6 (2021): 1907818. https://doi.org/10.1002/adma.201907818

[23] Chowdhury, Tomojit, Erick C. Sadler, and Thomas J. Kempa. "Progress and prospects in transition-metal dichalcogenide research beyond 2D." *Chemical Reviews* 120, no. 22 (2020): 12563–12591. https://doi.org/10.1021/acs.chemrev.0c00505

[24] Pi, Lejing, Liang Li, Kailang Liu, Qingfu Zhang, Huiqiao Li, and Tianyou Zhai. "Recent progress on 2D noble-transition-metal dichalcogenides." *Advanced Functional Materials* 29, no. 51 (2019): 1904932. https://doi.org/10.1002/adfm.201904932

[25] Zhang, Xixia, Sin Yong Teng, Adrian Chun Minh Loy, Bing Shen How, Wei Dong Leong, and Xutang Tao. "Transition metal dichalcogenides for the application of pollution reduction: A review." *Nanomaterials* 10, no. 6 (2020): 1012. https://doi.org/10.3390/nano10061012

[26] Wu, Xin, Huabin Zhang, Jian Zhang, and Xiong Wen Lou. "Recent advances on transition metal dichalcogenides for electrochemical energy conversion." *Advanced Materials* 33, no. 38 (2021): 2008376. https://doi.org/10.1002/adma.202008376

[27] Liang, Qijie, Qian Zhang, Xiaoxu Zhao, Meizhuang Liu, and Andrew TS Wee. "Defect engineering of two-dimensional transition-metal dichalcogenides: applications, challenges, and opportunities." *ACS Nano* 15, no. 2 (2021): 2165–2181. https://doi.org/10.1021/acsnano.0c09666

[28] Castellanos-Gomez, Andres, Xiangfeng Duan, Zhe Fei, Humberto Rodriguez Gutierrez, Yuan Huang, Xinyu Huang, Jorge Quereda, Qi Qian, Eli Sutter, and Peter Sutter. "Van der Waals heterostructures." *Nature*

Reviews Methods Primers 2, no. 1 (2022): 58. https://doi.org/10.1038/nature12385

[29] Liang, Shi-Jun, Bin Cheng, Xinyi Cui, and Feng Miao. "Van der Waals heterostructures for high-performance device applications: Challenges and opportunities." *Advanced Materials* 32, no. 27 (2020): 1903800. https://doi.org/10.1002/adma.201903800

[30] Wang, Peiqi, Chuancheng Jia, Yu Huang, and Xiangfeng Duan. "Van der Waals heterostructures by design: from 1D and 2D to 3D." *Matter* 4, no. 2 (2021): 552–581. https://doi.org/10.1016/j.matt.2020.12.015

[31] Di Bartolomeo, Antonio. "Emerging 2D materials and their van der Waals heterostructures." *Nanomaterials* 10, no. 3 (2020): 579. https://doi.org/10.3390/nano10030579

[32] Sierra, Juan F., Jaroslav Fabian, Roland K. Kawakami, Stephan Roche, and Sergio O. Valenzuela. "Van der Waals heterostructures for spintronics and optospintronics." *Nature Nanotechnology* 16, no. 8 (2021): 856–868. https://doi.org/10.1038/s41565-021-00936-x

[33] Zhao, Bei, Zhong Wan, Yuan Liu, Junqing Xu, Xiangdong Yang, Dingyi Shen, Zucheng Zhang et al. "High-order superlattices by rolling up van der Waals heterostructures." *Nature* 591, no. 7850 (2021): 385–390. https://doi.org/10.1038/s41586-021-03338-0

[34] Das, Sonali, Deepak Pandey, Jayan Thomas, and Tania Roy. "The role of graphene and other 2D materials in solar photovoltaics." *Advanced Materials* 31, no. 1 (2019): 1802722. https://doi.org/10.1002/adma.201802722

[35] Shinde, Pratik V., and Manoj Kumar Singh. "Synthesis, characterization, and properties of graphene analogs of 2D material." In Morgan Hywel, Chandra Sekhar Rout and Dattatray J. Late (Eds.), *Fundamentals and sensing applications of 2D materials*, pp. 91–143. Woodhead Publishing, 2019. https://doi.org/10.1016/B978-0-08-102577-2.00004-X

[36] Qiao, Hui, Huating Liu, Zongyu Huang, Rong Hu, Qian Ma, Jianxin Zhong, and Xiang Qi. "Tunable electronic and optical properties of 2D monoelemental materials beyond graphene for promising applications." *Energy & Environmental Materials* 4, no. 4 (2021): 522–543. https://doi.org/10.1002/eem2.12154

[37] Liu, Bo, and Kun Zhou. "Recent progress on graphene-analogous 2D nanomaterials: Properties, modeling and applications." *Progress in Materials Science* 100 (2019): 99–169. https://doi.org/10.1016/j.pmatsci.2018.09.004

[38] Akinwande, Deji, Cedric Huyghebaert, Ching-Hua Wang, Martha I. Serna, Stijn Goossens, Lain-Jong Li, H-S. Philip Wong, and Frank H.L. Koppens. "Graphene and two-dimensional materials for silicon technology." *Nature* 573, no. 7775 (2019): 507–518. https://doi.org/10.1038/s41586-019-1573-9

[39] Gonzalez-Ortiz, D., C. Salameh, Mikhael Bechelany, and P. Miele. "Nanostructured boron nitride–based materials: Synthesis and applications." *Materials Today Advances* 8 (2020): 100107. https://doi.org/10.1016/j.mtadv.2020.100107

[40] Roy, Soumyabrata, Xiang Zhang, Anand B. Puthirath, Ashokkumar Meiyazhagan, Sohini Bhattacharyya, Muhammad M. Rahman, Ganguli Babu et al. "Structure, properties and applications of two-dimensional hexagonal boron nitride." *Advanced Materials* 33, no. 44 (2021): 2101589. https://doi.org/10.1002/adma.202101589

[41] Cai, Qiran, Lu Hua Li, Srikanth Mateti, Amrito Bhattacharjee, Ye Fan, Shaoming Huang, and Ying Ian Chen. "Boron Nitride Nanosheets: Thickness-Related Properties and Applications." *Advanced Functional Materials* (2024): 2403669. https://doi.org/10.1002/adfm.202403669

[42] Li, Mengna, Gordon Huang, Xiujuan Chen, Jianan Yin, Peng Zhang, Yao Yao, Jian Shen, Yuwei Wu, and Jing Huang. "Perspectives on environmental applications of hexagonal boron nitride nanomaterials." *Nano Today* 44 (2022): 101486. https://doi.org/10.1016/j.nantod.2022.101486

[43] Chakraborty, Suman Kumar, Baisali Kundu, Biswajeet Nayak, Saroj Prasad Dash, and Prasana Kumar Sahoo. "Challenges and opportunities in 2D heterostructures for electronic and optoelectronic devices." *Iscience* 25, no. 3 (2022). https://doi.org/10.1016/j.isci.2022.103942

[44] Ahn, Ethan C. "2D materials for spintronic devices." *NPJ 2D Materials and Applications* 4, no. 1 (2020): 17. https://doi.org/10.1038/s41699-020-0152-0

[45] Liu, Yanping, Siyu Zhang, Jun He, Zhiming M. Wang, and Zongwen Liu. "Recent progress in the fabrication, properties, and devices of heterostructures based on 2D materials." *Nano-micro Letters* 11 (2019): 1–24. https://doi.org/10.1007/s40820-019-0245-5

[46] Zhang, Bingyu, Pengcheng Lu, Roozbeh Tabrizian, Philip X-L. Feng, and Yingying Wu. "2D Magnetic heterostructures: Spintronics and quantum future." *NPJ Spintronics* 2, no. 1 (2024): 6. https://doi.org/10.1038/s44306-024-00011-w

[47] Irvine, Stuart, and Peter Capper, (Eds). (2019). *Metalorganic Vapor Phase Epitaxy (MOVPE): Growth, Materials Properties, and Applications*. Wiley. https://doi.org/10.1002/9781119313021

[48] Pohl, Udo W., and Udo W. Pohl. (2020). *Epitaxy of Semiconductors: Physics and Fabrication of Heterostructures*. pp. 427–467. https://link.springer.com/book/10.1007/978-3-030-43869-2

[49] Chen, Zhaolong, Yue Qi, Xudong Chen, Yanfeng Zhang, and Zhongfan Liu. "Direct CVD growth of graphene on traditional glass: Methods and mechanisms." *Advanced Materials* 31, no. 9 (2019): 1803639. https://doi.org/10.1002/adma.201803639

[50] Sen, Soumya, Mamta Khosla, and Ashish Raman. "Design and investigation of electrostatic doped heterostructure vertical $Si_{(1-x)}Ge_x$/Si nanotube TFET.", *Microelectronics Journal* 153 (2024): 106417. ISSN1879-2391. https://doi.org/10.1016/j.mejo.2024.106417

[51] Song, Cen, Binghan Wu, and Gang Zhao. "Applications of novel semiconductor materials in chip design." *Journal of Industrial Engineering and Applied Science* 2, no. 4 (2024): 81–89. https://doi.org/10.5281/zenodo.12798364

[52] Parichenko, Alexandra, Shirong Huang, Jinbo Pang, Bergoi Ibarlucea, and Gianaurelio Cuniberti. "Recent advances in technologies toward the development of 2D materials-based electronic noses." *TrAC Trends in Analytical Chemistry* (2023): 117185. www.sciencedirect.com/science/article/pii/S0165993623002728

[53] Liu, Fu, and Zhanxi Fan. "Defect engineering of two-dimensional materials for advanced energy conversion and storage." *Chemical Society Reviews* 52, no. 5 (2023): 1723–1772. https://doi.org/10.1039/D2CS00931E

[54] Qi, Junlei, Zongxiao Wu, Wenbin Wang, Kai Bao, Lingzhi Wang, Jingkun Wu, Chengxuan Ke, Yue Xu, and Qiyuan He. "Fabrication and applications of van der Waals heterostructures." *International Journal of Extreme Manufacturing* 5, no. 2 (2023): 022007. https://doi.org/10.1088/2631-7990/acc8a1

Chapter 3

FA Overview on Distinct FETs

Device Physics and Properties

S. Maheswari and S. Agnes Shifani

3.1 INTRODUCTION TO FIELD EFFECT TRANSISTORS (FETS)

Field Effect Transistors (FET) can use an electric field to turn on and off a semiconductor channel and hence current flowing within it, and are often referred to as unipolar transistors because they operate with only a single carrier type. In essence, FETs work by making use of either holes (p channel) or electrons (n channel), but not both. There is a wide range of field effect transistors. Typically, field effect transistors have extremely high input impedance at low frequencies [1–5].

3.1.1 Origin of Field Effect Transistors (FETs)

Shortly after the 17-year patent expired, in 1947 [1], John Bardeen and Walter Houser Brattain, working for William Shockley at Bell Labs, identified and explained the transistor effect. Shockley first made an unsuccessful attempt to construct a functional FET by attempting to alter a semiconductor's conductivity. The physicists Julius Edgar Lilienfeld and Oskar Heil, who were born in Hungary, were the first to patent the idea of a field-effect transistor (FET) in 1925 and 1934, respectively. Although such an idea was attempted, they could not build a viable, useful semiconductor device using the idea. The first FET device to be successfully manufactured was the junction field effect transistor (JFET) [2], which was first patented by Heinrich Welker in 1945 [4,5]. A type of JFET was found by Japanese engineers Jun-ichi Nishizawa and Y. Watanabe in 1950.

3.1.2 What is a FET?

A semiconductor-based channel is a type of semiconductor device that consists of two electrodes plugged to the semiconductor channel. The two electrodes are the source and drain electrodes, which are connected to either end. The gate, a third electrode, controls the current that flows between

 DOI: 10.1201/9781003533535-3

the source and drain electrodes. Based on how they function, FETs are divided into two types: enhancement mode and depletion mode FETs. This is determined by whether the voltage applied at the gate terminal increases or decreases the current flow over the channel. The possibility that charges on nearby objects could draw charges into a semiconductor channel is the idea underlying the field effect transistor. This is basically how it works—the name came from the electric field effect [6–10].

Accordingly, there are multiple types of FETs, including metal semiconductor field effect transistors (MESFET), metal insulator field effect transistors (MISFET), and junction field effect transistors (JFET). Because an oxide (SiO_2) is utilized as the insulator, MISFET and MOSFET are also known by these names. Because the gate is separated from the channel through which current flows, it is often referred to as an insulated gate field effect transistor (IGFET) [11,12].

3.1.3 Distinct Types of Field Effect Transistors (FETs)

Bipolar transistors and JFETs appear to be essentially the same. A controlled input voltage is necessary for the operation of a FET. While a JFET is regulated by the input voltage, a BJT is controlled by the current. There are two kinds of FETs available [13,14]:

1) Junction Field Effect Transistor (JFET)
2) Metal Oxide Semiconductor FET (MOSFET)

The other FETS have unique characteristics which may useful in different applications. They are the Metal-Semiconductor FET (MESFET), Insulated-Gate Bipolar Transistor (IGBT), FinFET, and Tunnel FET (TFET).

MESFETs are similar to JFETs, except instead of a PN junction, they employ a Schottky barrier, or a metal-semiconductor junction. As a result, high-speed functioning is possible. Features from BJTs and MOSFETs are combined in IGBTs. They offer high input impedance and high current handling, with a gate configuration akin to a MOSFET and a conduction channel akin to a BJT. A kind of MOSFET known as a finFET has a three-dimensional structure in which the gate encircles a channel that resembles a fin. Leakage currents are decreased, and channel control is enhanced as a result. Quantum mechanical tunneling is used by TFETs to operate at minimal power. Their distinct structure enables carriers to tunnel through possible obstacles [15].

3.2 DEVICE PHYSICS OF JUNCTION FIELD EFFECT TRANSISTORS (JFETS)

The majority carriers, electrons or holes, are the basis of operation for JFETs. A silicon bar with PN connections on its sides is used to build these transistors. Here are some key ideas to remember in order to comprehend JFETs better. Using diffusion or alloying processes, the N-type bar is strongly doped to produce the gate of a JFET. These doped areas are called "gate" (G) [16].

The semiconductor bar provides the entry point for the majority of carriers, namely through the source. The drain is the other side of the story, allowing most bulk carriers to escape the semiconductor bar. The channel is the N type material area in which it is easier for the majority carriers to move from source to the drain. There are two varieties of JFETs that are frequently employed in the field of semiconductor devices: N-channel and P-channel JFETs. Each of these stands apart in a number of electronic applications [17,18].

3.2.1 Symbol and Essential Terms of JFET

The JFET is made up of three components. The JFET's "gate" element functions extremely similarly to the bipolar transistor's base. The "source" and "drain" components of the JFET stand in for the emitter and collector of a bipolar transistor, respectively. The charge carriers' path of transition within the device is referred to as the channel in all FETs. If the channel is composed of an n-type semiconductor material, it is referred to as an N-channel FET [10–12]. Similarly, a P-channel FET's channel material is composed of a p-type semiconductor. Figure 3.1 and Figure 3.2 depict the basic n-channel JFET and p-channel JFET designs, respectively.

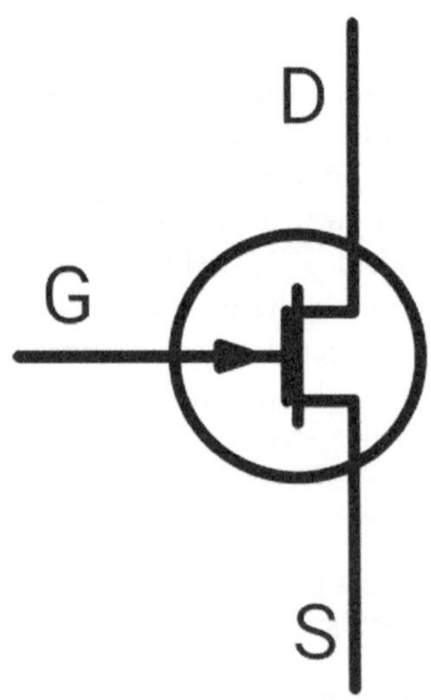

Figure 3.1 N-vhannel JFET [1].

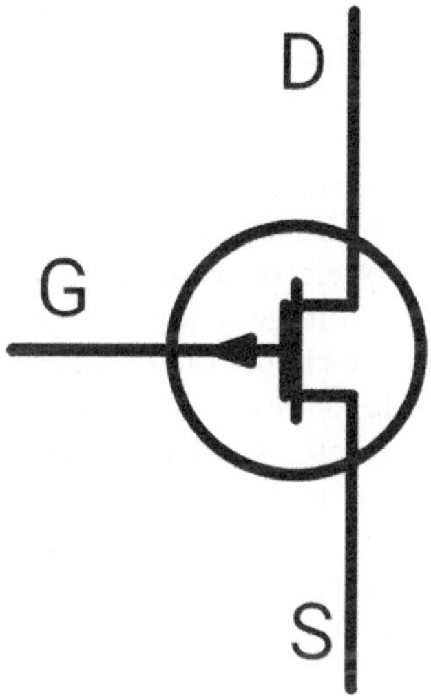

Figure 3.2 P-vhannel JFET.

The junction field effect transistor, also known as a JFET, is a semiconductor device with three terminals: This entails a source, a drain and a gate. The idea is that these may function based on the same principle as how an electric field controls the flow of electrons from the source to the drain. The gate is a crucial component in regulating the current flow via this semiconductor device. The terminal where the current enters is called the source, and the terminal where it exits is called the drain. The gate has control over the flow of electrons [13, 16].

3.2.2 Key properties and Operation behind FET

In field-effect transistors (FETs), the drain current flowing between the source and drain terminals is controlled by adjusting the voltage applied to the gate terminal. An N-channel FET's conductive channel forms between the source and drain when a positive voltage is supplied to the gate, attracting electrons. In contrast, a P-channel FET requires a negative gate voltage to establish the conductive path. When a voltage is applied to the gate of the semiconductor material, an electric field is created through the semiconductor material, increasing or decreasing the conductive channel in the semiconductor material. This modulation of the channel conductivity allows for precise control of the current flow [8,9].

This voltage-controlled system offers an incredibly effective way to manage current. Among the main characteristics of FETs are their fast switching speed, low power consumption, and high input impedance [7].

Because of their high input impedance, FETs require very little current from the stage before them, which makes them perfect for situations where power conservation is crucial, like in battery-powered devices. Because of their rapid switching rates, they are appropriate for digital circuits and high-frequency applications where short response times are required to handle sudden changes in signal states [6].

Both source and drain terminals are doped either with n-type or p-type material allowing the flow of electrons or holes from the source to the drain terminals; however, when a voltage is applied between the gate terminal, an electric field is induced in the channel region between the source and drain terminals. This electric field serves as a "valve" to modulate the conductivity of the channel and regulate the flow of current between the source and drain. The gate of an N-channel FET is pulled into the positive voltage, pulling free electrons into the channel increasing its conductivity, enhancing current flow between the source and drain. On a p-channel FET a negative gate voltage repels holes and increases conductivity [2,3].

3.2.2.1 Biasing of JFET

To properly bias the JFET, a variety of strategies are employed. Out of all the methods, the following three are commonly employed. They are Fixed Biasing, Self-Biasing, and Voltage Divider Biasing techniques [5].

In order to produce the right drain current IDS, a FET device is required to be put into DC bias on a gate source voltage (VGS). The value of IDS limits the drain current of a JFET. The extraordinarily high input impedance of the FET precludes the device from loading or changing the DC voltage on the gate which is either a voltage divider or a fixed battery voltage [5]. As a result, no gate current flows. Figure 3.3 illustrates the fixed-bias circuit for JFET.

An input battery is used to obtain fixed DC bias. With the help of this battery, you can be confident that the gate is always negative in relation to the source and that there is no current flowing through the gate terminal when $I_G = 0$. To bias the N-channel JFET, the battery supplies a voltage V_{GS}, but no current is subsequently pulled from the battery V_{GG}. A DC voltage drop of IG RG, or 0 volt, is equivalent to the DC voltage drop across resistor RG, which is provided so that any AC signal applied through capacitor C can expand across RG [6–9].

Applying KVL to the input circuits,
$V_{GS} + V_{GG} = 0$ for DC analysis with $I_G = 0$

$$V_{GS} = -V_{GG}$$

The fixed DC supply of V_{GS} is referred to as a "fixed bias circuit".

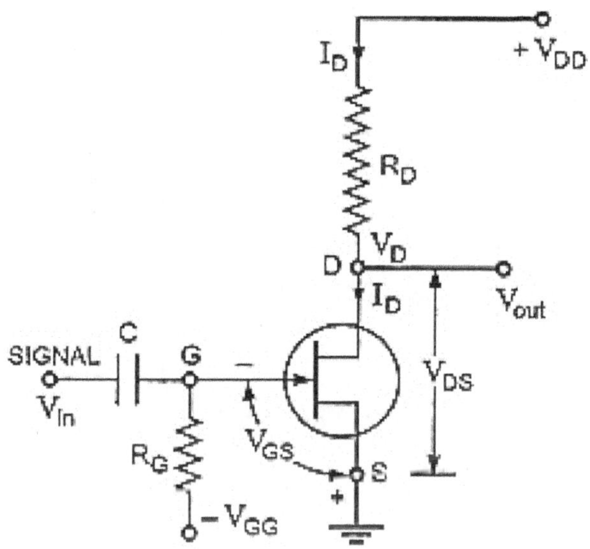

Figure 3.3 Fixed bias circuit for JFET.

$$I_{DQ} = I_{DSS}\left(1 - \frac{V_{GS}}{V_{GP}}\right)^2$$

The voltage drop across the drain resistor R_D that results from this current I_{DQ} is therefore indicated as $V_{DSQ} = V_{DD} - I_D R_D$

3.2.2.1.1 Self-Bias Circuits

Biasing a JFET is most commonly accomplished via self-bias circuits. Figure 3.4 illustrates the self-bias circuit for an N-channel JFET.

The JFET gate-source junction needs to be reverse biased at all times. Since the reverse-biased gate-source is not receiving any gate current, i.e.,

$I_G = 0$ and $V_G = I_G R_G = 0$ no gate current passes through it [10].

The voltage at the S with a drain current I_D is $V_S = I_D R_S$

The gate-source voltage is then $V_{GS} = -I_D R_S$ which is therefore dubbed self-biasing as voltage drop across resistance Rs produces the biasing voltage V_{GG} and no external source is needed for biasing [12].

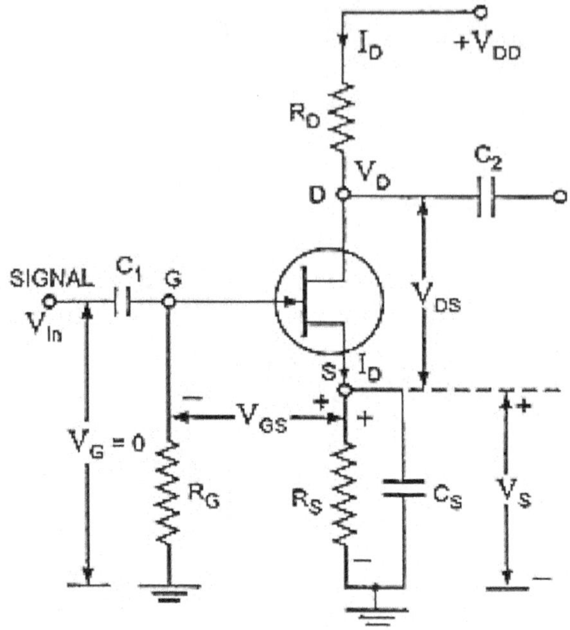

Figure 3.4 Self bias circuit for N channel JFET.

The operating point can readily be found using the equation:

$$V_{DS} = I_D\left(R_D + R_S\right) - V_{DD}$$

The JFET's quiescent operating point is shielded from changes in its properties, like transconductance, via self-biasing. The gate-source voltage, or V_{GS}, thereafter becomes more negative with every rise in the voltage drop across RS, which in turn lessens the increase in drain current [13].

3.2.2.1.2 Potential Divider Bias Circuits

Figure 3.5 illustrates the potential divider bias circuit for an N-channel JFET. A potential divider is formed by the resistors R_{G1} and R_{G2} across the drain supply V_{DD}. The required bias is supplied by the voltage V_2 across R_{G2} [12, 14].

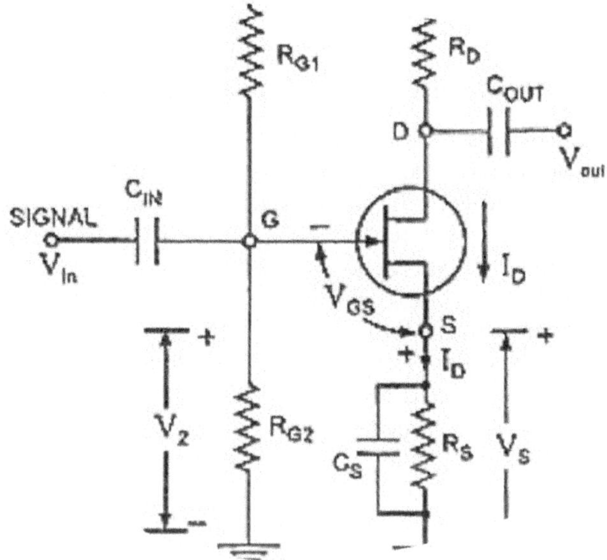

Figure 3.5 Potential divider Bias circuit for N channel JFET.

The additional gate resistor R_{Gl} from the gate to supply voltage allows for larger modifications to the DC bias point and the usage of bigger values of R_S.

Generally, $I_{DQ} = I_{DSS}\left(1-\frac{V_{GS}}{V_P}\right)^2$ yields the operating point of a JFET amplifier utilizing the Voltage-Divider Bias, $V_{DSQ} = V_{DD} - I_D (R_D + R_S)$ and $V_G - I_D R_S = V_{GSQ}$

3.2.3 JFET operating Characteristics

The advantage is that FETs do not have thermal runaway, unlike BJTs. Yet, with a fixed gate voltage, the ID levels are unpredictable due to large variations in the maximum and minimum transfer characteristics. Thus suitable limitations on quiescent drain currents ID and drain source voltage VDS are required and techniques using a source resistor and potential divider bias achieve this. This is plotted as a curve between drain current and drain source voltage defining the characteristics of a JFET. We define its characteristics as the drain current fluctuation as a function of the voltage applied at the drain source terminals at a constant gate source voltage. In essence, there are two categories of characteristics: These are commonly referred to as drain characteristics or transfer characteristics, with drain characteristics showing the relationship between drain current and drain-to-source voltage [12–15].

3.2.3.1 Output characteristics or Drain Characteristics

The JFET's drain characteristics in Figure 3.6 show that the drain current flows from the drain terminal to the source terminal because, in the absence of an external bias and with the channel completely open, there is no voltage between the gate and source terminals. However, no current flows when the drain source voltage (VDS) is zero because there is no attractive potential at the drain [10–13].

As a result, drain current ID is also 0. Low applied voltages VDS, essentially up to the knee point, show a simple semiconductor resistor behavior with the n-type bar and the drain current rising linearly with VDS. Between the knee point and the rightmost part of the knee (the channel ohmic zone), the JFET behaves like a regular resistor; it is in the channel ohmic zone.

The ohmic voltage drop between the source and channel regions reverse biases the gate junction as the drain current ID rises [14].

In the JFET's operation, the majority of the charge carriers are travelling from source to drain, so current flows from drain to source. Since there is no external reverse biasing, we know that the channel width must be larger because the depletion layer's width will not be changing initially. It allows a great deal of current through the channel.

As the VDS is increased steadily the gate channel connection eventually fails. Figure 3.7 shows JFET drain characteristics when $V_{GS} < 0v$ [12–13].

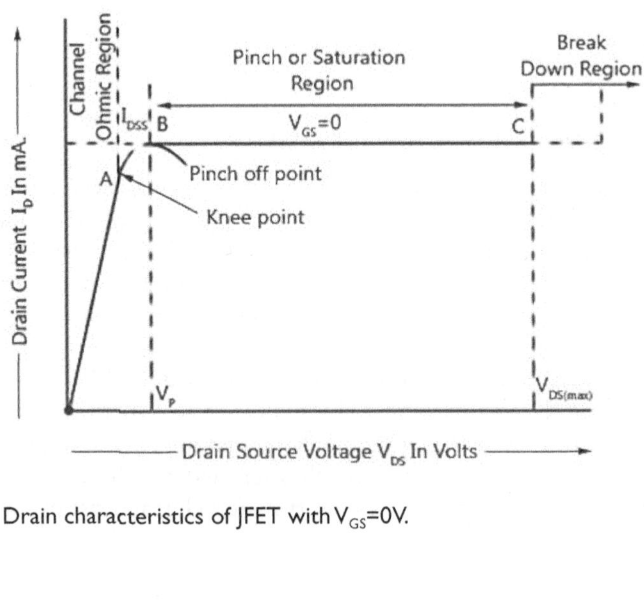

Figure 3.6 Drain characteristics of JFET with V_{GS}=0V.

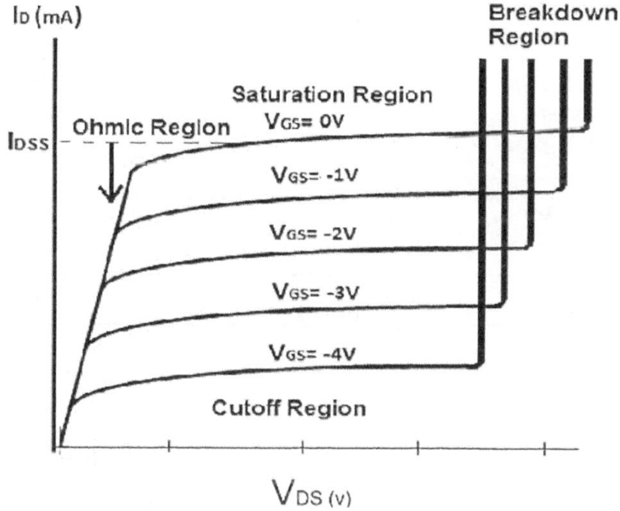

Figure 3.7 Drain characteristics of JFET with V_{GS}< 0V.

Current is now increasing extremely quickly and the JFET might be ruined. This occurs as a result of an avalanche effect caused by the charge carriers at the gate channel junction accelerating to a high velocity and constituting the saturation current.

- Knee Point: The characteristics curve suggests a linear relationship between the drain current and the drain-source voltage, however it turns into a curve after that.
- Channel ohmic region is that area in the characteristics curve to the left side of the knee point.
- Pinch-off point is the point on the curve above which, no matter how much the drain to source voltage is raised, the drain current doesn't increase anymore. Because the current is fully switched to constant at this voltage, it is referred to as the pinch off voltage.
- The drain-source saturation current is the current that flows from the drain to the source and either stabilizes or fully reaches a saturation situation [10, 12].

3.2.3.2 Transfer characteristics

The drain characteristics are simply the opposite of the transfer characteristics; in the latter case the gate source voltage is held non varying while the values of the drain current at various drain source voltage values are measured. Using such a fixed drain-source voltage, changes in the gate

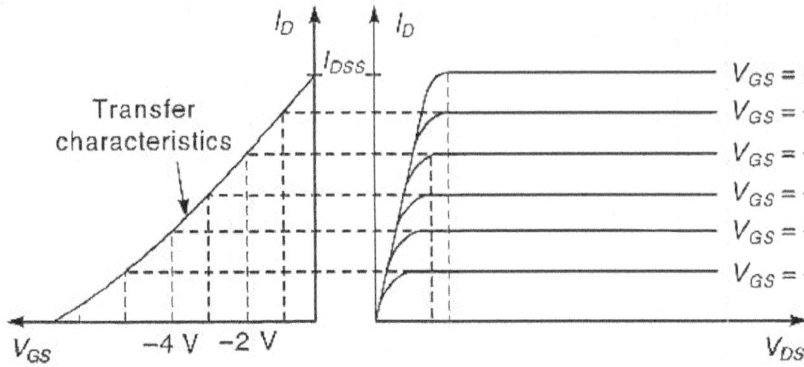

Figure 3.8 Transfer characteristics of JFET.

source voltage can be seen in the drain current, which will ascertain the transfer characteristics of the device [16, 17].

The experimental transfer characteristic of a JFET is obtained by maintaining the VDS constant while measuring the ID for different V_{GS} values. The relation between gate source voltage (V_{GS}) and drain current (ID) is depicted in Figure 3.8.

It has been seen that:

(i) Drain current falls to negative when gate-source bias is increased;
(ii) Drain current becomes I_{DSS} at $V_{GS} = 0$; and
(iii) Drain current becomes 0 when $V_{GS} = V_D$.

3.3 DEVICE PHYSICS AND DISTINCT TYPES OF METAL OXIDE SEMICONDUCTOR FET (MOSFET)

MOSFETs are like JFETs, in that they have two areas on either side called the source and drain terminals, and a channel composed of semiconductor material. In a MOSFET, the gate is isolated from the channel by an insulating layer. In this configuration, when a voltage is applied to the gate, an electric field is created, which, depending on the polarity of the voltage, either attracts or repels the charge carriers in the channel. This procedure manages the width of the channel as well as the current flow through between the source and drain terminals. Further, there are two subtypes of MOSFETs: enhancement mode (EMOSFET) and depletion mode MOSFETs with NMOS as N-channel and PMOS as P-channel MOSFETs. MOSFETs are essentially two types [10–15]. They are as follows:

- Depletion Type
- Enhancement Type

3.3.1 Depletion Type MOSFET

When it comes to depletion, a MOSFET transistor is similar to a switch that is "normally closed". To switch off the device, gate-source voltage V_{GS} must be given to depletion type transistors. The N-channel and P-channel MOSFET symbols are displayed in Figure 3.9 and Figure 3.10.

The fourth terminal, also known as the substrate, is connected to ground in the above symbols, but it is attached to the source terminal in discrete MOSFETs. The type of depletion is shown by the thick, continuous line that connects the source and drain terminals. The channel type, such as N- or

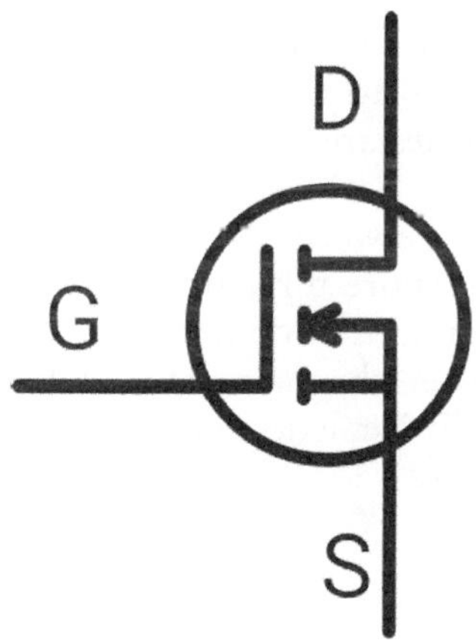

Figure 3.9 N-channel MOSFET.

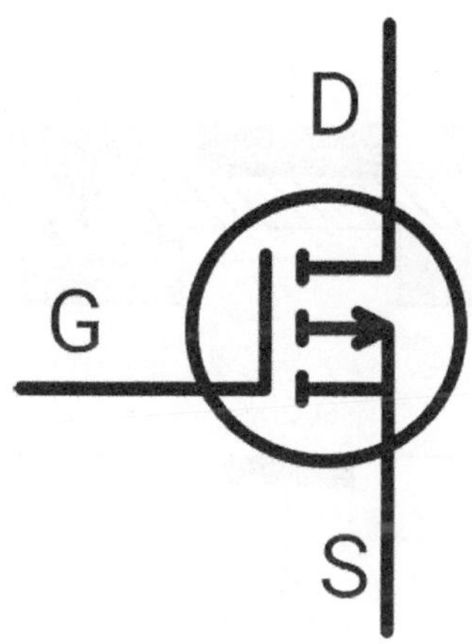

Figure 3.10 P-channel MOSFET.

P-channel, is indicated by the arrow symbol. These MOSFETs have a tiny layer of silicon underneath the gate terminal. The depletion mode MOSFET transistors are normally ON when the gate-source voltage (V_{GS}) is zero. The conductivity of the channel is lower in depletion MOSFETs than in enhancement MOSFETs [9–14].

3.3.1.1 N-Channel Depletion MOSFET

The N-channel depletion MOSFET's construction is depicted in Figure 3.11. The source and drain of this sort of depletion MOSFET are connected via a small N-type semiconductor strip. The majority of the charge carriers in this kind of MOSFET are electrons, and the substrate is a P-type semiconductor. These sources and drains are heavily medicated here.

Except for how it functions, the structure of the N-channel depletion mode MOSFET is identical to that of the enhancement mode N-channel MOSFET. N-type impurities are present in the gap between the drain and source terminals [8–13].

A potential difference between the source and drain terminals causes current to flow over the entire n-region of the substrate. Charge carriers, such as electrons, will be rejected and transported downward in the n-region beneath the dielectric layer when a negative voltage is supplied at the gate terminal of this MOSFET. Depletion of charge carriers will therefore take place inside the channel [12].

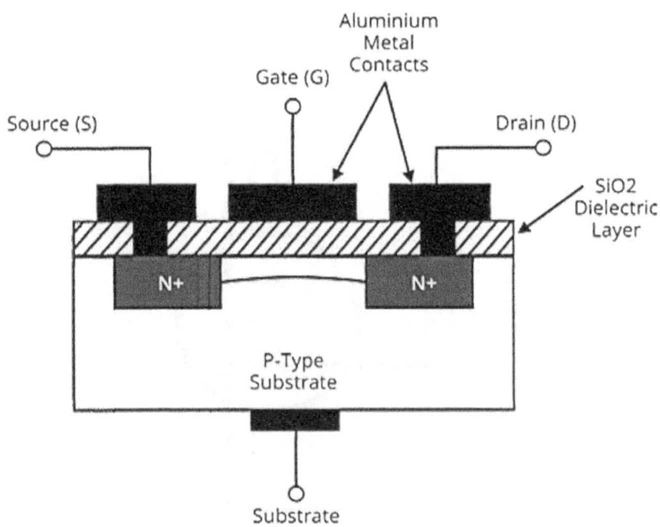

Figure 3.11 N-channel depletion MOSFET.

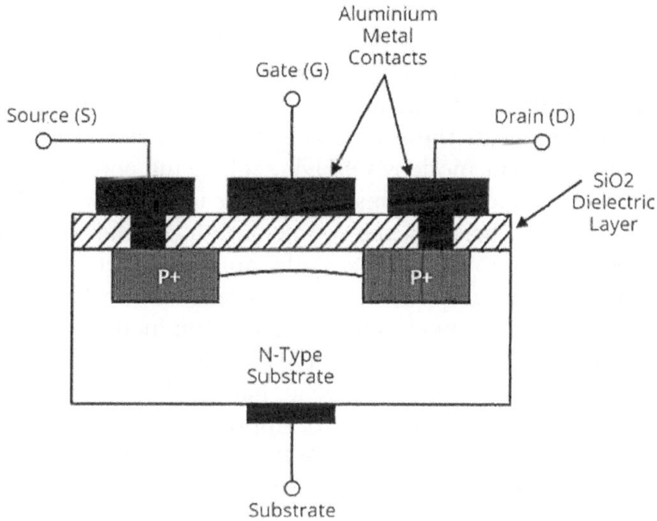

Figure 3.12 P-channel depletion MOSFET.

The overall channel conductivity consequently drops. In this instance, the moment the GATE terminal receives the same voltage, the drain current will decrease. It requires more increases in the negative voltage before the pinch-off mode is attained [6].

3.3.1.2 P-Channel Depletion MOSFET

The P-channel depletion MOSFET's drain and source are connected by a small P-type semiconductor strip. The source and drain are composed of P-type semiconductors, and the substrate is composed of N-type semiconductors. Most charge carriers are holes. Unlike a p-channel depletion MOSFET, an n-channel depletion mode MOSFET is very different. As shown in Figure 3.12, this MOSFET has an n-type substrate and a p-type channel, and it is very doped with p-type impurities at the channel between the source and drain regions [3–10].

Electrostatic forces will attract minority charge carriers, like electrons in the p-type region, to the MOSFET's gate terminal, where they will form fixed negative impurity ions when a positive voltage is applied. The channel's conductivity will decrease as a result of the development of a depletion zone inside the channel. The drain current can be controlled by applying a positive voltage at the gate terminal [3, 12].

3.3.2 Enhancement Type MOSFET

The enhancement mode MOSFET is comparable to a "Normally Open" switch in that it needs a gate-source voltage in order to turn on the device. Here, we can see that the enhancement mode type is represented by the broken line that connects the source and drain. Carriers are added to the channel in enhancement mode MOSFETs, enhancing conductivity through an increase in the oxide layer. This oxide layer is commonly referred to as the "Inversion layer". Between the source and drain, the channel forms in the substrate's opposite type; for example, an N-channel is produced with a P-type substrate, and a P-channel is produced with an N-type substrate. The P-type or N-type channel determines the conductivity of the channel because of holes or electrons, respectively [8–14].

Enhancement mode MOSFETs have a channel that is ordinarily off and needs to be turned on with the help of applying a positive voltage to the gate. The channel in depletion mode MOSFETs is normally getting on, and to turn it off, the gate must receive a negative signal.

Conductivity in a MOSFET type of transistor is dependent upon the semiconductor channel that is present between the source and drain terminals. Depending on how the MOSFET is configured, this semiconductor channel can be either P-channel or N-channel. A MOSFET is composed of the drain, source, and gate terminals. Applying a voltage across the gate and source creates an inversion layer or, in the case of a threshold voltage, a channel between the drain and source. The lowest voltage at which current must flow is known as the threshold voltage. No channel will form if the applied voltage is lower than the threshold voltage. As a result, the MOSFET cannot accept current. The cutoff area is the name given to this situation [6–12].

The MOSFET is composed of three layers: drain, gate, and source. The internal substrate connection of the MOSFET to the source is referred to as the "substrate". The metallic gate terminal of a MOSFET can be isolated from the semiconductor layer by a silicon dioxide or dielectric layer. Two of the components needed to create this EMOSFET are P-type and N-type semiconductors. A substrate gives structural support for the gadget. A thin coating of SiO, which is also a good electrical insulator, covers the space between the source and drain terminals. A metallic layer sits on top of the oxide layer to form the gate electrode [4–10].

There is no channel that connects the source and drain when V_{GS} is 0V. The drain current is zero because there aren't many thermally generated charge carriers which are minority, such as free electrons, in the p-type substrate. For this reason, the MOSFET will often be turned off.

Minority charge carriers from the p-substrate, are drawn to the gate (G), which is positive (+ve), where they combine through the holes beneath the

SiO_2 layer. Additional V_{GS} will provide the electrons with the potential to overcome bonding and produce additional charge carriers, or electrons that are deposited in the channel.

In this instance, the dielectric is employed to stop the electrons from passing through the layer of silicon dioxide [3, 5].

The accumulation between the source and drain terminals will result in the formation of an N-channel. As a result, the generated drain current can end up flowing all the way through the channel. The resistance of the channel, which is established by the charge carriers attracted to the gate's positive terminal, is directly connected to the drain current [6].

There are two types available: N-channel enhancement MOSFET and P-channel enhancement MOSFET. The source and drain terminals of the N-channel enhancement type are composed of a mildly doped p-substrate and two strongly doped n-type regions. Electrons make up the majority of the charge carriers in this kind of E-MOSFET. The source and drain terminals of the P-channel type are formed by a slightly doped N-substrate and two heavily doped p-type regions. The majority of the charge carriers in this kind of E-MOSFET are holes [7, 8].

3.3.3 VI Characteristics of MOSFET

The input/transfer and output characteristics of the MOSFET are specified below.

3.3.3.1 *Input or Transfer Characteristics*

In Figure 3.13, the input characteristics of drain current with the gate source voltage are shown. Threshold voltage is the st voltage required to initiate

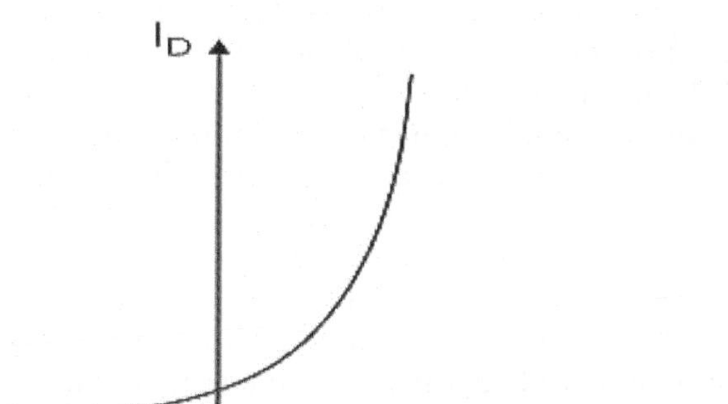

Figure 3.13 MOSFET VI characteristics: I_D versus V_{GS}.

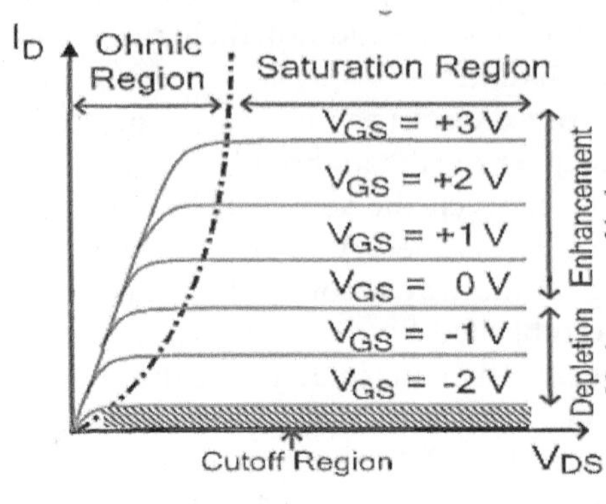

Figure 3.14 I_D versus VDS curve for MOSFET.

the drain current flow. The Vth is between 2 and 3 volts. If the gate source voltage is equal to or less than the threshold voltage, the current from the drain is zero or no current flows in the load; if the gate source voltage exceeds the threshold level then the current from the drain begins to rise. The drain is proportional to the magnitude level of the gate source voltage. The device is said to have transfer qualities since it is able to switch from the OFF state to the ON state.

As the V_{GS} voltage rises, the drain current grows exponentially before becoming constant. There is a minimal gate to source voltage, or threshold voltage, as demonstrated by the MOSFET VI characteristics, at which current merely begins to flow across the channel [3–6].

3.3.3.2 *Output Characteristics*

The MOSFET VI characteristics mentioned in Figure 3.14 show that there are three zones of operation: the cut-off, saturation, and ohmic region.

It functions as an off state in the cut-off region, an open switch in the ohmic region, and an amplifier in the saturation zone. Figure 3.5 shows how, as the VDS voltage rises, the current I_D first climbs and then gradually reaches a constant amount.

a. Cut-Off Region

Since there won't be any current passing through the MOSFET at the cut-off zone, it will be OFF. Because MOSFETs function as open switches in this situation, they are used when they must function as electronic switches [6].

b. Ohmic or Linear Region

In the ohmic or linear zone, the current ID rises as the VDS value does. When MOSFETs are configured to function in this domain, they can be used as amplifiers.

c. Saturation Region

The MOSFETs then move into the saturation area and their ID becomes constant when VDS impacts the pinch off voltage. Data may then move through the device like a closed switch until the IDS value reaches a saturation point [4].

As a result, this operating area is chosen in order to use MOSFETs for switching operations. If the drain to source current IDS through the device is plotted against the gate to source voltage VGS in the transfer characteristics, there is no current flowing through the device up until V_{GS} in the transfer characteristics exceeds the threshold voltage V_T for an empty channel connecting the source and drain terminals. The output characteristics (ID vs. VDS) are shown to be such that a rise in VDS will not cause current flow. When this occurs, the device functions as a closed switch, letting data pass through until the IDS value reaches a saturation point [6, 7].

Therefore, whenever MOSFETs are required to carry out switching responsibilities, this operating region is utilized. There won't be any current passing through the device until the gate-to-source voltage V_{GS} in the transfer characteristics drain-to-source current IDS against gate-to-source voltage V_{GS} exceeds the threshold voltage V_T. An empty channel that was intended to connect the source and drain terminals is the cause of this. The corresponding output characteristics (ID vs. VDS) demonstrate that, even a rise in VDS will not result in current flow. Consequently, this case is nothing but the MOSFET's cut-off region operation [2–6].

3.4 REVIEW BETWEEN TYPES OF FETS

Field-effect transistors (FETs) are a broad class of transistors that comprise a number of varieties. These include transistors whose current is controlled by a reverse-biased PN junction. The MOSFET, a more specialized type of FET, uses an insulating oxide layer to keep the gate and channel apart.

To conduct, the enhancement MOSFET requires a positive gate source voltage for the n-channel as well as a negative gate source voltage for P-channel operation. Depletion mode MOSFETs must be operated at zero gate source voltage and therefore must have a negative gate source voltage for N-channel operation and a positive gate source voltage for P-channel operation [12].

3.4.1 Basic Structure and Mechanism

A generic FET's gate is a PN junction, which means that altering the gate-source voltage directly regulates the current flowing through the channel. In a MOSFET, a small layer of silicon dioxide (or other insulating material) separates the gate, which comprises metal or a strongly doped polysilicon, from the channel. This permits a very high input impedance since the gate consumes very little current [4].

Although the gate construction and control mechanism can change, the voltage supplied to the gate directs the current flowing through the channel. In JFETs, the gate voltage regulates the depletion zone that impacts the channel conductivity. The current that flows between the drain and the source in a MOSFET is controlled by the voltage that is provided to the gate.

Applying a voltage to the gate of enhancement mode MOSFETs produces an inversion layer that permits current to flow. Applying a gate voltage depletes the channel in depletion mode MOSFETs, where the channel is commonly on [6].

Because the gate is isolated from the channel by a PN junction or other structures, FETs often have high input impedance but given that the oxide layer acts as a superior insulator and permits very little gate current, MOSFETs have an even higher input impedance while in comparison with JFETs [7].

Depending on the kind, power consumption varies. When compared to MOSFETs, JFETs could have larger gate leakage currents [9].

MOSFETs are well known for their low power consumption due to their high input impedance and low gate current, which makes them perfect for battery-operated devices and low-power applications. Depending on the type of FET, switching speeds can be very high. When compared to MOSFETs, JFETs typically have slower switching speeds. Because of their high input impedance and lack of charge storage effects, MOSFETs usually have very high switching rates. There are differences in thermal stability. Because JFETs have a positive temperature coefficient, which means that as the temperature rises so does their resistance, this helps prevent thermal runaway [4, 6].

3.4.2 Applications and Research Trend of Field Effect Transistors (FETs)

FETs are less temperature dependent and more immune to radiation. Fabrication of field effect devices is simpler (less number of steps) than bipolar devices. They occupy less area on the chip. Because of their high gain, low output impedance, and high input impedance, these devices are frequently utilized in electronic circuits [13].

As FETs are immune to radiation, they are used in amplifiers required for satellite communication in the GHz frequency range. Because of high input impedance, they are suitable for input stages of operational amplifiers, instrumentation amplifiers etc. [6].

Since MOSFETs have negative temperature coefficients of resistance, which aid in preventing thermal runaway, they often provide higher thermal stability, particularly when utilized in power applications. MOSFETs are generally more complicated to manufacture because of the requirement for exact control over the gate structures and oxide layer, which can raise manufacturing costs, particularly in nodes with advanced technology [15].

JFETs are commonly used in analog applications that call for precise signal processing because of their strong linearity and minimal distortion. Instrumentation, analog signal processing circuits, and high-fidelity audio equipment all use them. Even though MOSFETs have outstanding linearity, switching performance and efficiency are often what make them preferred over linearity in particular. In analog applications, they are typically found in circuits that demand high precision and low distortion. They are utilized in both analog and digital circuits [16].

FETs are used in many different applications, such as impedance matching, signal amplifiers, and analog switches. JFETs work especially well in input stages with high impedance. MOSFETs are frequently utilized in power switching, amplification, and digital circuits (such as those utilizing CMOS technology). In applications involving high-speed switching, power electronics, and integrated circuits, they are essential [17, 18].

JFETs are employed in situations where a modest switching speed is adequate. When their switching speed is suitable for the task, they are usually used in analog circuits such as mixers and RF amplifiers. For applications where high-speed switching is required, MOSFETs are recommended. They are extensively utilized in digital circuits, including memory devices, microprocessors, and logic gates. Because of their quick switching times, MOSFETs are also essential in power electronics for power supply and inverter switching [16].

When compared to MOSFETs, JFETs are typically easier and less expensive to utilize in circuits. They are frequently employed in more basic analog circuits where it is necessary to maintain low costs and circuit complexity. Because of their production method and requirement for exact oxide layers, MOSFETs can be more complicated and expensive. However, they are preferred for more sophisticated and advanced applications, such as digital integrated circuits and power electronics, because of their benefits in switching speed, power consumption, and efficiency. Because of their well-known low noise properties, JFETs are a good fit for delicate analog applications like instrumentation, RF receivers, and low-noise amplifiers. Despite having low noise characteristics, MOSFETs are typically employed in applications where power efficiency and switching performance are more important than noise concerns [12–17].

FET integration with other semiconductor components is being worked on to build more efficient and compact analog circuits. FETs can be used in CMOS technology to combine their high input impedance with the circuits' high density and low power consumption [12].

Researchers are looking into alternative materials for MOSFETs, like silicon carbide as well as gallium nitride (GaN), to enhance their performance in high-frequency and high-power applications. These materials offer advantages over traditional silicon-based MOSFETs, including higher breakdown voltages and better thermal conductivity [15].

In order to improve battery life and performance, MOSFETs are utilized in power management and switching circuits in consumer devices such as laptops, tablets, and smartphones. MOSFETs are utilized in infotainment, lighting, and engine controls in the car sector. They are appropriate for automotive electronics that need to be very reliable because of their efficiency and ruggedness. MOSFETs are utilized in wind turbine controllers and solar inverters, among other renewable energy systems, to aid in the effective conversion and administration of energy.

REFERENCES

[1] Dixit A. *MOSFET on the Horizon: What's New and What's Next.* MOSFET – Developments and Trends. IntechOpen, Apr. 16, 2024. doi: 10.5772/intechopen.1004452.

[2] Dhurandhar N, and Dwivedi P. A review article on fin field effect transistors technology. *CSVTU Research Journal on Engineering and Technology*, 2019, 8(1), 31–38. ISSN: 0974-8725.

[3] Qiao R, Zhang H, Zhao S, Yuan L, Jia R, Peng B, and Zhang Y. A state-of-art review on gallium oxide field-effect transistors. *Journal of Physics D: Applied Physics*, 2022, 55(38), p. 383003. DOI 10.1088/1361-6463/ac7c44

[4] Şahin M E, and Blaabjerg F. An overview on MOSFET drivers and converter applications. *Electric Power Components and Systems*, 2021, 49(8), 828–847. https://doi.org/10.1080/15325008.2021.2002477

[5] Luo H, Iannuzzo F, Reigosa P D, Blaabjerg F, Li W, and He X. Modern IGBT gate driving methods for enhancing reliability of high-power converters—an overview. *Microelectronics Reliability*, 2016, 58, 141–150. https://doi.org/10.1016/j.microrel.2015.12.022

[6] Puschkarsky K, Grasser T, Aichinger T, Gustin W, and Reisinger H. Review on SiC MOSFETs high-voltage device reliability focusing on threshold voltage instability. *IEEE Transactions on Electron Devices*, Nov. 2019, 66(11), 4604–4616. https://doi.org/10.1109/TED.2019.2938262.

[7] Ma C T and Gu Z H. Review on driving circuits for wide-bandgap semiconductor switching devices for mid-to high-power applications. *Micromachines*, 2021, 12(1): 65. https://doi.org/10.3390/mi12010065.

[8] Yoori S, Eul-Yong S, Yong-Young N, et al. Flexible bottom-gated organic field-effect transistors utilizing stamped polymer layers from the surface of water. *ACS Applied Materials & Interfaces*, 2020.

[9] Islam M M, Rivero M H, Rose G, et al. Low-power dynamic circuit design with steep-switching hybrid phase transition FETs (Hyper-FETs). *IEEE Transactions on Electron Devices*, 2022, 70(2): 819–825.

[10] Hyuk G S, Geon-Beom L, Eun-Su A, et al. High-performance field-effect transistor and logic gates based on GaS-MoS2 van der Waals heterostructure. *ACS Applied Materials & Interfaces*, 2020.

[11] Hu C, Andrew W, Chun M, et al. The effect of ring expansion in Thienobenzo [b] indacenodithiophene polymers for organic field-effect transistors. *Journal of the American Chemical Society*, 2019.

[12] Guo N, Xiao L, Gong F, et al. Light-driven WSe2-ZnO junction field-effect transistors for high-performance photodetection. *Advanced Science*, 2020, 7(1): 1901637.

[13] Srisonphan S, Jung Y S, Kim H K. Metal–oxide–semiconductor field-effect transistor with a vacuum channel. *Nature Nanotechnology*, 2012, 7(8): 504–508.

[14] Roccaforte F, Greco G, Fiorenza P, et al. An overview of normally off GaN-based high electron mobility transistors. *Materials*, 2019, 12(10): 1599.

[15] Torsi L, Magliulo M, Manoli K, et al. Organic field-effect transistor sensors: A tutorial review. *Chemical Society Reviews*, 2013, 42(22): 8612–8628.

[16] Pullano S A, Greco M, Fiorillo S A, et al. Design and fabrication of an EGFET based chemical sensor using transistor association technique. *2020 IEEE International Symposium on Medical Measurements and Applications (MeMeA)*. IEEE, 2020: 1–5.

[17] Hirayama Y. *Encyclopedia of Condensed Matter Physics* (Second Edition), Field Effect Transistors: MISFET, MESFET, and HEMT, 2005.

[18] Thompson, M T. Chapter 9 - Introduction to Field-Effect Transistors (FETs) and Amplifiers. In *Intuitive Analog Circuit Design*, Elsevier, 2013, pp. 297–339. doi: 10.1016/b978-0-12-405866-8.00009-7.

Chapter 4

Essential Foundations of CMOS Technology

Soumya Sen, Atanu Chowdhury, Ashish Raman, Prabhat Singh, and Subarna Mondal

4.1 INTRODUCTION

The rise of CMOS technology has revolutionized the field of microelectronics. This has made it possible to develop faster, smaller, and more energy efficient devices. CMOS technology has become the backbone of contemporary electronics. From consumer electronics to medical devices, the ceaseless quest for smaller, quick, and more economical semiconductor devices that power a wide range of sensing systems and applications has driven the adoption of semiconductor devices. This has led to exponential improvements in efficiency and energy consumption. However, as CMOS transistors approach the nanoscale regime, new challenges have arisen. These include leakage current, variability, and increased thermal management issues. A basic understanding of MOS transistor physics and CMOS circuit design principles is essential to overcome these challenges. This chapter provides a comprehensive introduction to CMOS technology, focusing on the production, operation, and applications of n-type and p-type MOS transistors. A deep understanding of CMOS technology is essential for developing innovative solutions to overcome the limitations of current devices. According to Swanson and Mendel, "The continued scale-up of MOS transistors requires a thorough understanding of CMOS technology." This chapter aims to provide readers with a comprehensive understanding of CMOS technology, enabling them to contribute to the ongoing advancement of microelectronics. By exploring the basic principles of CMOS technology, readers will gain a comprehensive understanding of its design. Knowledge of the optimization and the application of CMOS circuits will facilitate the development of the upcoming era of electronic devices. Advancing across diverse fields such as artificial intelligence, the Internet of Things, and quantum computing [1–4].

4.1.1 MOS Transistor Basics

At the nucleus of CMOS technology is the Metal-Oxide-Semiconductor Field-Effect Transistor (MOSFET). A MOSFET is a voltage-governed

 DOI: 10.1201/9781003533535-4

device that modulates the current between source and drain terminals based on the voltage rendered to the gate. As a high enough gate voltage is exercised, the MOS transistor forms an inversion layer (or channel) behind the gate oxide.

There are two main types of MOS transistors used in CMOS technology:

I. N-Channel MOSFET (NMOS): As illustrated in Figure 4.1(a), electrons act as the charge carriers. A positive gate voltage relative to the source attracts electrons to form a conductive channel within the source and drain.
II. P-Channel MOSFET (PMOS): In this case, holes act as the charge carriers. A negative gate voltage relative to the source creates a hole-rich inversion layer, allowing current to flow between the source and drain, as indicated in Figure 4.1(b).

The unique advantage of CMOS technology lies in combining both NMOS and PMOS transistors on a single substrate. This configuration allows for low static power consumption since inverters and logic gates made from CMOS only draw power during switching events.

4.1.1.1 Complementary Operation of NMOS and PMOS Transistors

CMOS circuits use the complementary behaviour of NMOS and PMOS transistors. In a typical CMOS inverter, as shown in Figure 4.1(c), the former conducts when the input is high, dragging output to ground (logic 0), while the latter conducts when the input is subdued, pulling to V_{DD} (logic 1). This complementary action enables CMOS circuits to achieve low

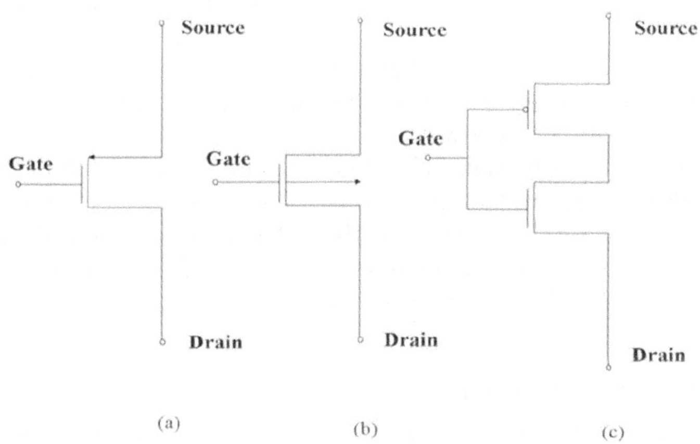

Figure 4.1 Schematic diagram of: (a) NMOS; (b) PMOS; (c) CMOS.

static power dissipation, as either of them is off at any given time when the circuit is in a steady state. As a result, CMOS circuits only consume dynamic power during switching transitions, making them highly energy-efficient. The power consumption in CMOS circuits can be approximated by equation 4.1,

$$P_{dynamic} = C_L V_{DD}^2 f \tag{4.1}$$

where $P_{dynamic}$, C_L, V_{DD}, and f, are the dynamic power consumption, load capacitance, supply voltage, and switching frequency, respectively [1–4]. A more elaborate explanation is provided in section 4.3.

The dominance of CMOS technology in the semiconductor industry is largely due to several key advantages it offers over other technologies, such as NMOS-only or bipolar junction transistors (BJT). These advantages include:

1. **Low Power Consumption:** CMOS circuits are known for their high energy efficiency, making them ideal for low-power applications. This results in minimum static strength dissipation, making CMOS ideal for battery-powered devices and systems requiring low electricity consumption.
2. **Exceptional Noise Resilience:** CMOS circuits are recognized for their sturdy noise margins, which make them exceedingly immune to signal loss because of external interference or parasitic capacitance. This feature ensures reliable performance, even under demanding conditions or high-speed applications.
3. **Enhanced Scalability:** CMOS generation adapts properly to smaller device sizes, permitting the industry to adhere to Moore's Law over many years. As transistors reduce in size, a wider variety can be accommodated on a unmarried chip, thereby boosting processing ability and functionality at the same time as preserving area and keeping costs fairly low.
4. **Integration Potential:** The CMOS era helps combine hundreds of thousands, and even billions, of transistors onto a single chip. This functionality supports the advent of compound structures, like microprocessors, memory devices, and System-on-Chip (SoC) architectures.

4.1.2 MOS transistor physics

This section provides an in-depth examination of the physical concepts that underlie MOS transistors, illustrated in Figure 4.2, which can be essential to the operation of CMOS technology. The physics of MOS transistors dictate

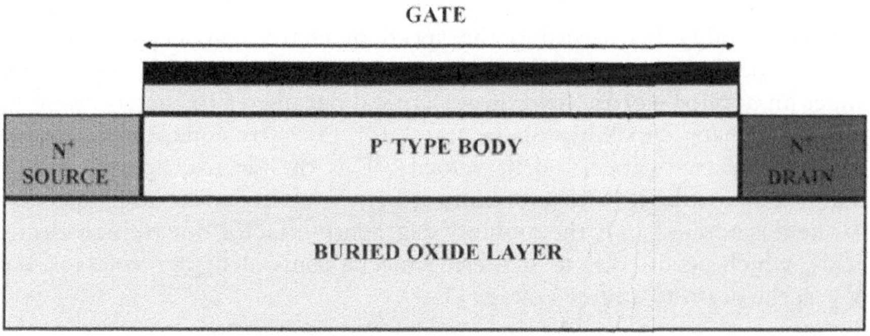

Figure 4.2 Internal structure of MOS transistor, showing all regions.

their electrical behaviour, which is fundamental to understanding their performance in digital and analog circuits. This section covers threshold voltage, mobility, sub threshold leakage, capacitance, and channel length modulation.

4.1.2.1 Threshold Voltage (V_{th})

The threshold voltage is the minimal gate-to-source voltage necessary to form an electrical pathway amid the source and drain of a MOSFET. It can be represented as in equation 4.2.

$$V_{th} = V_{FB} + 2\varphi_F + \frac{\sqrt{2q\varepsilon_S N_A 2\varphi_F}}{C_{OX}} \tag{4.2}$$

where V_{FB}, is the flat band voltage; φ_F and ε_S, are the Fermi potential representing the energy contrast between the Fermi and semiconductor's intrinsic energy level and the silicon substrate's permittivity; N_A is as the acceptor doping concentration in the p-type substrate; and $C_{OX} = \frac{\varepsilon_{ox}}{t_{ox}}$, is the gate oxide capacitance, where ε_{ox} is the permittivity of the gate oxide material and t_{ox} is the thickness of the oxide layer [1–4].

Considering the effect on device operation, by reducing the V_{th}, off-state leakage current increases, leading to higher power dissipation, whereas in the other case, leakage gets reduced but may lower the drive current in the on-state, affecting performance.

4.1.2.2 Carrier Mobility (μ) and Channel Length Modulation

Carrier mobility (μ) describes the speed of charge carriers (electrons in nMOS and holes in pMOS transistors) navigating through the channel under an applied electric field. It is a critical parameter for determining the current-carrying capability of the transistor, given by equation 4.3, where v_d is the charge carriers' drift velocity, E is the electric field across the channel, μ_0, is the low-field mobility (carrier mobility when no proper electric field is applied), θ is the mobility degradation factor due to high electric fields, which occurs due to scattering mechanisms at higher voltages, and V_{DS} is the drain-to-source voltage [1–4].

$$\mu = \frac{v_d}{E} = \mu_0 \left(1 + \theta V_{DS}\right) \tag{4.3}$$

In short-channel devices, channel length modulation is the phenomenon where the effective channel length is altered by the drain-source voltage, leading to a change in the saturation current. The drain current in saturation can be approximated as in equation 4.4, with K' as the process-dependent parameter, V_{GS} as the gate-source voltage, and λ as the channel length modulation coefficient, representing the change in effective channel length with V_{DS}.

$$I_{DS} = \frac{K'}{2} = \left(V_{GS} - V_{th}\right)^2 \left(1 + \lambda V_{DS}\right) \tag{4.4}$$

Channel length modulation causes the drain current to increase slightly with increasing V_{DS}, even in saturation, which can affect analog circuit performance, particularly in amplifiers [1–4].

4.1.2.3 Subthreshold Leakage Current

In digital CMOS circuits, subthreshold leakage (illustrated in equation 4.5) occurs when the transistor operates below the threshold voltage but still allows a small current to flow, between the drain and source. This current is exponentially related to the gate voltage in the subthreshold region, where I_0 is a pre-exponential factor dependent on the device structure and process technology, V_T $\left(V_T = \frac{KT}{q}\right)$, which depends on temperature, and n is the subthreshold slope factor, typically greater than 1, representing the effectiveness of gate control over the channel current.

$$I_{DS(Sub)} = I_0 e^{\frac{(V_{GS} - V_{th})}{nV_T}} \tag{4.5}$$

The subthreshold leakage can be affected by a higher doping concentration elevating V_{th}, reducing leakage but affecting on-state performance. Additionally, thinner gate oxides improve control over the channel but increase gate leakage, contributing to power dissipation [1–4].

4.1.2.4 Capacitance in MOS Transistors

Capacitance in MOSFETs, as shown in Figure 4.3, arises from the regions where charge is stored, such as the gate oxide, depletion region, and overlap regions. The significant capacitance in a MOS transistor is gate capacitance (C_{ox}), which is a critical factor determining the transistor's switching speed, as it governs the ability of the gate to supervise the channel. It is given by equation 4.6,

$$C_{ox} = \frac{\varepsilon_{ox}}{t_{ox}} \tag{4.6}$$

where ε_{ox}, is the permittivity of the gate oxide material and t_{ox} is the oxide thickness. A thinner oxide increases C_{ox}, allowing faster switching but at the cost of higher gate leakage currents. In addition to gate capacitance, the junction capacitance (C_j) arises from the pn-junctions formed amongst the source/drain and the substrate. This junction capacitance is given by equation 4.7,

$$C_j = \frac{\varepsilon_s A}{\sqrt{2qN_A\left(V_{bi} + V_R\right)}} \tag{4.7}$$

This depends on the depletion region width and reverse bias across the junction. Junction capacitance affects the transistor's dynamic performance, especially in analog applications, impacting switching frequency and power consumption. Both capacitances play a vital role in determining the overall speed and efficiency of MOS transistors, particularly in scaled-down devices where their influence becomes more pronounced [1–4].

4.1.2.5 Short Channel Effects and Scaling

As MOSFETs are scaled down to nano dimensions, the traditional long-channel model no longer accurately describes their behaviour. Instead, a

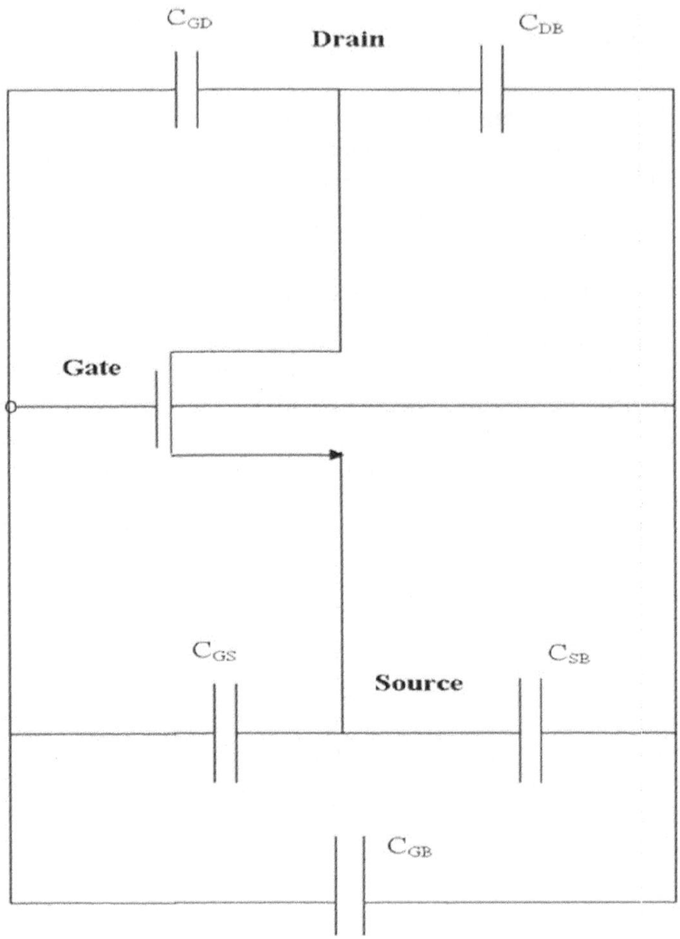

Figure 4.3 CMOS capacitances.

range of phenomena, known as short-channel effects (SCE), begin to dominate the device's operation. These effects arise from the reduction in the channel length, which alters the electric field distribution within the transistor and compromises gate control over the channel. Understanding these effects is crucial in modern CMOS design, particularly in submicron and nanometre technologies.

I. **Drain Induced Barrier Lowering:** Drain-Induced Barrier Lowering (DIBL), depicted in Figure 4.3, occurs when the electric field from the drain terminal influences the potential barrier at the source end of the channel. In long-channel devices, the gate voltage primarily controls the barrier, but in short-channel MOSFETs, the drain voltage also plays a role. As the drain voltage increases, it reduces the barrier height at the source, effectively lowering the threshold voltage, V_{th}. The impact of DIBL can be modelled as in equation 4.8,

$$\Delta V_{th} = S_{DIBL}.V_{DS} \tag{4.8}$$

where S_{DIBL} is the DIBL coefficient, which indicates how sensitive the threshold voltage is to changes in the drain-source voltage V_{DS}.

DIBL results in a lowering of the threshold voltage, increasing the subthreshold leakage current and degrading the transistor's off-state performance. In digital circuits, DIBL can cause excessive standby power consumption due to increased leakage current in the off state [1–4].

II. **Threshold Voltage Roll-Off:** In long-channel devices, the threshold voltage remains relatively constant. However, in short-channel MOSFETs, it decreases as the channel length reduces, leading to "threshold voltage roll-off". This effect occurs because the gate drops some influence over the channel, and the source and drain depletion regions encroach into the channel region. Threshold voltage roll-off can be expressed as a function of channel length L (equation 4.9),

$$V_{th} \propto \frac{1}{L} \tag{4.9}$$

As the channel length shortens, the electrostatic integrity of the channel weakens, and the threshold voltage decreases. This creates challenges in ensuring consistent transistor behaviour, especially in modern nanometre-scale technologies [1–4].

III. **Velocity Saturation:** In short-channel MOSFETs, the electric field across the channel becomes very strong as the channel length decreases. While the electric field reaches an acute value, known as the saturation electric field E_{sat}, the carrier speed increases non-linearly with the electric field. Instead, it saturates at a maximum value, prescribed as the saturation velocity v_{sat}, limiting the current. It is expressed by equation 4.10,

$$v_{sat} = \mu E_{sat} \tag{4.10}$$

where μ is the carrier mobility and E_{sat} is the saturation electric field. In long-channel MOSFETs, the current escalates quadratically with the gate voltage in the saturation region. However, in short-channel devices with velocity saturation, the current increases linearly with gate voltage, reducing the drive current and affecting the transistor's switching speed; adding to the same velocity saturation limits the transistor's performance at high drain-source voltages, reducing the potential speed improvements from scaling [1–4].

IV. **Channel Length Modulation:** Channel Length Modulation (CLM) is the process of shrinking the effective channel length as the drain voltage increases in short-channel MOSFETs. As the drain voltage rises, the depletion zone near the drain expands, shortening the effective length of the conducting channel. This causes a rise in the drain current, even in the saturation area. Equation 4.11 may be used to calculate the drain current at saturation while taking channel length modulation into account,

$$I_{DS(Sat)} = \frac{K'}{2}\left(V_{GS} - V_{th}\right)^2 \left(1 + \lambda V_{DS}\right) \tag{4.11}$$

K′ is a process-dependent constant, V_{GS} is the gate-to-source voltage, V_{DS} is the drain-to-source voltage, and λ is the channel length modulation coefficient, which indicates how much the effective channel length changes with VDS. CLM allows the drain current to rise slightly with rising drain voltage, even when the transistor is meant to be saturated. This action lowers the output resistance of the transistor, which can diminish the gain of analogue circuits such as amplifiers [1–4].

V. **Hot Carrier Injection:** Hot Carrier Injection happens when strong electric fields near the drain allow carriers (electrons or holes) to accumulate a lot of kinetic energy. Some of these "hot" carriers can be induced into the gate oxide, causing oxide damage and long-term performance deterioration in the transistor. The chance of hot carrier injection grows as:

A. **High drain voltage (V_{DS}):** The electric field near the drain increases, accelerating the carriers to higher velocities.
B. **High gate voltage (V_{GS}):** More carriers are present in the channel, increasing the likelihood of hot carriers interacting with the oxide.

Hot carrier injection leads to long-term reliability issues, such as threshold voltage shifts and degradation of carrier mobility. This effect is more prominent in short-channel devices and can cause a gradual deterioration in device performance, particularly in high-speed circuits [6–10].

4.2 MOS TRANSISTOR FABRICATION PROCESS

The MOS transistor fabrication process, is a highly intricate procedure involving multiple steps to create functioning MOSFET devices on a silicon wafer. The process begins with *wafer preparation*, followed by *oxide growth*, *doping*, *lithography*, and *metallization*, each step ensuring the precise formation of the transistor's core components: the gate, source, drain, and channel [11–14].

4.2.1 Wafer Preparation and Oxide Formation

The fabrication process begins with the selection of a high-quality silicon wafer, which serves as the substrate for building MOS transistors. Initially, the wafer undergoes cleaning to remove any surface contaminants, and the crystal orientation is precisely controlled for optimal performance. After cleaning, *thermal oxidation* is performed, where the wafer is exposed to high-temperature oxygen or steam, resulting in the formation of a thin layer of silicon dioxide (SiO_2) on the surface. This oxide layer acts as the *gate oxide* and insulates the gate from the underlying silicon channel. The oxide thickness (t_{ox}) plays a critical role in the transistor's electrical characteristics, particularly the threshold voltage V_T, which can be modelled as in equation 4.12,

$$V_T = V_{FB} + 2\varphi_F + \frac{Q_{OX}}{C_{OX}} \tag{4.12}$$

where V_{FB} is the flat-band voltage, φ_F is the Fermi potential, Q_{ox} represents the charge in the oxide, and C_{ox} is the capacitance of the gate oxide [1–4].

4.2.2 Doping and Source/Drain Formation

Next, the process of *ion implantation* is used to introduce dopants (typically boron for p-type and phosphorus or arsenic for n-type) into specific regions of the silicon substrate to create the source and drain regions. These doped areas form the *p-n junctions* essential for MOSFET operation. The doping concentration is controlled to modify the electrical properties of the MOSFET, such as carrier mobility and junction depth. A *photoresist* is exercised over the wafer, and *lithography* is used to generate patterns, exposing only the areas to be doped. The dopant atoms are accelerated and implanted into the wafer, followed by *annealing* to activate the dopants and repair the crystal structure. The junction depth (X_j) and the doping profile affect the *threshold voltage* and *subthreshold slope*, and can be modelled using equation 4.13,

$$X_j = \sqrt{\frac{2\varepsilon_s V_{bi}}{qN_A}} \tag{4.13}$$

where ε_s is the permittivity of silicon, V_{bi} is the built-in potential, q is the electronic charge, and N_A is the doping concentration of the substrate [1–4].

4.2.3 Gate Electrode Formation and Metallization

After the source and drain areas have been specified, the gate electrode is formed. A stratum of polysilicon is placed over the oxide, allowing it to function as a conductive gate fabric. Photolithography is used once more to define the gate location, and the excess polysilicon is etched away, leaving the gate electrode aligned over the channel. The next step is metallization, in which a metal layer, typically aluminium or copper, is implemented to create interconnects that link the transistors with other elements inside the included circuit. This steel layer establishes connections between the gate, source, drain terminals, and the outside circuits, permitting manipulation over the current inside the MOSFET. The gate capacitance performs a vital position in influencing switching velocity and may be represented by employing equation 4.14:

$$C_{OX} = \frac{\varepsilon_{ox}}{t_{ox}} A \tag{4.14}$$

where ε_{ox} is the permittivity of the gate oxide, t_{ox} is the oxide thickness, and A is the area of the gate [1–4].

After the layers are deposited and patterned, the wafer undergoes passivation and packaging processes to protect the device and prepare it for integration into larger systems. The final result is an efficient MOS transistor, able to switch and amplify electronic indicators, which serves as the cornerstone of contemporary digital circuits [15–20].

4.3 CMOS CIRCUIT DESIGN PRINCIPLES

CMOS circuits are built using a combination of p-type and n-type MOS transistors, enabling the design of digital logic circuits characterized by low power consumption and high noise margins. CMOS generation is favoured for its minimal static power dissipation, durability, and scalability, all of which are critical for modern-day electronic gadgets. This section explores the fundamental principles of CMOS circuit design, covering topics such as logic gates, power consumption, delay, noise margins, and optimization techniques using mathematical models.

4.3.1 Basic CMOS Inverter Design

The CMOS inverter, illustrated in Figure 4.4, is the essential building block for all CMOS circuits. It consists of a p-type MOSFET (PMOS) and an n-type MOSFET (NMOS) connected in series between the supply voltage (VDD) and ground. The input voltage V_{in} is carried out to each gate, while the output voltage V_{out} is obtained from the drain nodes of the transistors.

1. **Operating Principle:** When the input exceeds $V_{in} = V_{DD}$, the NMOS transistor is turned ON (carrying out), while the PMOS transistor is grown to become OFF (non-undertaking), resulting in the output being dragged to the ground, which equates to a logical 0. When the input voltage is curtailed ($V_{in} = 0V$), the pull-up network turns on, and NMOS turns off, pushing the output to V_{DD} (logical 1). This complementary action consumes less static power since there is no direct link between V_{DD} and ground during steady-state operation.
2. **Voltage Transfer Characteristics**: The relationship between V_{in} and V_{out} for a CMOS inverter is outlined by the **voltage transfer characteristic** (VTC) curve, illustrated in Figure 4.5, which is governed by the following regions of operation:
   a) Region 1: NMOS OFF, PMOS ON (high output)
   b) Region 2: Both transistors in transition (output switching)
   c) Region 3: NMOS ON, PMOS OFF (low output)

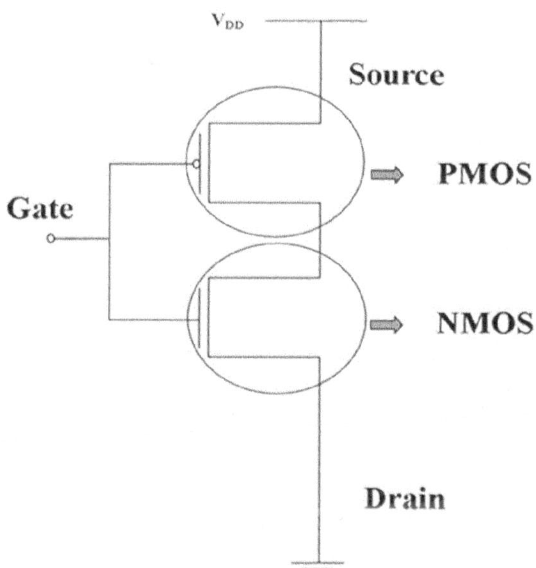

Figure 4.4 Schematic diagram of CMOS inverter.

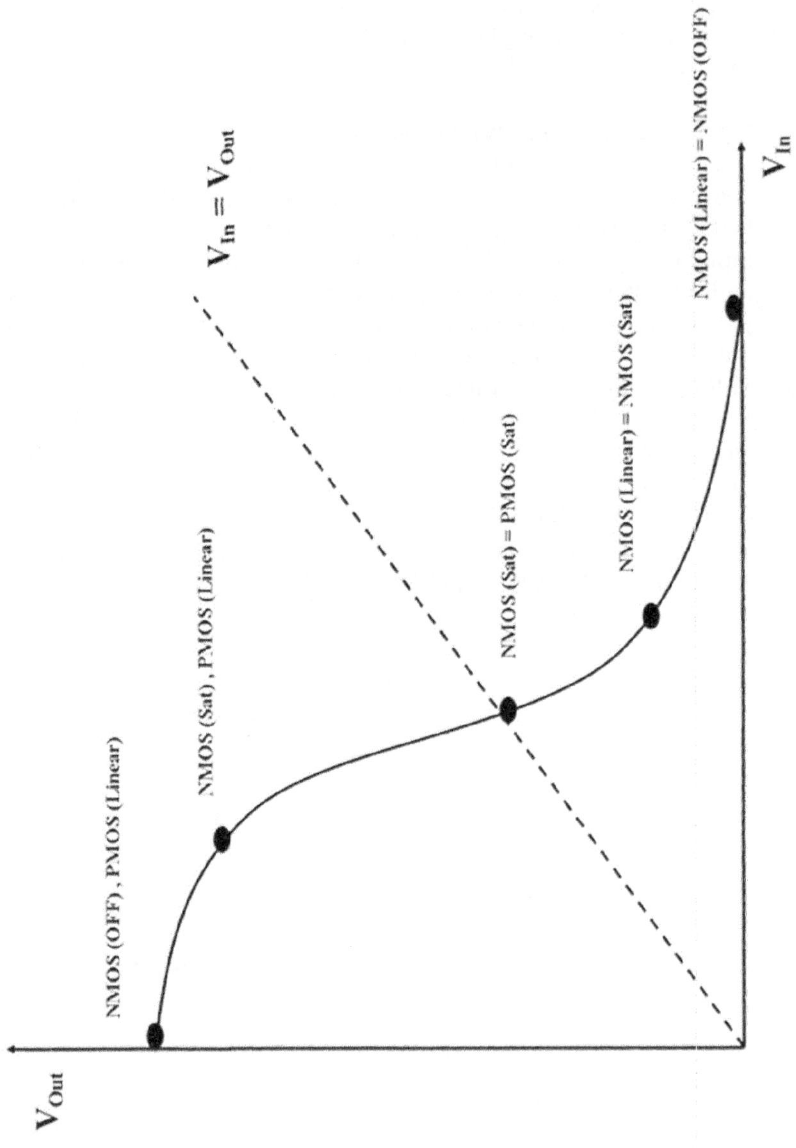

Figure 4.5 Voltage transfer characteristics curve for CMOS inverter.

The **switching threshold** V_{th} of the CMOS inverter, where $V_{in}=V_{out}$, is derived by equating the currents of the NMOS and PMOS in saturation. Assuming equal mobility for simplicity, the threshold can be approximated as in equation 4.15,

$$V_{th} - \frac{V_{DD}}{2} \tag{4.15}$$

This results in a symmetric switching behaviour, where the inverter transitions smoothly between logic levels [1–4].

3. **Power Consumption in CMOS circuits:** The entire power dissipation in CMOS circuits is separated into two categories: dynamic and static power. The primary factor in CMOS power consumption is dynamic power, which occurs during switching events, whereas static power is triggered by leakage currents in the off-state.
   a) **Dynamic Power:** The dynamic power consumption is associated with the charging and discharging of capacitances in the circuit during transitions between logic states. The dynamic power P_{dyn} is given by equation 4.16,

   $$P_{dyn} = \alpha C_L V_{DD}^2 f \tag{4.16}$$

   where α is the activity factor, representing the fraction of time the circuit is switching, C_L is the load capacitance, V_{DD} is the supply voltage, and f is the clock frequency [1–4].
   b) **Static Power:** Static power consumption P_{static} mainly arises from leakage currents that occur when the transistors are intended to be in the off state. This power consumption can be represented by equation 4.17,

   $$P_{static} = I_{leak} V_{DD} \tag{4.17}$$

   where I_{leak} is the leakage current, which increases with scaling and reduced threshold voltages. Minimizing leakage requires careful design techniques such as multi-threshold CMOS (MTCMOS) and power gating [1–4].
   c) **Delay and Performance Metrics:** Propagation postponement in CMOS circuits is a critical overall performance parameter, indicating the time required for a sign to journey through a gate. This cut-off is affected by the charging and discharging of the burden capacitance, which depends on the size of the transistors and the intrinsic capacitance of the tool. The propagation postpone τ_p of a CMOS gate may be approximated using equation 4.18,

$$\tau_p = \frac{C_L V_{DD}}{I_{drive}} \tag{4.18}$$

where C_L is the load capacitance, V_{DD} is the supply voltage, and I_{drive} is the drive current of the MOSFET in saturation. Increasing the drive current by sizing up the transistors (increasing their width W) or reducing the load capacitance is effective in minimising delay. However, larger transistors increase capacitance, creating a trade-off between speed and power consumption [1–4].

d) **Noise Margin and Robustness:** The noise margin indicates how much voltage version a CMOS circuit can tolerate without impacting the output logic tiers. CMOS common sense circuits function key noise margins:

1. Noise Margin Low (NML): This is the difference between the maximum voltage interpreted as a low input and the maximum output voltage whilst the circuit produces a low output.
2. Noise Margin High (NMH): This represents the difference between the minimum voltage considered an elevated input and the minimum output voltage whilst the circuit generates a high output.

The noise margins can be mathematically described using the subsequent equations 4.19 and 4.20, respectively:

$$NML = V_{IL} - V_{OL} \tag{4.19}$$

$$NMH = V_{OH} - V_{IH} \tag{4.20}$$

where V_{IL} and V_{IH} are the input logic levels, and V_{OL} and V_{OH} are the output logic levels for low and high, respectively. The noise margins are essential for ensuring the reliable operation of CMOS circuits in the presence of electrical noise and variations in fabrication [1–4].

e) **Optimization Strategies in CMOS Circuit Design:** Designing optimized CMOS circuits entails balancing overall performance (pace), power consumption, and place (PPA). Key techniques include:

1. Sizing: Adjusting transistor sizes is a crucial approach for optimizing overall performance. Larger transistors supply higher power currents, which can decrease cut-off, however, in addition they boom vicinity and strength consumption.
2. Voltage Scaling: Lowering the supply voltage (V_{DD}) reduces dynamic electricity intake, however this will cause expanded delay and reduced noise margins.

3. Clock Gating: Techniques like clock gating, which turns off the clock in inactive circuit sections, and strength gating, which cuts power to idle areas, are effective in minimizing both dynamic and leakage energy.
4. Multi-Vt Design: Using transistors with exclusive threshold voltages allows for a trade-off between speed (lower V_T) and leakage (better V_T).

As CMOS technology evolves, addressing power consumption and thermal management challenges while exploring emerging technologies like FinFETs, Gate- All-Around FETs, and 2D materials is essential. Through effective optimization strategies and innovations, the future of lower power design remains bright, allowing for continued advances in high-performance and scalable electronic systems [21–24].

4.4 EMERGING TECHNOLOGIES IN CMOS DESIGN

As CMOS technology scales down to meet the demands for higher performance and integration density, challenges related to power consumption, thermal management, and new device architectures arise. In this section, we will explore advanced design techniques, the impact of power and thermal issues, and the emerging technologies that are shaping the future of low power circuits. The focus will be on mathematical models that capture these critical aspects and provide insight into optimizing CMOS design for modern applications [25–27].

1. **FinFET Technology**: One of the most promising emerging technologies is FinFET (Fin Field-Effect Transistor), depicted in Figure 4.6, which addresses the short-channel effects and leakage problems of traditional planar MOSFETs. These devices use a three-dimensional fin structure to control the channel electrostatically, providing better performance at lower power. The current equation for a FinFET in saturation can be modelled using equation 4.21:

$$I_D = \mu C_{OX} \frac{W}{L} \left(V_{GS} - V_{th}\right)^2 \tag{4.21}$$

where μ is the carrier mobility, C_{ox} is the oxide capacitance per unit area, W is the effective channel width (dependent to the height and number of fins), L is the channel length, V_{GS} is the gate-source voltage, and V_{th} is the threshold voltage. FinFETs provide better electrostatic control over the channel, resulting in less short-channel effects and lower leakage currents than planar transistors [1–4, 28].

Figure 4.6 FIN-field effect transistor (FINFET).

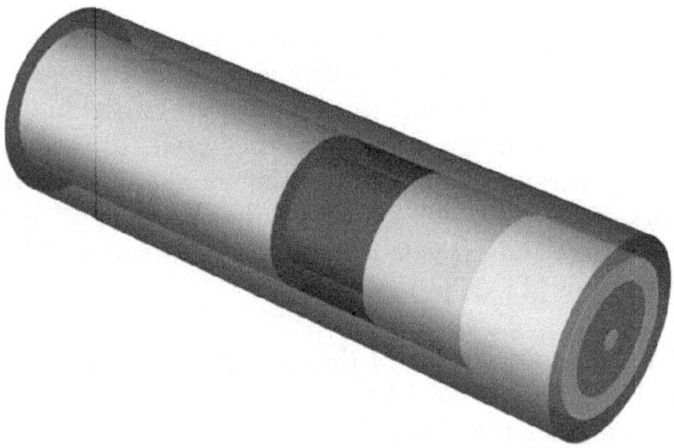

Figure. 4.7 Gate all around field effect transistor.

2. **Gate-All-Around (GAA) FETs**: Another promising technology is Gate-All-Around FETs, illustrated in Figure 4.7, where the gate completely surrounds the channel, providing even better electrostatic control than FinFETs. This further reduces leakage and short-channel effects, making GAA FETs suitable for continued scaling [29, 30].
3. **Graphene and 2D Materials**: Beyond traditional semiconductors, graphene and 2D materials such as transition metal dichalcogenides (TMDs) are being investigated for their potential in CMOS circuits. Graphene's great electron mobility, mixed with the unique bandgap traits of transition metal dichalcogenides (TMDs), positions those substances as promising candidates for next-generation low-power,

high-performance gadgets. The contemporary equation for a transistor utilizing 2D substances may be modelled similarly to that of conventional MOSFETs, even though it requires modifications to account for material-unique parameters which include mobility and bandgap [31–40].

4.5 CONCLUSION

This chapter has provided a comprehensive review of CMOS technology, addressing essential standards inclusive of the physics of MOS transistors, fabrication methods, and 2D-order effects. We examined the advantages of CMOS circuits, specifically their power efficiency and scalability, which have installed them as cornerstones of present-day electronics. Key demanding situations, consisting of power intake and thermal control, have been discussed, in conjunction with answers like dynamic voltage scaling and advanced cooling strategies to address the growing complexity of enormously incorporated CMOS designs.

Looking forward, CMOS technology is poised for continued evolution via innovations in regions consisting of 3D integrated circuits (ICs), FinFET, and post-silicon materials like graphene and different 2D semiconductors. As scaling approaches its physical limits, opportunity architectures, together with Nanosheet FETs and Gate-All-Around FETs, in conjunction with heterogeneous integration, will become increasingly more crucial. Emerging fields consisting of quantum computing, neuromorphic structures, and flexible electronics can even force the expansion of CMOS into new applications, ensuring its ongoing relevance in future technologies.

Future advancements in CMOS technology will focus on addressing power and thermal challenges, particularly as devices become more compact and power densities continue to increase. Techniques like close to-threshold computing and multi-threshold transistors will assist mitigate power dissipation, even as modern cooling strategies, which include liquid and microfluidic cooling systems, might also emerge as critical for dealing with heating in densely integrated chips. As more state-of-the-art structures-on-chip (SoC) are evolved for programs in AI, IoT, and independent structures, strength performance and thermal control will remain critical areas for innovation.

Furthermore, the integration of recent materials, together with high-k dielectrics, low-k interconnects, and 2D semiconductors like graphene, will propel the CMOS era into its subsequent developmental phase. These substances, along with evolving transistor architectures and GAAFETs and Nanosheet FETs, will facilitate ongoing scaling and overall performance improvements. As CMOS advances, it will play an important function in the improvement of future technologies, from wearable electronics and clever fabrics to self-sufficient motors and quantum computing.

REFERENCES

[1] Tsividis, Y. (2002). *Mixed Analog-digital VLSI devices and technology*. World Scientific.

[2] Khoury, John, and Mihai Banu. "Yannis Tsividis' early contributions to MOS filters." *IEEE Solid-State Circuits Magazine* 6, no. 4 (2014): 36–40.

[3] Taur, Yuan, and Tak H. Ning. *Fundamentals of modern VLSI devices*. Cambridge University Press, 2021.

[4] Taur, Yuan, Y-J. Mii, David J. Frank, H-S. Wong, Douglas A. Buchanan, Shalom J. Wind, Stephen A. Rishton, G. A. Sai-Halasz, and Edward J. Nowak. "CMOS scaling into the 21st century: 0.1 μm and beyond." *IBM Journal of Research and Development* 39, no. 1.2 (1995): 245–260.

[5] Ranuárez, Juan C., M. Jamal Deen, and Chih-Hung Chen. "A review of gate tunneling current in MOS devices." *Microelectronics Reliability* 46, no. 12 (2006): 1939–1956.

[6] Sze, Simon M., Yiming Li, and Kwok K. Ng. *Physics of semiconductor devices*. Wiley, 2021.

[7] Wong, H. "On the CMOS device downsizing, more Moore, more than Moore, and more-than-Moore for more Moore." In *2021 IEEE 32nd International Conference on Microelectronics (MIEL)*, pp. 9–15. IEEE, 2021.

[8] Hill, Ian, Parvez Chanawala, Rohit Singh, S. Arash Sheikholeslam, and André Ivanov. "CMOS reliability from past to future: A survey of requirements, trends, and prediction methods." *IEEE Transactions on Device and Materials Reliability* 22, no. 1 (2021): 1–18.

[9] Rawat, Udit, Jackson D. Anderson, and Dana Weinstein. "Design and applications of integrated transducers in commercial CMOS technology." *Frontiers in Mechanical Engineering* 8 (2022): 902421.

[10] Sharma, Vijay Kumar. "A survey of leakage reduction techniques in CMOS digital circuits for nanoscale regime." *Australian Journal of Electrical and Electronics Engineering* 18, no. 4 (2021): 217–236.

[11] Teng, Qiao, Yongkang Hu, Ran Cheng, Yongyu Wu, Guodong Zhou, and Dawei Gao. "Reliability challenges in CMOS technology: A manufacturing process perspective." *Microelectronic Engineering* (2023): 112086.

[12] Zhao, Yuda, Marco Gobbi, Luis E. Hueso, and Paolo Samorì. "Molecular approach to engineer two-dimensional devices for CMOS and beyond-CMOS applications." *Chemical Reviews* 122, no. 1 (2021): 50–131.

[13] Ren, Liwang, Zhiwei Li, Yawei Lv, X. Li, D. Zhang, and W. Li. "Efficient modulation of MoS." *Applied Physics Letters* 120, no. 5 (2022).

[14] Netchitailo, Vladimir S. "Cosmology and classical physics." *Journal of High Energy Physics, Gravitation and Cosmology* 8, no. 4 (2022): 1037–1072.

[15] Kuo, M, Yi-Shan, Shen-Yang Lee, Chia-Chin Lee, Shou-Wei Li, and Tien-Sheng Chao. "CMOS-compatible fabrication of low-power ferroelectric tunnel junction for neural network applications." *IEEE Transactions on Electron Devices* 68, no. 2 (2021): 879–884.

[16] Wong, Hei. "Abridging CMOS technology." *Nanomaterials* 12, no. 23 (2022): 4245.

[17] Patel, Rashmit, Yash Agrawal, and Rutu Parekh. "Single-electron transistor: Review in perspective of theory, modelling, design and fabrication." *Microsystem Technologies* 27, no. 5 (2021): 1863–1875.

[18] Prasanna, J. Lakshmi, M. Ravi Kumar, Ch Priyanka, and Chella Santhosh. "Evaluation of device fabrication from FET to CFET: A review." *Journal of Nano-and Electronic Physics* 13, no. 6 (2021).

[19] Liu, T. C., & Hu, C. (2011). *MOS transistor physics and simulation*. Wiley.

[20] Agarwal, Arun, Gourav Misra, and Kabita Agarwal. "Semiconductor III–V nanowires: Synthesis, fabrication and characterization of nanodevices." *Journal of The Institution of Engineers (India): Series B* (2022): 1–11.

[21] Sen, Soumya, Mamta Khosla, and Ashish Raman. "Design and investigation of electrostatic doped heterostructure vertical Si (1-x) Gex/Si nanotube TFET." *Microelectronics Journal* (2024): 106417.

[22] Sen, Soumya, Mamta Khosla, and Ashish Raman. "Design and Challenges in TFET." In: Dharmendra Singh Yadav, Shiromani Balmukund Rahi, Sukeshni Tirkey. (eds) *Advanced Field-Effect Transistors*, pp. 23–45. CRC Press, 2023.

[23] Sen, Soumya, Ashish Raman, and Mamta Khosla. "A literature survey on tunnel field effect transistors." *AIJR Proceedings* (2021): 506–512.

[24] Singh, Prabhat, and Dharmendra Singh Yadav. "Impact of temperature on analog/RF, linearity and reliability performance metrics of tunnel FET with ultra-thin source region." *Applied Physics A* 127, no. 9 (2021): 671.

[25] Singh, P., and D. S.Yadav. "Design and Investigation of F-shaped tunnel FET with enhanced analog/RF parameters." *Silicon* 14 (2022): 6245–6260.

[26] Singh, P., and D. S. Yadav. "Analysis of Channel Doping Variation on Transfer Characteristics to High Frequency Performance of F-TFET." 91–104.

[27] Sen, S., and M. Singh. . "Analysing the Behaviour of 14 nm, 10 nm, 7 nm FinFET and Predicting the Superiority among the Lot." In: Dhar, S., Mukhopadhyay, S.C., Sur, S.N., Liu, CM. (eds) *Advances in Communication, Devices and Networking. Lecture Notes in Electrical Engineering*, vol 776. Springer, Singapore, 2022.

[28] Dey, P., S. Sen, R. Ranjan, and A. Raman. . "A Survey on Heterostructure Tunnel Field Effect Transistors (H-TFET)." In: Rawat, S., Kumar, A., Raman, A., Kumar, S., Pathak, P. (eds) *Proceedings of Third International Conference on Computational Electronics for Wireless Communications. ICCWC 2023. Lecture Notes in Networks and Systems*, vol 962. Springer, Singapore, 2025.

[29] Sen, S., M. Khosla, and Raman, A. "Performance characterization and analytical modelling of electrostatic doped heterostructure nanotube tunnel field effect transistor." *Semiconductors* 58 (2024): 547–557.

[30] Singh, S. and B. Raj. "Design and analysis of a heterojunction Vertical t-Shaped tunnel field effect transistor." *Journal of Electronic Materials* 48 (2019): 6253–6260.

[31] Singh, Avtar, and Chandan Kumar Pandey. "Improved DC performances of gate-all-around si-nanotube tunnel FETs using gate-source overlap." *Silicon* 14, no. 4 (2022): 1463–1470.

[32] Bhol, Krutideepa, Biswajit Jena, and Umakanta Nanda. "Journey of Mosfet from planar to gate all around: A review." *Recent Patents on Nanotechnology* 16, no. 4 (2022): 326–332.
[33] Sen, Soumya, Mamta Khosla, and Ashish Raman. "Deciphering Phenomenal Performance with Gate-All-Around Strained Heterostructure Nanotube Tunneling FET (TFET) for better Drive Currents and Steep Subthreshold Slope." In *2024 International Conference on Integrated Circuits, Communication, and Computing Systems (ICIC3S)*, vol. 1, pp. 1–4. IEEE, 2024.
[34] Zurzolo, Chiara. "Tunneling nanotubes: Reshaping connectivity." *Current Opinion in Cell Biology* 71 (2021): 139–147.
[35] Xu, Rui, Lei Du, David Adekoya, Gaixia Zhang, Shanqing Zhang, Shuhui Sun, and Yong Lei. "Well-defined nanostructures for electrochemical energy conversion and storage." *Advanced Energy Materials* 11, no. 15 (2021): 2001537.
[36] Hossain, Nayem, Md Hosne Mobarak, Mariam Akter Mimona, Md Aminul Islam, Amran Hossain, Fatema Tuz Zohura, and Mohammad Asaduzzaman Chowdhury. "Advances and significances of nanoparticles in semiconductor applications–A review." *Results in Engineering* 19 (2023): 101347.
[37] Qin, Laixiang, Chunlai Li, Yiqun Wei, Ziang Xie, and Jin He. "Gate electrostatic controllability enhancement in nanotube gate all around field effect transistor." *AIP Advances* 13, no. 6 (2023).
[38] Singh, Shailendra and Balwinder Raj. "Modeling and simulation analysis of SiGe heterojunction Double Gate Vertical t-shaped tunnel FET." *Superlattices and Microstructures* 142 (2020): 106496, ISSN 0749-6036.
[39] Singh, S., and B. Raj. "Analytical and compact modeling analysis of a SiGe Hetero-Material Vertical L-Shaped TFET." *Silicon* 14 (2022): 2135–2145.
[40] Singh, P., D. P. Samajdar and D. S. Yadav. "Doping and Dopingless Tunnel Field Effect Transistor." *2021 6th International Conference for Convergence in Technology (I2CT)*, Maharashtra, India, 2021, pp. 1–7.

Chapter 5

Wide Band Gap Semiconductors and their Applications

S. Ponmudi, K. Baskaran, and C. Uma Rani

5.1 INTRODUCTION

Today, the constant rise in global energy utilization is one of the most pressing issues in our society, since the worldwide use of energy had risen exponentially to 13,147.3 million tonnes of oil equivalent (Mtoe) in 2015 from 8,588.9 Mtoe in 1995 [1]. It is estimated that the future worldwide primary energy requirement for the year 2040 will be 17,487 Mtoe, with energy utilization projected to increase by approximately 33% from 2015 to 2040 [2]. Power electronics plays a pivotal role in linking sources of energy production with consumers by facilitating the control and organization of electric power. Prior to the discovery of semiconductor devices like the transistor, most electronic devices were based on vacuum tubes, such as vacuum diodes, which have two electrodes (cathode and anode), and triodes, which have three electrodes (cathode, anode, and grid), etc. Electrons are emitted from a heated cathode in a vacuum tube, and their flow is controlled by adjusting the voltage between the electrodes. Vacuum tubes are commonly referred to as valves because electrons can move in only one direction, from the cathode to the anode. The main disadvantages of these vacuum tube devices are that they are huge, use high power, typically operate at higher voltages, and have a restricted life span and less consistency. Interestingly, the development of contemporary solid-state semiconductors began in the 1930s when it was discovered that certain semiconductors and their junctions could regulate both the amount and direction of charge carrier flow. Semiconductor devices require no external heating or large evacuated spaces. They are compact, consume very small power, work at low voltages, and offer extended life span and elevated reliability. Initially, elemental semiconductors such as germanium (Ge) and silicon (Si) were used in semiconductor devices. Later, compound semiconductors like CdS, GaAs, and InP were developed and introduced. These types of semiconductors generally have smaller band gap values. The gap between the top of the valence band (E_V) and the bottom of the conduction band (E_C) is called the energy band gap (E_g). Depending on the material, this gap may be large, small, or even zero.

DOI: 10.1201/9781003533535-5

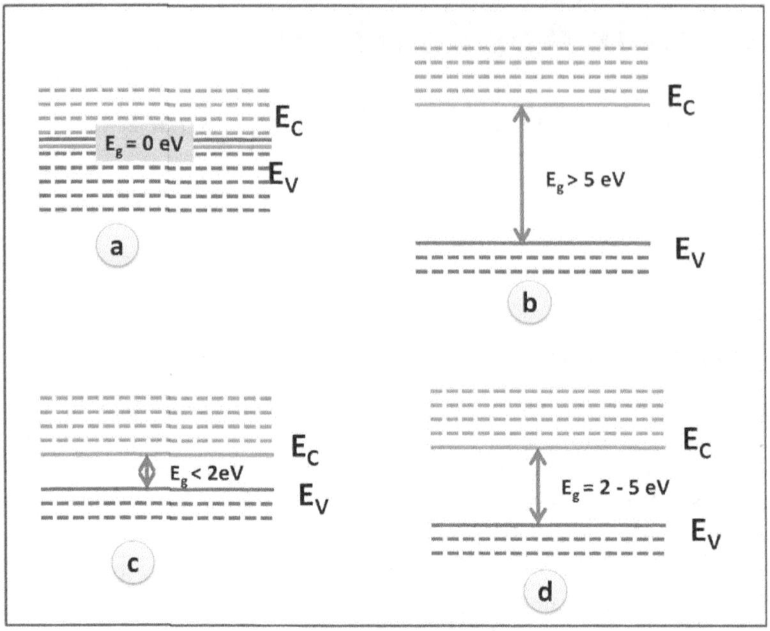

Figure 5.1 Energy band gap (Eg) for (a) conductor, (b) insulators, (c) narrow band gap semiconductor, and (d) wide band gap semiconductor.

Generally, metals (conductors) have zero energy band gap, insulators have band gap values greater than 5 eV, and semiconductors typically have band gap values between those of conductors and insulators. Normal semiconductors like Si and Ge are known to have a band gap of less than 2 eV and are therefore referred to as narrow band gap semiconductors (NBGs), while semiconductors with band gap values greater than 2 eV are called wide band gap semiconductors (WBGs). These various scenarios are illustrated in Figure 5.1.

WBGs are both scientifically and technologically significant, gaining prominence as essential materials for high-performance optoelectronic and electronic devices due to their large band gap energy (*Eg*). Large *Eg* allows them to absorb or emit ultraviolet (UV) light, making them particularly suitable for optoelectronic applications. WBGs generally possess high chemical and mechanical stability (Chemical stability refers to the resistance of a material to chemical reactions, while mechanical stability indicates resistance to physical wear and tear), making them suitable for operation in harsh environments, including in extreme temperatures, high radiation, and corrosive conditions. The combination of higher values of thermal conductivity, breakdown voltage, and electron mobility allow WBG devices

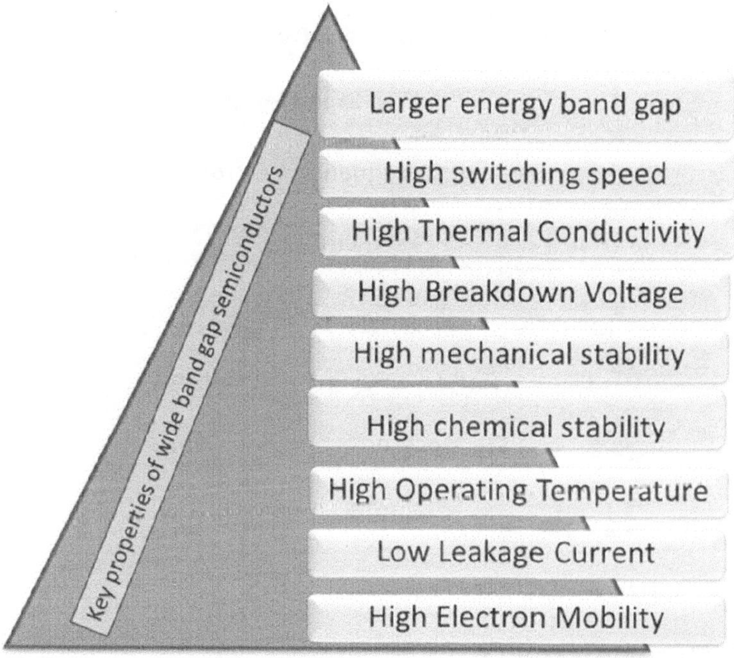

Figure 5.2 Key properties of wide band gap semiconductors.

to achieve high power density (generally, power density is the amount of power a device can handle per unit volume or area), which contributes to more compact and efficient electronic systems.

These inherent properties as shown in Figure 5.2 collectively enable WBG materials to execute efficiently in harsh environment applications, driving advancements across various technology sectors. Several noteworthy WBGs, including gallium nitride (GaN), silicon carbide (SiC), indium oxide (In_2O_3), zinc oxide (ZnO), aluminum nitride (AlN), gallium oxide (Ga_2O_3), indium-gallium-zinc-oxide (IGZO), and diamond, exhibit various fascinating properties and are widely used in a broad range of technologically important devices. Their band gap values range between 3.3 eV and 6 eV.

5.2 APPLICATIONS OF WIDE BAND GAP SEMICONDUCTORS

There is an escalating importance in WBGs due to their fascinating technological applications, such as in optoelectronics, power electronics, high-frequency electronics, spintronics, gas sensors, battery management systems, electrochromic smart windows, and quantum computing (Figure 5.3).

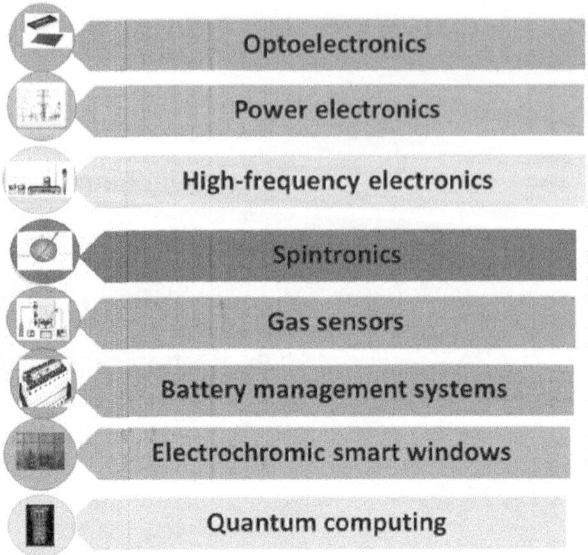

Figure 5.3 Applications of wide band gap semiconductors.

5.2.1 Optoelectronics

In the beginning, optoelectronics was defined as the interaction of light with matter in solid, liquid, and gaseous forms, as well as the devices that depend on these interactions. Later, this definition was broadened by many authors, who focused primarily on the ways in which light interacts with semiconductors [3]. In general, optoelectronic devices (OED) are electronic devices which convert electric charge (C) into light, as in lasers and light emitting diodes (LEDs), or convert light into electric current (I), as in photovoltaic (PV) devices and photodetectors (PDs). It is important to mention here that photodetectors (PDs), photovoltaic cells (PVCs), liquid crystal displays (LCDs), touch screens, light-emitting diodes (LEDs), and flexible organic light-emitting diodes (OLEDs) all rely on electrodes with over 80% transmittance ($T > 80\%$) and low sheet resistance (R_S) for optimal performance [4]. WBGs are transformative in optoelectronics, providing high efficiency, durability, and performance across a range of devices. Their ability to emit and detect light across different wavelengths, combined with their thermal and mechanical stability, makes them essential for advancing lighting, communication, and display technologies.

It is well known that the energy of photons emitted by a semiconductor is directly related to its band gap and can be expressed by the equation:

$$E = \frac{hc}{\lambda} \tag{5.1}$$

where, h is Planck's constant, c is velocity of light, and λ is wavelength of the emitted photon.

A larger band gap leads to the emission of shorter-wavelength photons, crucial for generating light in the blue, violet, and ultraviolet (UV) regions of the electromagnetic spectrum. This is why materials with wide band gaps/ultra-wide band gaps are crucial for producing laser diodes as well as light-emitting diodes that work in the visible to UV range. Further, WBGs are at the forefront of materials that meet these requirements, including group IV, III–V, and II–VI elements such as GaN (3.4 eV), SiC (3.2 eV), and ZnO (3.4 eV). With even larger band gaps exceeding 4 eV, ultra-wide band gap semiconductors (UWBGs) include III-nitrides (aluminum nitride (AlN), boron nitride (BN)), ternary III-nitrides (aluminum gallium nitride (AlGaN), indium aluminum nitride (InAlN)), and diamond, as well as gallium (III) oxide (Ga_2O_3). These materials vary in their technological maturity, with Ga_2O_3 swiftly advancing and poised to facilitate new ultraviolet (UV) and deep-ultraviolet (d-UV) optoelectronic devices [5]. The blue LED is a familiar and most important wide band gap semiconductors. However, despite the intense global interest in using wide band gap semiconductors for blue LEDs, unfortunately it took time before technological breakthroughs enabled the manufacture of GaN p-n junctions. After 1990, numerous research groups became actively involved in wide bandgap semiconductors (WBGs), leading to the first successful operation of blue LEDs [6]. SiC is another important indirect wide band gap semiconductor, and before modern developments in nitride-based semiconductors, SiC p-n junctions were also utilized to fabricate blue LEDs. Although zinc selenide (ZnSe) has an E_g of about 2.7 eV and is extensively recognized as an outstanding WBGs for manufacturing blue-green LEDs and LDs, it was identified that the life span of these optical devices (ODs) was too short for commercialization, leading to a decline in research in this area as nitride semiconductors showed greater promise.

Metal halide perovskite solar cells (MHPSCs) are recognized as a leading technology in the next generation of solar cells. Of late, organic-inorganic perovskite (OIP) materials have gaining attention owing to their outstanding opto-electronic properties, such as high carrier mobility, extended diffusion length, and strong light-absorption. The efficiency of power conversion of perovskite solar cells (PSCs) has considerably improved to 22.1% from 3.8%, in just a few years. This rapid improvement is a strong indicator of the potential for PSCs in the photovoltaic industry. The notable rise in efficiency suggests that perovskite solar cells hold great promise for developing high-efficiency and low-cost solar energy solutions, which could play a vital role in making solar energy more accessible and commercially viable. It is important to mention that nickel oxide is used as hole transport layers (HTLs) in perovskite solar cells owing to its huge band gap, comparatively deep valence band (5.4 eV), encouraging alignment of energy level with photoactive layers, low cost, and excellent environmental stability [7].

As the research and development continue, WBGs are expected to drive further innovations and improvements in optoelectronic applications.

5.2.2 Power Electronics

Wide band gap semiconductors (WBGs) are becoming more and more fashionable in power electronics since they offer superior performance compared to conventional silicon-based devices, e.g., enhancing switching frequencies and thermal conductivity and reducing power losses. WBG materials such as GaN and SiC can handle higher voltages and currents than silicon. This makes them suitable for high-power applications including renewable energy systems (RES), electric vehicles (EVs), and industrial machinery. Silicon carbide has an electric breakdown field strength nearly 10 times higher than silicon, allowing silicon carbide power electronic devices (PEDs) to be made one tenth as thin [8]. This higher breakdown strength also means that SiC devices can handle larger current densities, resulting in smaller overall sizes. In particular, for power electronic devices, SiC rectifiers outperform their Si counterparts in many aspects, including higher voltage handling, better thermal performance, and greater efficiency at higher current densities. Schottky Barrier Diodes (SBD), Positive-Intrinsic-Negative (PIN) diodes, and Junction Barrier Schottky (JBS) diodes are the three main types of SiC power rectifiers, each suited to different power applications depending on the trade-offs between switching speed, voltage rating, and efficiency. Due to SiC's larger critical dielectric field, SiC devices can achieve elevated blocking voltages with the similar drift layer thickness compared to their silicon counterparts. Additionally, SiC's high thermal conductivity provides a significant advantage, allowing for operation at high density of current and reducing the size of cooling systems. The ability of SiC devices to operate at high temperatures makes them ideal for various applications in aerospace and space missions [9].

High-electron mobility transistors (HEMTs) made by GaN/AlGaN are being explored in potential modern high-power devices (HPDs) applications. GaN/AlGaN HEMTs enable "ultrahigh-power-density (UHPD) operation" with lesser power loss, owing to the high carrier mobility (μ) in "two-dimensional (2D) electron gas channels" and the huge breakdown voltage resulting from their great critical electric fields [10, 11]. These special characteristics make them ideal for power electronics applications, as they help reduce on-resistance and minimize power losses during high-frequency switches. In general, most GaN/AlGaN HEMTs demonstrated thus far exhibit depletion mode (normally-on) characteristics, with gate threshold voltages below –4 V. However, enhancement mode (normally-off) characteristics are preferred for power electronics applications to avoid device failure during a short circuit. Some depletion mode GaN/AlGaN HEMTs have been reported with AlGaN layers just a few nanometers thick to reduce "two-dimensional electron gas (2DEG) density". However, the on-resistances of these devices unintentionally increased as the 2DEG density decreased. Interestingly, Wataru Saito et al. [10] successfully increased the threshold voltage for normally-off recessed-gate structure AlGaN/GaN HEMTs to –0.14 V, while without the recessed-gate structure, the threshold voltage was approximately –4 V. The recessed-gate structure effectively shifts the gate threshold voltage without increasing the on-resistance, due to selective decrease of two-dimensional electron gas density under the gate electrode. This is an efficient method for addressing this issue.

In commonly used metal-oxide-semiconductor field-effect transistors (MOSFETs) for power electronics, "the specific on-resistance (R_{ON}) is inversely proportional to the cube of the electric breakdown field strength" [8]. As a result, the R_{ON} of SiC MOSFETs can be lower than those of silicon devices. Additionally, since SiC's thermal conductivity is nearly three times greater than that of silicon, systems with SiC power electronics are able to utilize simpler cooling solutions. Thus, SiC holds great promise for the future, offering a compact size, lower losses, high efficiency, and fast heat dissipation [6]. GaN, like SiC, has outstanding physical properties such as a high saturation velocity and a high critical electric field. These characteristics make GaN a strong candidate for manufacturing power devices with a low specific on-resistance (R_{ON}), high breakdown voltage (B_V), and high switching frequency. GaN theoretically offers even better performance than SiC owing to its higher electron mobility and critical electric field. This suggests that GaN devices could outperform SiC in terms of efficiency and speed. Despite these promising properties, GaN technology lags behind SiC in certain areas. It exhibits processing difficulties and, crucially, the absence of high-class freestanding GaN substrates, which limits the development of vertical GaN devices essential for high-power applications [12].

5.2.3 High-Frequency Electronics

Interestingly, WBG materials have higher electric field breakdown strength, enabling devices to handle higher voltages without breakdown, which improves performance in high-frequency power electronics. Additionally, WBGs offer higher electron mobility, leading to faster switching speeds and greater efficiency in radio frequency (RF) and microwave frequencies. Due to these material characteristics, WBG semiconductors are extensively used in 5G communications, power amplifiers, and radars. In particular, HEMTs made from WBG materials such as GaN offer high power density, efficiency, and frequency capability, making them widely used in RF and microwave amplifiers.

The fifth generation of mobile networks (5G technology) offers significant advancements over its predecessor, 4G. It is designed to meet the growing demands for faster, more efficient, and more reliable wireless communication, enabling a wide range of new applications and services. 5G networks can deliver data speeds of 10 Gbps, which is 100 times quicker than 4G. This allows for near-instantaneous downloads and uploads, making activities like streaming high-definition (HD) videos and downloading big data much quicker. WBG semiconductors are crucial in the development of 5G technology, where high frequency and high power are essential.

Wide band gap semiconductors, such as GaN and SiC, are increasingly critical in the development of advanced radar systems. Their unique material properties offer significant advantages in radar applications, particularly in terms of power efficiency, frequency operation, and thermal management. It is noteworthy to mention here that WBG semiconductors are more durable and reliable in harsh environments, for example, high altitudes, intensive radiation areas, and extreme temperatures, making them ideal for military and space-based radar systems. WBG semiconductors improve the performance of phased-array radars (PAR) and Synthetic Aperture Radar (SAR) systems. PAR represents a significant advancement in radar technology, offering unparalleled speed, accuracy, and flexibility. Their application across military, aviation, telecommunications, and meteorology highlights their importance in modern technology. As these systems continue to evolve, they will play a crucial role in enhancing situational awareness, improving safety, and enabling new technological innovations. SAR is a powerful and versatile technology with wide-ranging applications across Earth observation, military intelligence, environmental monitoring, and beyond. Its ability to provide high-resolution images under any conditions makes it indispensable in scenarios where traditional imaging methods fall short. As WBGs technology advances, SAR systems are expected to become even more sophisticated, offering greater capabilities and expanding their use in both existing and new fields.

5.2.4 Spintronics

Wide band gap semiconductors are increasingly being explored for use in spintronics, a cutting-edge field of electronics that exploits the intrinsic spin of electrons, along with their charge, to store, process, and transfer information. Spintronics offers the potential for energy-efficient devices with non-volatile memory capabilities. WBGs can be doped with magnetic ions to induce magnetic properties within the material. This enables the creation of dilute magnetic semiconductors (DMS), which combine both semiconductor and magnetic properties, crucial for the development of spintronic devices such as spin transistors and magnetic tunnel junctions. The ferromagnetization in dilute magnetic semiconductors (DMS) arises due to the alignment of the magnetic moments of doped ions, which are typically magnetic elements, embedded within a non-magnetic wide band gap semiconductor host material. This alignment is maintained by various spin coupling mechanisms that induce the magnetic ordering within the material.

The phenomenon of room temperature ferromagnetism (RTFM) observed in undoped WBG metal oxide semiconductors such as Cr_2O_3, Cu_2O, SnO_2, Al_2O_3, and ZnO has been extensively attributed to intrinsic defects within these materials. These intrinsic defects, such as metal and/or oxygen vacancies, play a crucial role in inducing and stabilizing ferromagnetic ordering, even in the absence of traditional magnetic dopants (like transition metals). On the other hand, RTFM can also be achieved through extrinsic defects introduced by doping WBGs with transition metals (TM). These doped systems exhibit a significant modification of structural, electrical, optical, and magnetic properties, primarily due to the incorporation of TM ions, which introduce localized magnetic moments into the host material. When WBG semiconductors (such as Cr_2O_3, ZnO, SnO_2, CeO_2, etc.) are doped with TM like Ti, Ni, Cr, Ag, Cu, or Mn, the TM ions bring unpaired d-electrons that generate localized magnetic moments. These magnetic ions introduce spin states into the host material's electronic structure, making ferromagnetic ordering possible via exchange interactions between the doped ions. Doping is crucial for tuning the magnetic properties of the material. For instance, in Cr-doped ZnO or Mn-doped SnO_2 thin films, the dopant ions' magnetic spins interact with the semiconductor's electronic carriers (electrons or holes) via exchange interactions such as the double exchange or superexchange mechanism. The $Al_2Cr_2Cu_2O_8$ thin film, formed by a solid solution of chromia (Cr_2O_3), alumina (Al_2O_3), and cupric oxide (CuO), displaying RTFM after being prepared via RF magnetron sputtering, is a notable result. This combination of materials in the form of a ternary metal oxide solid solution is highly significant due to several key factors related to both the material composition and the synthesis technique [13]. Both Al_2O_3 and Cr_2O_3 are wide band gap semiconductors with insulating behavior under normal conditions. However, the addition of CuO, which is a narrower band gap p-type semiconductor, can modify the overall electronic structure of the ternary oxide system. The Cu^{2+} ions in CuO, Al^{3+} ions in Al_2O_3, along with Cr^{3+} ions from Cr_2O_3, can introduce magnetic moments into the system owing to the existence of unpaired d-electrons. The presence of Cr^{3+}, Al^{3+}, and Cu^{2+} ions can generate localized spins, and these spins can couple ferromagnetically, leading to the observed RTFM in the $Al_2Cr_2Cu_2O_8$ thin film.

Spin Field-Effect Transistors (SpinFETs) represent a promising frontier in semiconductor technology, offering the potential for efficient, non-volatile, and faster electronic devices. Unlike traditional field-effect transistors (FETs), which control electron flow by manipulating their charge, SpinFETs use the spin state of electrons to control the current flow. The spin can be either "up" or "down," and the transistor operates based on the manipulation and detection of these spins' states. While there are significant challenges to overcome in material development, device integration, and spin coherence, ongoing research and technological advancements hold the promise of making SpinFETs a key component in future electronic and quantum systems. With the continuing development of new SpinFETs using WBGs materials, they could play a transformative role in computing, memory and beyond.

5.2.5 Gas Sensors

Gas sensors play an essential role across various fields, including environmental monitoring, public security, automobile safety systems, domestic safety, spacecraft, air conditioning in aircrafts, and industrial settings. In recent times, WBG metal oxide-based semiconducting sensors have garnered significant interest from researchers due to their numerous advantages over other gas sensing technologies. These sensors stand out for their easy fabrication procedure, economical, extended operational life, and reliable performance in real-time applications [14]. Their high sensitivity, stability, and ability to operate effectively at elevated temperatures make them particularly suitable for a broad range of industrial, environmental, and safety applications, further enhancing their appeal for gas detection solutions. The fundamental mechanism of WBG metal oxide sensors operates via a variation in electrical resistance, which occurs on the WBG material surface due to the chemical interaction between the material and the gas molecules [14]. When gas molecules are absorbed onto the surface of the metal oxide, they either donate or accept electrons, altering the material's electrical properties. These sensors generally perform two key functions: the receptor function, which identifies or recognizes the specific gas molecules; and the transducer function, which converts the resulting chemical interaction into an electrical output signal that can be measured. This dual-function system enables precise and efficient gas detection. The performance of sensors primarily depends on their operating temperature and chemical composition. As the demand for detecting toxic and polluting gases at room temperature grows, it becomes essential to enhance the sensors' sensitivity, selectivity, and stability for effective room-temperature operation. To achieve this, various approaches have been explored, such as incorporating suitable dopants into WBGs, which can significantly improve these key properties. By optimizing the materials and doping techniques, researchers aim to enhance the sensor's capability to accurately identify specific gases under challenging conditions, while maintaining long-term stability and reliability. Recently, mixed WBGs have gained popularity among researchers due to their potential for "tailoring" sensitivity toward a particular gas. By adjusting the relative composition of the ingredient compounds, it is possible to fine-tune the gas sensing properties, while the availability of a double receptor-transducer function in the same material system further enhances performance. As a

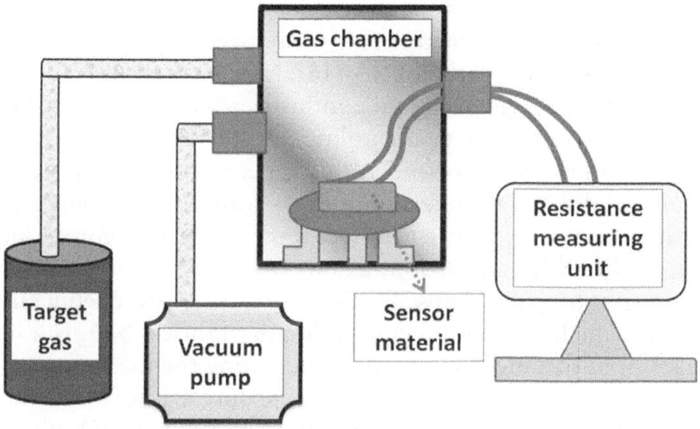

Figure 5.4 Schematic diagram of the gas sensing setup.

result, many binary metal oxide semiconducting gas sensors have been successfully developed. These advancements have paved the way for improved gas detection capabilities. A gas sensing setup, illustrating the typical components and processes involved in gas detection, is shown in Figure 5.4.

TiO_2–CeO_2 powders and thin films with a nanocrystalline structure exhibit a distribution of small particle size ranging from 19.2 to 22.4 nm. The structure of the TiO_2–CeO_2 film shows promising surface properties for sensing performance. TiO_2–CeO_2 demonstrates a stable, rapid, and consistent response to CO gas at 200°C, confirming that the WBG TiO_2–CeO_2 thin film gas sensor, with low power consumption and high thermal stability has strong potential for industrial applications [15].

Gao et al. [16] developed WBG tungsten trioxide–chromium trioxide (WO_3–Cr_2O_3) thin films using the sol–gel process for the detection of acetone. Several diseases are mainly related to characteristic gases; for example, acetone is a byproduct of ketone body metabolism. Medical studies show that acetone levels in the breath of healthy persons are below 0.8 ppm, while in diabetic patients, they exceed 1.8 ppm [17]. Thus, quantitative acetone analysis in the breath of human is highly beneficial for diabetes management. It was reported that WO_3 films doped with 5 mol% Cr_2O_3 and heated at 500°C for one hour demonstrated the highest sensitivity and selectivity for acetone gas, with a sensitivity of 8.91 to 20 ppm acetone. These sensors exhibited a high sensitivity, fast response and superior selectivity at low acetone concentrations, making them promising candidates for non-invasive diabetes diagnosis based on breath analysis (electronic noses).

Tin (IV) oxide (SnO_2), indium (III) oxide (In_2O_3) and indium tin oxide (In_2O_3-SnO_2) are famous WBG metal oxide semiconductors, with their optical band gaps generally falling within the range of 3 to 4 eV. Tin oxide (SnO_2) thin films with nanocrystalline structure were produced with varying concentrations of a precursor through spray pyrolysis for nitrogen dioxide (NO_2) gas monitoring by Kamble et al. [18]. NO_2 is commonly formed during combustion in thermal power plants, industries/factories and vehicles. This gas contributes significantly to air pollution and forms ozone (O_3), a main component of smog. While ozone in the upper atmosphere is beneficial for blocking ultraviolet radiation, ground-level ozone is hazardous, especially in urban environments. Exposure to lower-level ozone formation due to NO_2 can lead to reduced lung function, aggravating conditions like asthma and increasing respiratory problems. Furthermore, NO_2 can chemically convert to nitric acid, which corrodes metals and fades fabrics. NO_2 is a hazardous gas with a 3 ppm threshold limit value, indicating the maximum concentration that can be tolerated for repetitive exposure without causing adverse health effects. It was clearly noticed that the synthesized SnO_2 film exhibited a response to NO_2 gas concentrations up to 5 ppm. A sensor response of approximately 4.31 was recorded for 5 ppm of NO_2, with the response increasing as the gas concentration rose.

Francioso et al. [19] reported on the sensing performance of In_2O_3-SnO_2 films for detecting NO_2, comparing them to those of single oxide sensors In_2O_3 and SnO_2. These films exhibited high responses to NO_2 with variations in sensing properties observed based on surface morphology and operating temperature. Overall, the In_2O_3-SnO_2 -based sensor demonstrated superior performance in terms of response and sensitivity.

Indium-doped zinc oxide (IZO) thin films were fabricated onto glass substrates by Badadhe and Mulla [20] using the spray pyrolysis technique. They were utilized for the detection and monitoring of hydrogen sulfide (H_2S). H_2S is a toxic, malodorous gas commonly produced during the manufacturing of natural gas, coal, coal oil, and gasoline, and in sewage systems. Human exposure to H_2S levels above 250 ppm can indeed cause significant neurobehavioral toxicity and potentially be fatal [21]. Thus, detecting the traces of this hazardous gas are crucial. The IZO film showed a response to H_2S starting at 100 ppm, with the response reaching up to 13,000 for 1000 ppm H_2S at 250°C.

Thin film of Al_2O_3:Cr_2O_3 with a rhombohedral phase was prepared using RF magnetron sputtering. It is a WBG material (E_g = 3.5 – 4.9 eV) that exhibits higher sensing performance towards ammonia gas at room temperature [22]. Ammonia, a toxic gas, is widely used in different industries (automotive, fertilizer manufacturing, agriculture, and food manufacturing). Due to its toxic nature, accidental inhalation of ammonia in factories can cause severe health effects, with a threshold limit of approximately 25 ppm for humans. Although ammonia has a strong pungent odor, it becomes detectable by smell only when its concentration exceeds 50 ppm. To prevent life-threatening risks to workers in industries using ammonia, early detection of leaks is crucial. Al_2O_3:Cr_2O_3 (1:1) film annealed at 1000°C demonstrates outstanding sensitivity to NH_3 at room temperature, detecting concentrations as low as 1 ppm. The response value increased with rising ammonia concentrations, reaching 99.7% at 100 ppm.

In the meantime, ternary metal oxide WBG semiconductors have also been considered for further developments of sensing behaviors, utilizing the benefits of the three different phases of ternary systems. In particular, Al_2O_3:Cr_2O_3:CuO thin films fabricated by RF magnetron sputtering with band gap value of 3.62 eV show a higher gas sensing response to NH_3 at room temperature [23].

5.2.6 Battery Management Systems

With the ongoing rise in global temperatures, the need for alternative, eco-friendly transportation solutions has become increasingly urgent. Electric and hybrid-electric vehicles have emerged as efficient and environmentally friendly options over the past decade. These vehicles primarily rely on lithium-based batteries as their power source. However, battery packs composed of these cells require constant monitoring due to their hazardous and potentially explosive nature, especially when exposed to adverse conditions. Proper maintenance of these batteries is therefore crucial. Selecting the appropriate battery chemistry is critical for optimizing overall system performance.

At present, lithium-ion (Li-ion) chemistry is the preferred option due to its high energy density, power output, and efficient charge/discharge cycles, especially in pulsed energy systems. However, Li-ion batteries can fail if overdischarged/overcharged or operated outside their safest temperature limit. This makes battery management systems (BMS) essential, not only for enhancing the performance, life span, and safety of batteries in electric vehicles (EVs) but also for their role in renewable energy storage systems.

The function of a BMS can vary depending on perspective. For some, it is mainly about controlling key battery parameters, such as current, voltage, and internal and ambient temperatures. These systems typically relay data to protection mechanisms, which then trigger alarms or disconnect the battery from the charger if any of the monitored parameters exceed safe operating limits. It is well known that this BMS is typically an electronic control unit responsible for regulating and monitoring battery performance during both charging and discharging cycles. Additionally, the BMS communicates with other electronic systems to exchange critical data related to battery parameters. These parameters include voltage, capacity, temperature, power consumption, state of charge (SOC), state of health (SOH), charging cycles, and other relevant characteristics. The BMS uses this data to calculate the SOC and SOH of the battery [24].

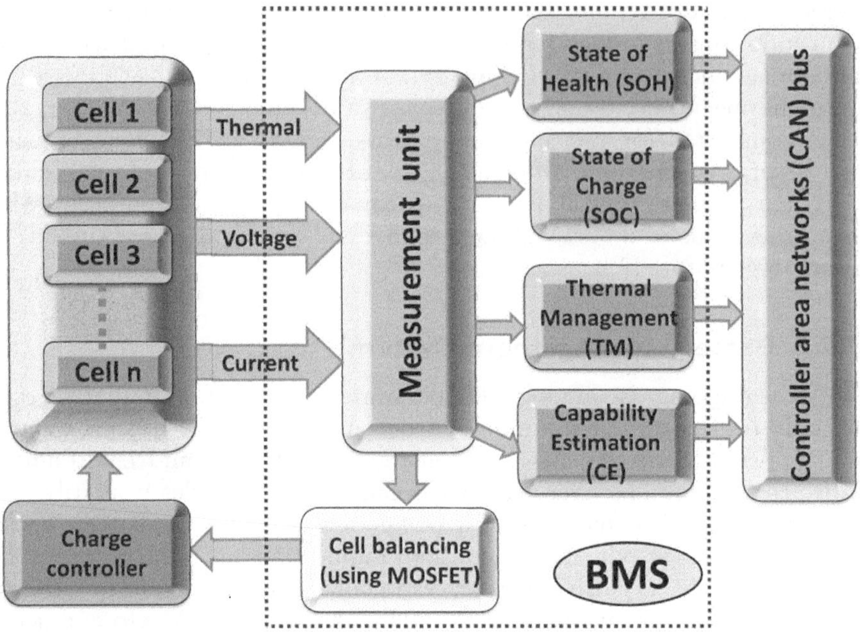

Figure 5.5 Schematic of a battery management system (BMS) connected to a battery pack.

The SOH is a key metric indicating the current capacity of a battery in comparison to its original capacity at the beginning of its life, providing valuable insight into predicting the battery's end of life. The SOC represents the ratio of the available charge at a given moment to the maximum charge the battery can store. Maintaining the battery between defined minimum and maximum SOC levels is crucial to avoid overcharging or deep discharging, as specified in the battery profile.

The key function of a BMS is to ensure the optimal utilization of the energy stored in the battery cells while protecting them from potential damage. Deep discharge, caused by excessive power draw, and overcharge, due to rapid charging, are harmful conditions the BMS safeguards against. Figure 5.5 illustrates a schematic of a BMS connected to a battery pack, ensuring the optimal and secure operation of the battery.

The BMS also integrates with a Controller Area Network (CAN bus), a vigorous vehicle communication standard that permits microcontrollers and devices to interact. This system enables multiple battery cells to function together as one, coordinating data measurements such as voltage, current, and temperature for efficient and safe operation.

Another key function of a BMS is cell balancing, which ensures that all battery cells have the same charge and discharge requirements. By balancing the cells, the BMS maintains them at equal voltage levels, thereby maximizing the capacity utilization of the battery pack. This balancing can be achieved using power MOSFETs within the BMS. A power MOSFET is a specialized transistor designed for handling various power loads, making it ideal for regulating current and voltage output in battery systems. By controlling the flow of current and voltage, the power MOSFET ensures stable and safe operation of the battery pack, helping to prevent imbalances that can reduce performance or cause damage. Power MOSFETs are widely used due to their simplicity in operation, ease of maintenance, and versatility in many battery applications.

The superior properties of power MOSFETs made from WBG materials can enhance various aspects of BMS, including efficiency, thermal management, and reliability. It is already known that WBG materials can operate at higher switching frequencies compared to traditional silicon-based components, which leads to reduced switching losses and improved overall efficiency in the power conversion stages of BMS, such as DC-DC converters. DC-DC converters are electronic circuits designed to convert direct current (DC) from one voltage level to another, either stepping up (boosting) or stepping down (bucking) the input voltage to meet the specific needs of a device or system. The low on-resistance of WBG devices results in reduced conduction losses. This is particularly beneficial in high-current applications, where efficiency improvements can significantly impact battery performance and energy consumption. Interestingly, WBG materials are more robust and less susceptible to degradation under high-voltage and high-temperature conditions. This enhances the reliability and longevity of the BMS, contributing to longer battery life and reduced maintenance requirements.

5.2.7 Electrochromic Smart Window

In an era of increasing technological importance, electrochromic smart windows (ECSW) are gaining significant interest due to their ability to manage light and heat dynamics within both domestic and commercial buildings. The primary goal of developing these smart-window products is to promote a greener environment. ECSW adjust the amount of light and temperature entering a building, contributing to energy efficiency and maintaining a comfortable living environment with minimal power consumption. The incorporation of WBG semiconductors into electrochromic smart windows holds the potential to revolutionize how buildings manage light and heat.

Figure 5.6 illustrates a typical schematic of a WBG oxide-based ECSW, which consists of five superimposed thin layers: electrode 1, ion storage layer, electrolyte, EC layer, and electrode 2. These layers are positioned between two transparent substrates, commonly made of glass, which is particularly suited for building applications.

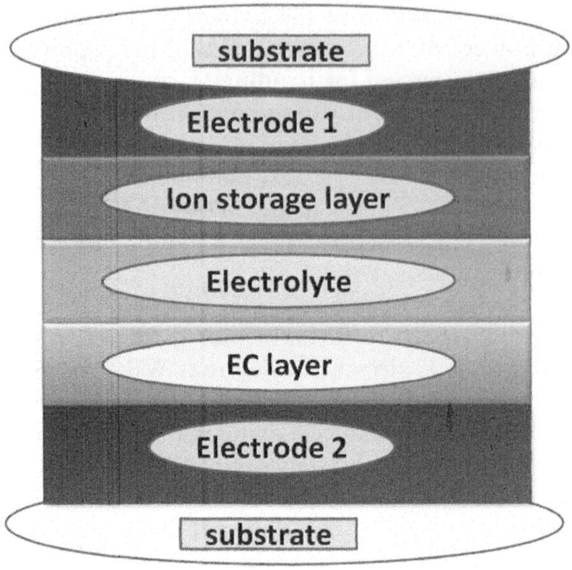

Figure 5.6 Schematic of a wide band gap oxide-based electrochromic smart window.

When an external voltage is applied, optical activity is induced in the electrochromic films, altering their optical transmittance by creating color compounds. This change in optical properties occurs as ions are inserted into or extracted from the EC layer. Voltage pulses with opposite polarity can return the device to its original state, allowing for dynamic control of light and heat passing through the window. This makes electrochromic smart windows highly useful for energy-efficient building designs. Materials capable of changing color upon application of electrical potential are commonly called electrochromic materials [25].

Some metal oxides inherently exhibit electrochromic properties due to their ability to exist in multiple oxidation states, and these are referred to as electrochromic oxides (ECOs). ECOs are classified into two types based on their electrochemical behavior:

Cathodic Electrochromic Oxides: These materials darken when they gain electrons during reduction, usually via the insertion of cations like Li^+ or H^+. Examples of cathodic ECOs include titanium oxide, niobium oxide, molybdenum oxide, tungsten oxide and tantalum oxide.

Anodic Electrochromic Oxides: These materials change color when they lose electrons during oxidation. Examples of anodic ECOs include iridium oxide, iron oxide, manganese oxide, nickel oxide, cobalt oxide, chromium oxide and rhodium oxide. In addition, vanadium-based oxides are considered hybrid materials because they exhibit both anodic and cathodic electrochromic behaviors, making them versatile in various electrochromic applications.

It is important to highlight those electrochromic systems (ECOS) used in applications such as displays, energy-efficient buildings and smart windows require critical characteristics like excellent reversibility, high optical contrast, high coloration efficiency, and long-term stability. Titanium dioxide (TiO_2), identified as a cathodic electrochromic material, undergoes color changes during charge intercalation and deintercalation: it turns deep blue when a charge is intercalated and returns to white upon deintercalation. During the electrochemical reactions, the formation of Ti^{3+} improves the conductivity of the lithiated domains, which is crucial for the material's electrochromic performance. The properties of TiO_2, such as coloration efficiency, switching kinetics and reversibility, are heavily influenced by the rate of ion insertion and the material's ion storage capacity, making these factors key determinants of its performance in electrochromic applications.

Dhandayuthapani et al. [26] investigated the preparation of anatase-phase titanium dioxide (TiO_2) thin films using the nebulized spray deposition technique, with a focus on their electrochromic properties. The study highlighted the influence of substrate temperature on the electrochromic behavior of TiO_2 films. The E_g for the films prepared at substrate temperatures of 350, 300, 250, and 200°C were estimated to be 3.24, 3.24, 3.21, and 3.2 eV, respectively. These values align with the typical indirect band gap of anatase-phase TiO_2, which is approximately 3.2 eV (in general, bulk TiO_2 can exist in three primary crystalline phases under normal pressure: rutile, anatase, and brookite, as well as in the amorphous form). Among the samples, the films prepared with the lower substrate temperature of 200°C exhibited superior electrochromic performance. This lower temperature likely contributed to optimal film structure and morphology, resulting in enhanced electrochromic properties.

The electrochromic behavior of TiO_2 films was analyzed through the "intercalation and de-intercalation of Li^+ ions using an electrolyte solution of 1 M $LiClO_4$ dissolved in propylene carbonate (PC)" [26]. When a negative potential is applied, lithium ions (Li^+) and electrons (e-) are intercalated into the TiO_2 electrode, reducing the Ti^{4+} ions to Ti^{3+}, resulting in the formation of a deep dark blue Li_xTiO_2 phase. Conversely, applying a positive potential reverses the process, i.e., de-intercalation of Li^+ ions and electrons restores TiO_2 to its original state, with the oxidation of Ti^{3+} back to Ti^{4+}. This reversible electrochromic process, characterized by the color change, is described by the following relationships:

$$TiO_2 + xLi^+ + xe^- \leftrightarrow Li_xTiO_2 \tag{5.2}$$

These films exhibited impressive electrochromic performance, with 260.7 mC cm^{-2} coloration efficiency at 6000 Å, rapid switching kinetics (bleaching time of 1.7 s and coloration time of 2.37 s), excellent reversibility (99%), and constant stability over 1000 cycles. Such results indicate that TiO_2 thin films deposited via nebulized spray deposition are highly promising for electrochromic applications, offering advantages like high efficiency, rapid response, and durability over extended use.

The extensively used cathodic EC material is tungsten oxide (WO_3), known for its high electrochromic performance. Eren et al. [27] investigated the deposition of WO_3 films on flexible substrates by RF magnetron sputtering. Key EC parameters such as coloration efficiency, electrochromic contrast and switching time were measured as a function of thickness of the WO_3 film. This study revealed that an electrochromic device using WO_3 film with a film thickness of 465 nm exhibited rapid switching times, with a bleaching time (t_x) ~5.8 s and a coloration time (t_c) ~1.6 s. "The direct band gaps (E_g) of the colored states of WO_3 films with different thicknesses (223, 465, and 650 nm) were estimated to be 2.21, 2.64, and 2.48 eV, respectively" [27]. Interestingly, the band gap increased as the film thickness increased from 223 nm to 465 nm but decreased slightly when the thickness increased to 650 nm. This trend indicates that the optical and electronic properties of WO_3 films are thickness-dependent.

The coloration/bleaching mechanism of WO_3 films involves the intercalation of lithium ions (Li^+) and electrons (e^-) into the WO_3 matrix during coloration, forming tungsten bronze (Li_xWO_3). Upon applying reverse voltages, the deintercalation of Li^+ and electrons returns the WO_3 to its original, bleached state. This process is described by the following electrochemical reaction

$$WO_3 + xLi^+ + xe^- \leftrightarrow Li_xWO_3 \tag{5.3}$$

Molybdenum trioxide (MoO_3) films have also been recognized as a capable electrochromic WBG semiconductor material for smart windows due to their beneficial properties. These films are characterized by exceptional electrochemical stability, low cost, versatile fabrication methods, and high cyclic stability. MoO_3 films show electrochromism through the reversible intercalation and deintercalation of electrons and ions; in their colored state, they appear brown due to the presence of Mo^{6+} in MoO_3. Through a subsequent electrochemical process, these films can revert to a transparent bleached state with Mo^{4+} in MoO_2 [28].

$$MoO_3 + xLi^+ + xe^- \leftrightarrow Li_xMoO_3 \tag{5.4}$$

MoO_3 exists in two primary crystalline phases: α-MoO_3 and β-MoO_3, with α-MoO_3 being particularly notable due to its layered structure. This layered nature of α-MoO_3 allows it to simply accommodate a huge number of positive ions, making it highly conducive to band gap manipulation. The E_g of α-MoO_3 lies in the range of 3.0–3.1 eV, which offers significant potential when MoO_3 is used as a coating material. By coating materials like TiO_2 with MoO_3, the band structure can be modified, enhancing the ion intercalation capabilities of TiO_2, which could lead to improved electrochromic (EC) properties. This combination could further enhance the overall performance of EC devices, especially in terms of efficiency and ion storage capacity. MoO_3 films have distinct optical characteristics, particularly in their colored state, where they appear gray. This gray color is attributed to a smooth spectrum of absorption that spans the visible range. Since the MoO_3 absorption band aligns closely with the sensitivity of the human eye, it becomes an excellent material choice for building lighting systems and other energy-efficient applications [29].

NiO is an anodic electrochromic material capable of changing from colorless to brown when a positive voltage is applied. Devices utilizing NiO as an electrochromic component demonstrate excellent color change capabilities and efficient energy storage performance. This makes NiO a valuable candidate for various applications in electrochromic devices (ECDs), where both visual aesthetics and energy efficiency are important. Sahu et al. [30] prepared NiO films using electron-beam evaporation. The structure of the nickel oxide thin films exhibits a preferred orientation at (111) plane, which is advantageous for EC properties. Films with this (111) orientation display tetrahedron-island structures, enhancing their performance. At a wavelength of 5500 Å, the best-performing film demonstrates higher transmittance modulation after 800 cycles with coloration efficiency of 34 C/cm^2.

Dong et al. [31] synthesized NiO thin films using the sputtering technique for an electrochromic study. The optical study, carried out under different applied potentials, demonstrated an exceptional contrast between the bleached and colored states. While the coloration mechanism in NiO was previously thought to be a surface phenomenon, this study demonstrates for the first time that ion insertion and extraction lead to changes in the lattice constant, proving that electrochromic coloration involves the bulk material's unit cells. The E_g of NiO in its initial, colored, and bleached position were calculated using the Tauc plots, revealing values of 3.900 ± 0.008 eV for the initial state, 3.874 ± 0.007 eV for the colored state, and 3.930 ± 0.008 eV for the bleached state. These results highlight the variations in NiO's optical band gap during its electrochromic transitions, indicating changes in its electronic structure throughout the coloration and bleaching processes. The possible electrochromic mechanism for NiO thin film is,

$$NiO + xLi^{+} + xe^{-} \leftrightarrow Li_{x}NiO \tag{5.5}$$

Vanadium pentoxide (V_2O_5) is a highly attractive material for electrochromic applications due to its impressive performance in both anodic and cathodic states. Its orthorhombic layered structure gives it a high capacity for lithium ion (Li^+) storage, making it a strong candidate for use as a counter electrode in EC devices. This layered structure enables V_2O_5 to intercalate and deintercalate Li^+ ions, which is essential for efficient charge transfer in electrochromic systems.

The electrochromic behavior of vanadium pentoxide (V_2O_5) can be explained by the reversible intercalation and deintercalation of ions (typically lithium ions, Li^+) and electrons. This process involves a redox reaction where V^{5+} in V_2O_5 is reduced to V^{4+} during ion intercalation, leading to a colored state, and oxidized back to V^{5+} during deintercalation, returning to a bleached state. The general electrochromic equation for V_2O_5 can be represented as:

$$V_2O_5 + xLi^+ + xe^- \leftrightarrow Li_xV_2O_5 \tag{5.6}$$

However, V_2O_5 films face significant challenges, including low ionic conductivity, poor cyclic stability, and low reversibility during charge–discharge cycles. As the material undergoes repeated cycling, structural distortions occur, leading to ion imprisonment within the crystalline lattice. This entrapment of ions increases the electrical resistance of the material and gradually reduces its Li^+ storage capacity, limiting its long-term effectiveness for energy storage and electrochromic applications. To mitigate these issues, research has shown that incorporating titanium dioxide (TiO_2) into the V_2O_5 matrix can significantly improve its electrochemical properties. The addition of TiO_2 enhances the stability of V_2O_5 films by reducing structural distortions and improving the material's overall reversibility and ionic conductivity. TiO_2 acts as a stabilizing agent that prevents the reorganization of the V_2O_5 structure during cycling, leading to better electrochemical stability and longer cycle life in devices. This makes the V_2O_5-TiO_2 composite a more durable and efficient material for use in electrochromic devices and energy storage systems [32].

Moura et al. [32] prepared V_2O_5-TiO_2 thin films on FTO transparent electrodes using the sol-gel process. These films underwent heat treatment for 30 minutes at 350°C. The optimal performance was achieved with the film containing 7.5 mol% TiO_2, which exhibited the maximum ion storage capacity of approximately 106 mC cm^{-2}. Additionally, at 633 nm, these films exhibited an optical modulation of 35% following the coloration and bleaching processes. These findings highlight the potential of V_2O_5-TiO_2 thin films for use in electrochromic applications, benefiting from their enhanced electrochemical and optical properties.

5.2.8 Quantum Computing

Quantum computing (QC) is an innovative technology that fundamentally differs from classical computing, allowing quantum systems to tackle problems beyond the reach of traditional computers. It has gained increasing attention for its transformative capabilities, particularly in organizing strongly entangled states and efficiently managing difficult mathematical operations. While quantum computing is advancing rapidly, compared to classical technology, it remains in its early stages. "Research continues to focus on mapping quantum algorithms to practical use cases, determining where quantum computing can enhance classical computation or introduce faster ways of performing tasks" [33].

In classical computers, information is processed using bits, each of which represents either a 0 or 1, like an on/off switch. These systems rely on manipulating millions of bits in sequence to perform tasks. On the contrary, quantum computers use "qubits", which can exist in both 0 and 1 states simultaneously. Furthermore, qubits have the ability to become entangled, where the state of one qubit is intrinsically correlated with that of another, allowing quantum devices to solve certain problems much faster by processing multiple possibilities at once. This makes quantum computers potentially far more powerful than classical ones for specific, complex tasks, though they are still in the early stages of development.

In nanoscale semiconductor quantum dot (QDs) structures, "the complete quantum confinement of electrons and holes leads to a discrete energy spectrum for charge carriers, resembling the energy levels in atoms" [34]. These unique spectral characteristics of QDs open up numerous potential applications, ranging from "optical memory, laser emitters, single-electron transistors, charge storage devices, field-effect transistors and quantum computers" [34].

"Luminescent quantum dots (QDs) are a captivating class of nanomaterials with distinctive electronic and optical properties that make them highly promising for a broad range of advanced applications in electronics, photonics and quantum computing (QC)" [35]. These semiconductor nanoparticles reveal size-dependent optical characteristics as a result of quantum confinement effects, which allow their emission wavelengths to be finely tuned. This size tunability, combined with their high photoluminescence quantum yields (PLQY), gives QDs a significant advantage over traditional materials. Luminescent metal halide QDs are nanoscale semiconducting materials that exhibit distinctive optical property, which arise from the phenomenon of "quantum confinement". Generally, these QDs are made up of cations such as cesium, cadmium, or lead and halide anions like iodine, bromine, or chlorine, arranged in a crystalline lattice. Their size typically ranges from one to ten nanometers, and it directly influences their optical behavior, including the band gap and emission wavelength. Recently, there has been increasing interest in using luminescent metal halide QDs for quantum computing due to their tunable emission spectra and potential integration into quantum information systems, making them promising candidates for advancing novel quantum technologies [35]. Machine learning (ML) is revolutionizing perovskite quantum dot (PQD) research by enabling more efficient materials discovery and device optimization. Beyond uncovering hidden patterns in large datasets, ML offers predictive modeling to forecast material properties, automated experimental design to streamline synthesis, and defect detection for quality control. It also aids in optimizing device performance by analyzing operational data and improving material synthesis through real-time process control. By leveraging ML, scientists can accelerate the discovery of PQD materials with optimal characteristics for specific applications, saving both time and cost compared to traditional experimental approaches. Additionally, ML techniques are being used to fine-tune the production processes of PQD-based devices, improving their performance and stability.

PQDs hold significant promise as key components in quantum computing, including quantum memory devices and quantum gates. Their distinctive electronic and optical properties make them ideal for "encoding and manipulating quantum information" [35]. Current developments have shown the viability of incorporating PQDs into quantum computing systems, emphasizing their capacity to transform the field by offering scalable, efficient, and tunable components for quantum information processing. As research progresses, PQDs could play a critical role in enabling the next generation of quantum technologies.

Thus, the integration of WBG materials and PQDs with quantum computing could lead to more robust, efficient, and scalable quantum systems. The high electron mobility and fast switching speeds of WBG materials can significantly enhance control and readout circuits, enabling faster processing and reducing noise in the classical systems that interface with qubits. These materials are ideal for constructing high-speed, low-noise electronics for signal processing and control, providing precise management of qubits and other quantum operations.

Although significant advancements have been made, numerous challenges persist in the development of PQDs for quantum computing. Key issues include improving the "coherence times" of PQD-based qubits, which is crucial for maintaining quantum states over longer durations, and developing scalable fabrication techniques for large-scale implementation. Overcoming these challenges is essential for PQDs to realize their full potential in quantum computing and to enable their integration into practical quantum architectures.

5.3 CONCLUSIONS

In conclusion, wide band gap semiconductors are transforming electronics by offering superior efficiency, reliability, and performance. Their high thermal conductivity and capability to work at high temperatures and voltages enable more compact, durable, and energy-efficient devices. Despite challenges like higher costs and complex manufacturing, ongoing advancements promise broader adoption across a range of applications, including power electronics, telecommunications, and emerging technologies. The continued progress of WBG materials is set to drive significant innovations and improvements in electronic systems, contributing to enhanced energy efficiency and performance.

REFERENCES

[1] K. Dong, X. Dong, and Q. Jiang, "How renewable energy consumption lower global CO_2 emissions? Evidence from countries with different income levels," *World Econ.*, vol. 43, no. 6, pp. 1665–1698, 2020.

[2] T. Ahmad and D. Zhang, "A critical review of comparative global historical energy consumption and future demand: The story told so far," *Energy Reports*, vol. 6, pp. 1973–1991, 2020.

[3] J. Wilson and J. F. B. Hawkes, Optoelectronics: An Introduction. in *Prentice Hall international series in optoelectronics*, John Wilson and John Hawkes, Eds., Prentice Hall Europe, 1998.

[4] A. Bablich, S. Kataria, and M. C. Lemme, "Graphene and two-dimensional materials for optoelectronic applications," *Electron.*, vol. 5, no. 1, 2016.

[5] J. B. Varley, B. Shen, and M. Higashiwaki, "Wide bandgap semiconductor materials and devices," *J. Appl. Phys.*, vol. 131, no. 23, pp. 1–13, 2022.

[6] A. Yoshikawa, H. Matsunami, and Y. Nanishi, "Development and Applications of Wide Bandgap Semiconductors," in *Wide Bandgap Semiconductors: Fundamental Properties and Modern Photonic and Electronic Devices*, K. Takahashi, A. Yoshikawa, and A. Sandhu, Eds., Springer Berlin Heidelberg, 2007, pp. 1–24.

[7] Z. Hu *et al.*, "Sol-gel-processed yttrium-doped NiO as hole transport layer in inverted perovskite solar cells for enhanced performance," *Appl. Surf. Sci.*, vol. 441, pp. 258–264, 2018.

[8] H. Matsunami, "State-of-the-art wide band-gap semiconductors for power electronic devices," *Int. Meet. Futur. Electron Devices,* vol. 2004., pp. 21–22, 2004.

[9] J. Millán, P. Godignon, X. Perpiñà, A. Pérez-Tomás, and J. Rebollo, "A Survey of wide bandgap power semiconductor devices," *IEEE Trans. Power Electron.*, vol. 29, no. 5, pp. 2155–2163, 2014.

[10] W. Saito, Y. Takada, M. Kuraguchi, K. Tsuda, and I. Omura, "Recessed-gate structure approach toward normally off high-voltage AlGaN/GaN HEMT for power electronics applications," *IEEE Trans. Electron Devices*, vol. 53, no. 2, pp. 356–362, 2006.

[11] W. Saito *et al.*, "High breakdown voltage AlGaN-GaN power-HEMT design and high current density switching behavior," *IEEE Trans. Electron Devices*, vol. 50, no. 12, pp. 2528–2531, 2003.

[12] F. Roccaforte *et al.*, "Emerging trends in wide band gap semiconductors (SiC and GaN) technology for power devices," *Microelectron. Eng.*, vol. 187–188, pp. 66–77, 2018.

[13] S. Ponmudi, R. Sivakumar, and C. Sanjeeviraja, "Tunable room-temperature ferromagnetism in RF magnetron sputtered $Al_2Cr_2Cu_2O_8$ thin film," *Ceram. Int.*, vol. 50, no. 2, pp. 4190–4203, 2024.

[14] S. Ponmudi, R. Sivakumar, C. Sanjeeviraja, C. Gopalakrishnan, and K. Jeyadheepan, "Tuning the morphology of Cr_2O_3:CuO (50:50) thin films by RF magnetron sputtering for room temperature sensing application," *Appl. Surf. Sci.*, vol. 466, pp. 703–714, 2019.

[15] M. R. Mohammadi and D. J. Fray, "Nanostructured TiO_2-CeO_2 mixed oxides by an aqueous sol-gel process: Effect of Ce:Ti molar ratio on physical and sensing properties," *Sensors Actuators B Chem.*, vol. 150, no. 2, pp. 631–640, 2010.

[16] P. Gao, H. Ji, Y. Zhou, and X. Li, "Selective acetone gas sensors using porous WO_3–Cr_2O_3 thin films prepared by sol–gel method," *Thin Solid Films*, vol. 520, no. 7, pp. 3100–3106, 2012.

[17] A. Rydosz, K. Wincza, and S. Gruszczynski, "Microsystem in LTCC technology for the detection of acetone in healthy and diabetes breath," in *2016 IEEE ANDESCON*, 2016, pp. 1–4.

[18] D. L. Kamble, N. S. Harale, V. L. Patil, P. S. Patil, and L. D. Kadam, "Characterization and NO_2 gas sensing properties of spray pyrolyzed SnO_2 thin films," *J. Anal. Appl. Pyrolysis*, vol. 127, no. 2, pp. 38–46, 2017.

[19] L. Francioso, A. Forleo, S. Capone, M. Epifani, A. M. Taurino, and P. Siciliano, "Nanostructured In_2O_3-SnO_2 sol-gel thin film as material for NO_2 detection," *Sensors Actuators B Chem.*, vol. 114, no. 2, pp. 646–655, 2006.

[20] S. S. Badadhe and I. S. Mulla, "H_2S gas sensitive indium-doped ZnO thin films: Preparation and characterization," *Sensors Actuators B Chem.*, vol. 143, no. 1, pp. 164–170, 2009.

[21] P. S. Shewale and Y. S. Yu, "H_2S gas sensing properties of undoped and Ti doped ZnO thin films deposited by chemical spray pyrolysis," *J. Alloys Compd.*, vol. 684, pp. 428–437, 2016.

[22] S. Ponmudi, R. Sivakumar, C. Sanjeeviraja, C. Gopalakrishnan, and T. Okamoto, "Development of room temperature sensor based on high quality rhombohedral Al_2O_3:Cr_2O_3 (1:1) thin film with bone like morphological feature for ultrasensitive detection of NH_3 gas," *J. Mater. Sci. Mater. Electron.*, vol. 31, no. 13, pp. 10123–10141, 2020.

[23] S. Ponmudi, R. Sivakumar, C. Sanjeeviraja, C. Gopalakrishnan, and K. Jeyadheepan, "Al_2O_3:Cr_2O_3:CuO (1:1:1) thin film prepared by radio frequency magnetron sputtering technique: A promising material for high sensitive room temperature ammonia sensor," *Mater. Res. Express*, vol. 6, no. 6, p. 066422, Mar. 2019.

[24] M. A. Rosen and A. Farsi, "Chapter 6 – Battery system design," in *Battery Technology*, M. A. Rosen and A. Farsi, Eds., Academic Press, 2023, pp. 161–198.

[25] B. Wang *et al.*, "An overview of recent progress in the development of flexible electrochromic devices," *Nano Mater. Sci.*, vol. 5, no. 4, pp. 369–391, 2023.

[26] T. Dhandayuthapani, R. Sivakumar, R. Ilangovan, C. Gopalakrishnan, C. Sanjeeviraja, and A. Sivanantharaja, "High coloration efficiency, high reversibility and fast switching response of nebulized spray deposited anatase TiO_2 thin films for electrochromic applications," *Electrochim. Acta*, vol. 255, pp. 358–368, Nov. 2017.

[27] E. Eren, G. Y. Karaca, U. Koc, L. Oksuz, and A. U. Oksuz, "Electrochromic characteristics of radio frequency plasma sputtered WO_3 thin films onto flexible polyethylene terephthalate substrates," *Thin Solid Films*, vol. 634, pp. 40–50, 2017.

[28] C. T. Lee *et al.*, "Synthesis and electrochromic properties of molybdenum oxide films," *Surf. Coatings Technol.*, vol. 363, pp. 426–429, 2019.

[29] Y. Wang, B. Chen, Y. Zhu, L. Fu, Y. Wu, and T. van Ree, "13 – Metal oxides in energy-saving smart windows," in *Metal Oxides in Energy Technologies*, Y. Wu, Ed., Elsevier, 2018, pp. 341–360.

[30] D. R. Sahu, Y.-H. Lee, T.-J. Wu, S.-C. Wang, and J.-L. Huang, "Synthesis and electrochromic property improvement of NiO films for device applications," *Thin Solid Films*, vol. 707, p. 138097, 2020.

[31] D. Dong *et al.*, "Electrochromic and colorimetric properties of anodic NiO thin films: Uncovering electrochromic mechanism of NiO," *Electrochim. Acta*, vol. 335, p. 135648, 2020.

[32] E. A. Moura *et al.*, "Electrochemical properties of thin films of V_2O_5 doped with TiO_2," *J. Phys. Chem. Solids*, vol. 119, pp. 1–8, 2018.

[33] H. Doga *et al.*, "How can quantum computing be applied in clinical trial design and optimization?," *Trends Pharmacol. Sci.*, vol. 45, no. 10, pp. 880–891, 2024.

[34] R. Bose and H. T. Johnson, "Coulomb interaction energy in optical and quantum computing applications of self-assembled quantum dots," *Microelectron. Eng.*, vol. 75, no. 1, pp. 43–53, 2004.

[35] K. Deepthi Jayan and K. Babu, "Luminescent perovskite quantum dots: Progress in fabrication, modelling and machine learning approaches for advanced photonic and quantum computing applications," *J. Lumin.*, vol. 277, p. 120906, 2025.

Chapter 6

Comprehensive Review of Carbon Nanotubes

Synthesis, Properties, Functionalization, Characterization, and Applications

S. Agnes Shifani and S. Maheswari

6.1 NANOTECHNOLOGY

6.1.1 Introduction

Nanotechnology involves the manipulation of matter at the nanoscale, typically with at least one dimension measuring less than 100 nanometers (nm). When describing the properties of matter, surface area and the applications of quantum mechanics come into play at this nanoscale [1]. This concept of nanotechnology encompasses all research and technological endeavors related to these distinctive features. The terms "nanotechnologies" and "nanoscale technologies" are sometimes used in the plural when referring to studies and applications that share elements of scale. The fullerene nanogears of the fundamental system are shown in Figure 6.1. An early understanding of nanotechnology focused on molecular nanotechnology, the modern name for the specific technological goal of accurately manipulating atoms and molecules to manufacture objects on a macroscale.

6.1.2 Evaluation of Nanotechnology

Although not commonly known, Norio Taniguchi coined the term "nanotechnology" in 1974. K. Eric Drexler later re-coined the term "nanotechnology" in his 1986 book *Engines of Creation: The Coming Era of Nanotechnology*. In it, he introduced the concept of a nanoscale "assembler" that could replicate itself and other objects of arbitrary complexity, with control at the atom level. Feynman's theories served as an inspiration for Drexler. In order to increase public awareness and understanding of the implications and concepts of nanotechnology, Drexler co-founded The Foresight Institute in 1986.

Nanotechnology emerged in the 1980s as a result of Drexler's theoretical and public work that created and advanced a conceptual framework in addition to well-known experimental accomplishments that raised awareness of the field's potential. Figure 6.2 illustrates the various sizes of nanomaterials.

 DOI: 10.1201/9781003533535-6

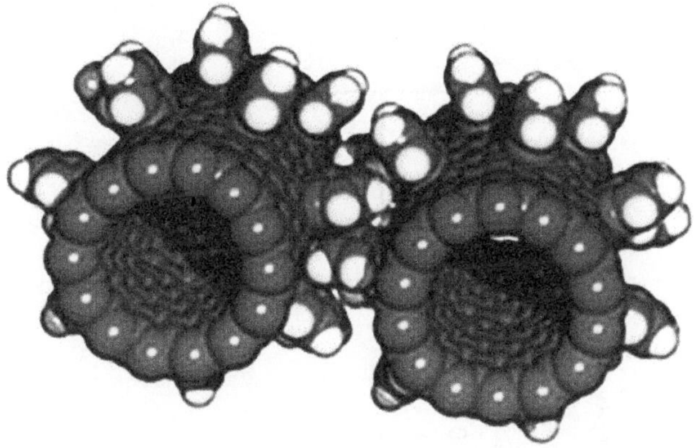

Figure 6.1 Fullerene nanogears.

6.1.3 Comparison of Microtechnology and Nanotechnology

Although they both require the reduction of materials to produce new goods, microtechnology and nanotechnology operate on distinct scales:

- Microtechnology
  Involves both physical and chemical processes as well as the formation or modification of structures that are one millionth of a meter in size (one micrometer). Microtechnology systems comprise optical, mechanical, electrical, and computer components [1].
- Nanotechnology
  Involves working at the nanoscale, approximately one to 100 nanometers, with atoms and molecules. Nanotechnology is distinct from microtechnology in that it deals with phenomena not observed in larger structures.

6.2 CARBON NANOTUBES

6.2.1 Evaluation of Carbon Materials

Carbon has been known to humans from ancient times. It is one of the elements that the cosmos contains the greatest amounts of. It is found in the atmospheres of most planets, comets, stars, and the sun. Carbon compounds (including soot, charcoal, graphite, and diamond) have been utilized by humans reaching back almost to the dawn of mankind. However, it was not until the latter part of the 17th century that carbon was recognized as an element.

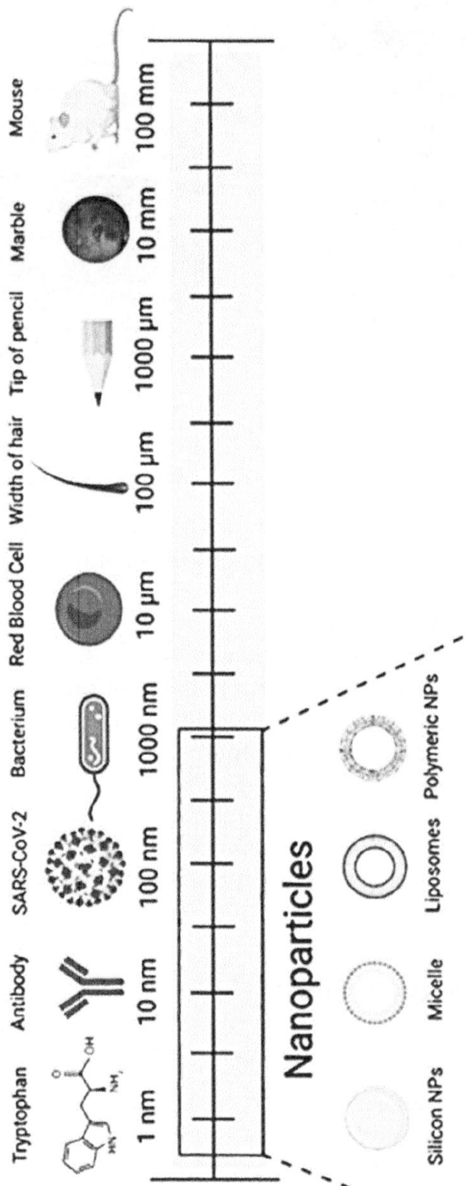

Figure 6.2 Comparison of nanomaterials sizes.

Table 6.1 Selected properties of carbon [2]

Atomic number	6
Atomic weight	12.011
Melting point	3550°C

The Sumerians and Egyptians used wood chars, also called charcoal, to reduce Cu, Zn, and Sn in order to make bronze circa 3750 BC. The ancient Egyptians used charcoal as a medicine about 1500 BC. It appears that the main use of charcoal has been for the adsorption of unpleasant gases from the digestive system and from wounds while they are healing.

More recently, the discovery of fullerenes and carbon nanotubes provided a whole new perspective on carbon compounds based on flat hexagonal layers, such as graphite. Like many important discoveries in science, fullerenes, also referred to as buckyballs, were discovered by accident. Scientists analyzed carbon clusters in the condensing plasma with a mass spectrometer and discovered, shockingly, that massive amounts of C60 clusters consistently occurred.

Carbon nanotubes (CNTs) were subsequently found in 1991 by Japanese electron microscopist Sumio Iijima of NEC [8]. He was looking into the residue left over from the cathode's arc-evaporation generation of fullerenes. NEC and IBM announced the discovery of CNTs composed of a single graphene layer two years later, in 1993. The two types of carbon nanotubes that were then identified were single-wall carbon nanotubes (SWCNTs) and multi-walled carbon nanotubes (MWCNTs). With two, five, and seven coaxial cylinders, MWCNTs were the first CNTs to be identified in 1991, according to this nomenclature.

6.2.2 Structure of Carbon Materials

Carbon, the most versatile element in the periodic table, can form bonds with a large number of elements, and these bonds can vary in strength and quantity. It has come to our attention that there is an additional family of carbon. There are many different types of carbon, including nanoparticles, amorphous films, carbon nanotubes, crystals, and manufactured carbons with slightly chaotic microstructures (as shown in Figure 6.3(a) and 6.3(b)). Among the crystalline forms of carbon are diamond, graphite, carbynes, and fullerenes [3].

Four of the six electrons in an atom of carbon are located in its valence bands. In the 1s2 orbital, there are two tightly bound core electrons. Additionally, the electrons occupying the $2s^2$ and $2p^2$ orbitals are referred to as valence electrons. A carbon atom's electrical state is represented by the notation 1s2, 2s2, and 2p2. The valence electrons are created by the valence electrons in the crystalline phase and are essential for the creation

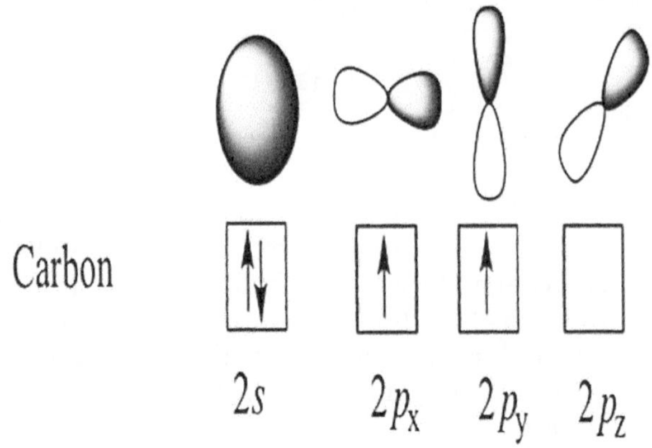

Figure 6.3 (a) Atomic orbital and sp hybridization of carbon; (b)hybrid orbitals of carbon.

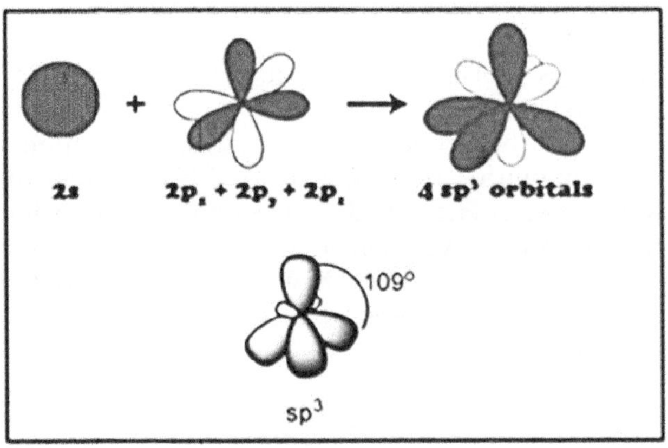

Figure 6.4 sp^3 hybrid orbitals.

of covalent bonds in carbon-based materials. In carbon, there is very little variation in orbital energy between the 2s and 2p orbitals compared to the energy generated during bond formation.

6.2.2.1 Tetrahedral Hybridization (sp^3)

After one 2s orbital hybridizes with the three 2p orbitals in sp3-hybridization, four sp3 orbitals are created, which are separated by an angle of 109.5°. As demonstrated in Figure 6.4, the unique tetrahedral shapes of the bonds are produced by the sp3 hybridization. Diamonds are among the hardest and most brilliant stones, with three-dimensional structures. Strong σ bonds are

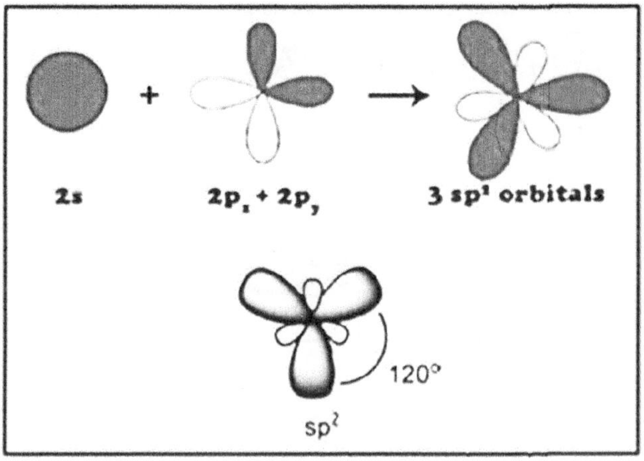

Figure 6.5 sp^2 hybrid orbitals.

formed with adjacent atoms by the tetrahedrally directed sp3 hybrid orbitals, which are the source of the structure. The diamond lattice is composed of two interpenetrating face center cubic (fcc) lattices with a lattice spacing of 0.357 nm [4].

6.2.2.2 Trigonal Hybridization (sp^2)

The hybridization of one 2s and two 2p orbitals, specifically px and py, is referred to as the sp2 hybridization. When combined, they result in a planar assembly (see Figure 6.5) in which the hybrid orbitals form a σ bond at an angle of about 120 degrees on average. Consequently, the fourth electron is located in a pz orbital normal to the σ bonding plane, while the other three are assigned to trigonally directed sp2 hybrid orbitals.

6.2.2.3 sp Hybridization

One s- and one p-orbital are joined together. The linear geometry of the sp hybridized orbitals, which are 180° apart, is shown in Figure 6.4. This superposition has no effect on the other two states, 2py and 2pz. For example, in the acetylene molecule (C_2H_2), this hybridization results in a strong covalent σ bond between the two carbon atoms due to their overlapping sp1 orbitals (see Figure 6.6).

Three forms of hybridization are possible for any carbon atom: sp (as in C_2H_2), sp2 (as in graphite), or sp3 (as in CH_4). This suggests a bivalence, which is actually only observed in a very small number of structures (carbyenes); carbon is tetravalent in the great bulk of its compounds. All three types of hybridization are listed in Table 6.2.

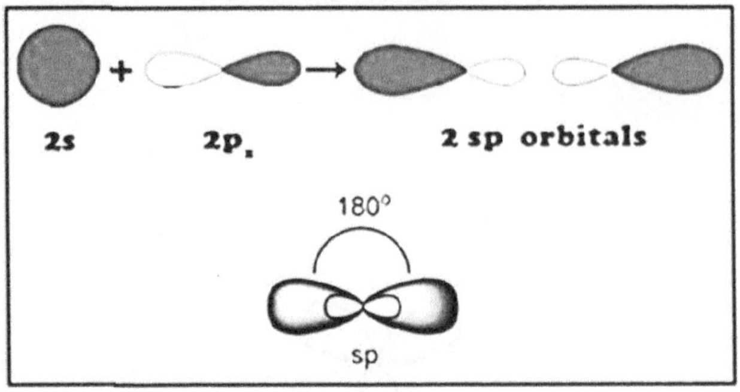

Figure 6.6 sp hybrid orbitals.

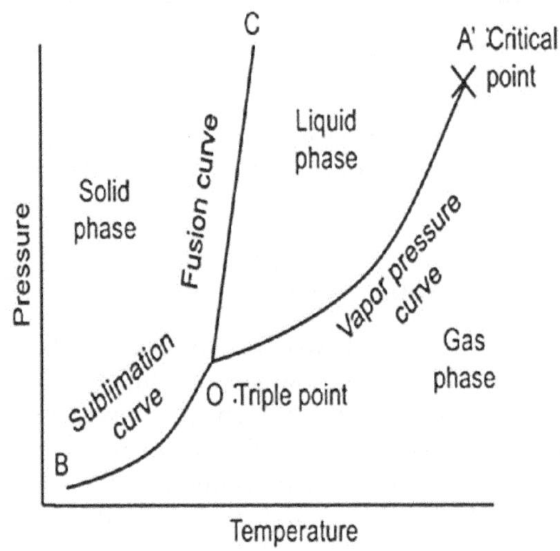

Figure 6.7 Equilibrium phase boundaries.

Table 6.2 Types of hybridization [5]

Types	*sp3*	*sp2*	*sp*
Used orbitals	1s & 3p	1s & 2p	1s & 1p
Unused p-orbitals	Nil	1	2
Angle	109.5	120	180
Geometry	Tetrahedral	Trigonal planar	Linear

Table 6.3 Crystalline representation of carbon

	Diamond	*Graphite*	*C 60*	*CNT*
Electrical Conductivity	Insulator	Conductor	Semi-conductor	Conductor to semiconductor
Hybridization	sp^3	sp^2	sp^2	sp^2

Because of its versatility in producing various forms of hybridization, carbon can exist in a wide range of allotropes, such as the well-known graphite, diamond, and amorphous carbon, as shown by the carbon phase diagram in Figure 6.7. Other regions of the phase diagram show the stability ranges for liquid carbon, hexagonal carbynes, and hexagonal diamonds.

6.2.3 Allotropes of Carbon

Many stable forms of carbon exist, such as 0wD fullerenes, 1D conducting and semiconducting carbon nanotubes, 2D semimetallic graphite, and 3D semiconducting diamonds. Carbon is an intriguing element. In contrast to other carbon forms, the carbynes have not been fully described. A few features of crystalline forms of carbon are listed in Table 6.3.

6.2.3.1 *Graphene*

From an experimental standpoint, graphene is the newest allotope—it has only been experimentally quantifiable since 2004. The longest known allotope in history is 3D graphite [6]. Graphene layers are stacked on top of one another to form bulk graphite. This suggests a bivalence, which is actually only observed in a very small number of structures (carbyenes); carbon is tetravalent in the great bulk of its compounds (Figure 6.8).

6.2.3.2 *Graphite*

Graphite is a well-known example of crystalline carbon. This stuff is really, really soft. Graphite is found at the interface between igneous (such as gneiss) and stratified rock. Given the high temperatures (between 300 and 1200° C) caused by the incoming magma, it is likely that coal in the stratified strata started to graphitize. The ideal crystal structure of graphite is made up of hexagonal carbon network sheets. These carbon sheets are stacked in the Bernal stacking, or ABAB pattern, as shown in Figure 6.9. In this design, different planes are shifted relative to one other.

Because of the strong in-plane C-C bond and small nearest neighbor gap between carbon atoms, graphite's in-plane lattice constant is extremely

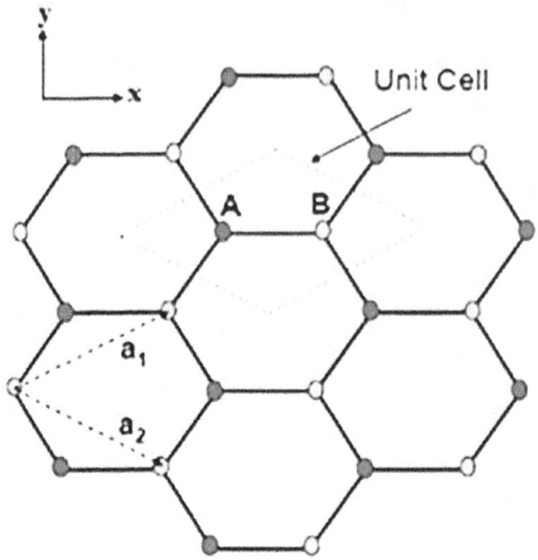

Figure 6.8 The crystal structure of graphene in real space.

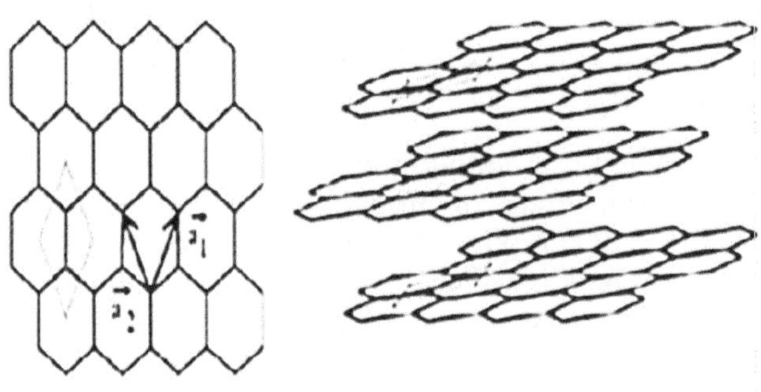

Figure 6.9 The structure of graphite: (a) the structure of one layer, showing the lattice vectors a_1 and a_2; (b) three stacked layers.

resilient to outside perturbations. Since graphene is held together by a weak van der Waals force, one effect of its low C–C value is that impurity species, except for boron, are unlikely to replace covalently bound in-plane lattice sites. Instead, they are likely to occupy an interstitial space between the planes of the graphene layer. The difficulties experienced with inserting substances other than boron into individual single-walled carbon nanotubes can be explained by the fact that the same considerations also apply to carbon nanotubes.

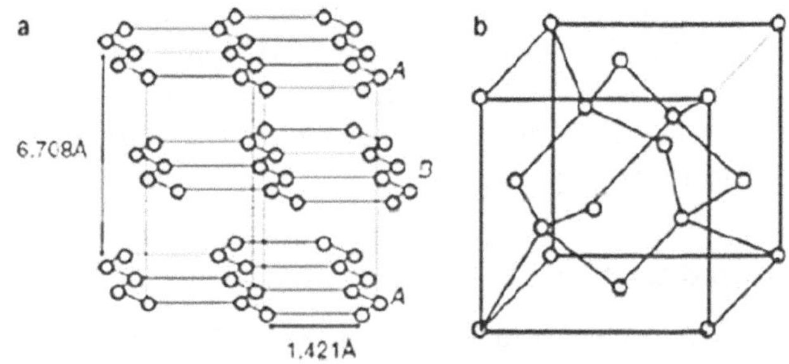

Figure 6.10 Crystal structure: (a) graphite; (b) diamond.

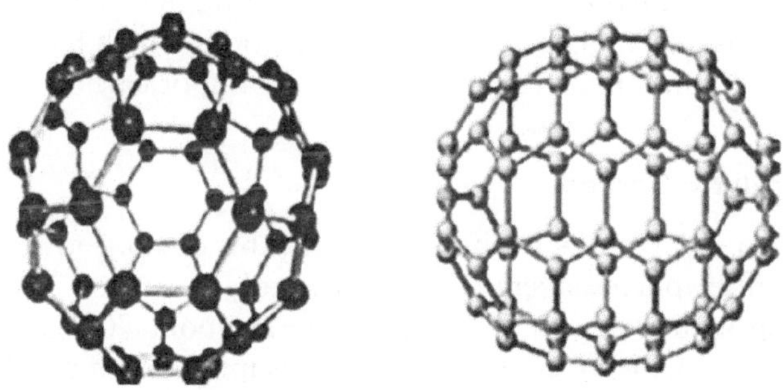

Figure 6.11 Model of the first fullerenes discovered: C_{60} and C_{70}.

6.2.3.3 *Diamond*

The word "diamond," which comes from the Greek words diaphanes (translucent) and adamas (invincible). Diamonds were first found in India around 4000 BC. Each carbon atom in a diamond has four immediate neighbors, as illustrated in Figure 6.10, which are located at the corners of a tetrahedron that is exclusive to that specific atom.

6.2.3.4 *Fullerenes*

It was decided to name the C60 molecule "buckminsterfullerene" (or "buckyball"). The name honors architect Buckminster Fuller, whose designs resemble the structure of fullerenes [7]. With an (almost) infinite number of hexagons, Euler's theorem for polyhedra suggests that 12 pentagons are generally required. The flat graphene sheet only includes hexagonal rings; pentagons and heptagons are the only shapes that can produce curved surfaces.

Seldom do fullerenes consisting of fewer than 60 carbon atoms exist. C20 is a theoretically feasible structure, but it is extremely unlikely because, as Figure 6.11 illustrates, two pentagons do not fit together well structurally.

Because geodesic molecules are hollow, atoms can be held within them just like atoms in a molecular cage. The strange abilities of buckyballs have been noted by the medical community. Indeed, some scientists believe that in the future, buckyballs could be used to target and destroy viruses, bacteria, or cancer cells, while protecting the rest of the body from the harmful effects of potent drugs.

6.2.3.5 Activated Carbon

With a radius of 0.5–2 nm and an area of 300–2000 m^2/g, activated carbon is a superior adsorbent for a variety of substances. Organic precursors like carbon fibers, coal, wood, turf, or coconut peel are heated gradually to create it. To improve the quantity and size of pores in the carbon matrix, raw materials are first carbonized at temperatures ranging from 600 to 650°C. The next step is a partial gasification or activation, which happens between 800 and 900 °C and involves the use of air, steam, or carbon dioxide.

6.2.3.6 Pyrolytic Carbon

Depending on the manufacturing conditions, this could lead to a fairly parallel order of graphene layers. Common pyrolytic carbon is produced by thermolyzing hydrocarbons at low pressure at about 700°C. The product arranges the graphene layers so that they precipitate on a plain surface. Its thermal and electrical conductivity anisotropy is about the same as that of natural graphite. The nearly completely parallel highly ordered pyrolytic graphite is created when precipitated carbon is heated to high temperatures and then subjected to shear pressures.

6.3 STRUCTURE OF CARBON NANOTUBES

A kind of carbon composed of rolling carbon atoms organized in a hexagonal lattice is called a carbon nanotube (CNT). A CNT is made up of one or more graphene sheets that have been coiled into a cylinder. SWCNTs are made of a single graphene sheet curled into a cylinder, as opposed to MWCNTs, which are made of several graphene sheets folded into concentric cylinders.

6.3.1 Characteristics of Carbon

A crucial method for describing CNTs and the composite materials they are used in is SEM (scanning electron microscope) analysis. MWCNTs and SWCNTs generally appear as fibers in SEM images [8]. From SEM images the nanotubes are plainly visible and it can be seen that the MWCNTs (B) and SWCNTs (A) have tubular architectures. However, transmission

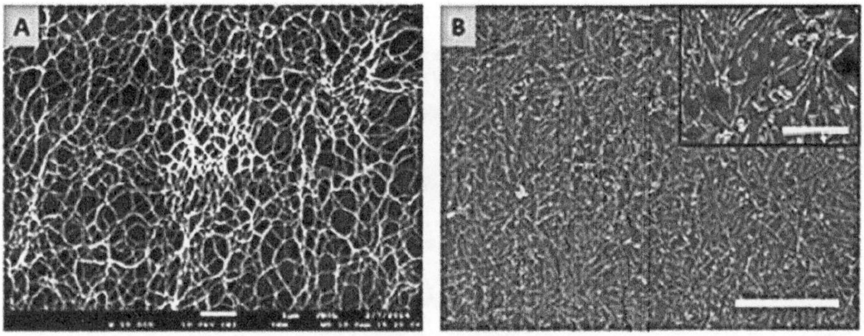

Figure 6.12 SEM analysis of SWCNT and MWCNT.

electron microscopy (TEM) analysis displays high-resolution images that show the CNT layers. SWCNT and MWCNT TEM images are shown in Figure 6.12. Generally speaking, CNT composites have some nodules on the nanotubes with dark layer shading.

6.3.2 Structure of Carbon

6.3.2.1 Zigzag Carbon Nanotubes

The pair (n, m), the bond length a, and the lattice translational indices could all be used to characterize the geometry of a CNT. Generally speaking, the term is used to define a CNT's diameter, d:

$$d = \frac{\sqrt{3}a}{\pi}\sqrt{n^2 + m^2 + nm}$$

m represents null because the topic of discussion is limited to zigzag CNTs (Figure 6.13). Here, the diameter is easily obtained using the formula.

$$d = \frac{\sqrt{3}an}{\pi}$$

The torsional U_t, bending U_b, and stretching U_a energies are the bond energies that exist between carbon atoms. When there are slight deviations from balance, these energies can look like this:

$$U_a = \frac{1}{2}K\left(r - r_0\right)^2 \quad (a)$$

Figure 6.13 Structure of zigzag CNTs.

$$U_b = \frac{1}{2}C\left(\odot - \odot_0\right)^2 \quad (b)$$

$$U_t = \frac{1}{2}C_t\left(\Phi - \Phi_0\right)^2 \quad (c)$$

6.3.2.2 Armchair Carbon Nanotubes

Armchair CNTs are produced when m = n and the CNTs are metallic. The chiral angle of an armchair CNT is 30 degrees. Zigzag CNTs are produced when the chiral angle is zero. The result of all other lattice vectors is chiral carbon nanotubes. Carbon nanotubes with chiral and zigzag shapes are semiconducting. It has been found that the inherent conductivity of CNTs is determined by their structure and electronic nature, which can be either metallic or semiconducting depending on the angle between the graphite lattices.

6.3.2.3 *Chiral Carbon Nanotubes*

Chiral carbon nanotubes are carbon nanotubes with a chiral angle between 0° and 30°. A graphene sheet is rolled in a specific way to get this angle. A CNT's chirality can be used to characterize its structural and electrical characteristics. The chirality of a carbon nanotube is determined by a chiral index, which is made up of the two numbers (n, m).

6.3.3 Properties of CNT

Elasticity Property

One amazing feature of CNTs is their flexibility. The term "elastic modulus" (or "modulus of elasticity") describes how elastic CNTs with one or more walls are. With the use of this apparatus, scientists can monitor and study the molecular vibrations caused by the heat pressures applied to the tubes' two corners.

Electromechanical Properties

A suspended carbon nanotube (CNT) can be worked with and its electrical conductivity measured using an AFM probe. This makes it possible to evaluate how the electrical characteristics of the CNT are impacted by mechanical deformation.

Chemical Properties

It was found that molecules with fully bonded sp^2 hybridization were attached to the carbon nanotube (CNT) through complex covalent bonding. It is evident that the molecules were able to be adsorbed onto CNTs due to the development of non-covalent forces, which could potentially affect the molecules' physical properties and potential applications.

Optical Properties

A rapid and secure method for representing a moderate quantity of carbon nanotubes is to use spectroscopy techniques. From a mechanical perspective, this depiction is very desirable, as it illustrates how many parameters of the nanotube union can be changed, either purposefully or accidentally, to change the quality of the nanotube.

6.3.4 Structures of Single-Walled Carbon Nanotubes

SWCNTs could be compared to a single, long graphene sheet that has been wrapped up into a cylinder. SWCNTs have a radius of roughly 0.5 nm and are frequently thousands of times longer. SWCNTs are the greatest choice for miniaturizing electronics in order to replace the tiny electromechanical systems that currently form the basis of modern electronics because of their distinct electrical properties when compared to MWCNT nanotube versions [9].

6.3.5 Structures of Multi-Walled Carbon Nanotubes

These nanotubes have a special structure because their walls are composed of several carbon sheets; each is one atom thick. The characteristic properties of MWCNTs are mainly caused by the fact that they are made up of many layers of graphene nanotubes stacked within each other.

MWCNTs typically range in size from 5 to 50 nm. However, the structure of MWCNTs is not as well understood as it may be because of the complexity and diversity. The advantageous material properties might be reduced in places where there is structural flaw. Table 6.4 compares SWCNTs and MWCNTs.

Prices for SWCNTs are higher than those for MWCNTs, a reflection of the difficulty of producing SWCNTs on a larger scale.

6.3.6 Defects in Carbon Nanotubes

One of the most interesting applications of nanotechnology is molecular electronics, and carbon nanotubes are a great candidate due to their unique electrical characteristics and quasi-one-dimensional structure. With regard to carbon nanotubes, it's most likely not as ideal as originally believed.

6.3.6.1 Elbow-like Defects

Imperfections have a major effect on the mechanical and vibrational properties of graphene and carbon nanotubes (CNTs) [11]. To investigate the effects of different parameters, including chirality, aspect ratio, boundary conditions, and the types and number of defects on the longitudinal and transverse vibrational behavior of CNTs and graphene sheets, molecular dynamics simulations were conducted on pristine and defective zigzag and armchair CNTs with clamped-free (C-F) boundary conditions, as well as pristine and defective armchair graphene sheets with different boundary conditions (Figure 6.14).

6.3.6.2 Adjacent Pentagon–Heptagon Pairs

Figure 6.15 shows an intramolecular junction (12, 0)-(9, 0) as well as the two hexagons needed for the necessary modification. Labels are used to indicate the sites whose local vibrational density of states (lVDOS) are shown in Figure 6.15.

Table 6.4 Comparison of SWCNT and MWCNT [10]

Single-walled carbon nanotubes (SWCNTs), also called graphene nanotubes (GNTs)	**Graphene**	**Multi-walled carbon nanotubes (MWCNTs)**	**Nanographite structures** (incl. few-layer "graphene", graphene flakes (graphene nanoplatelets), graphene oxide (GO), reduced graphene oxide (RGO) and others)	**Graphite structures** (incl. carbon black)
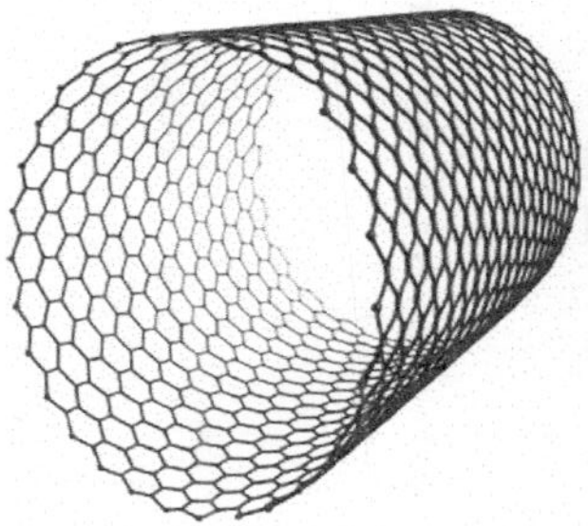	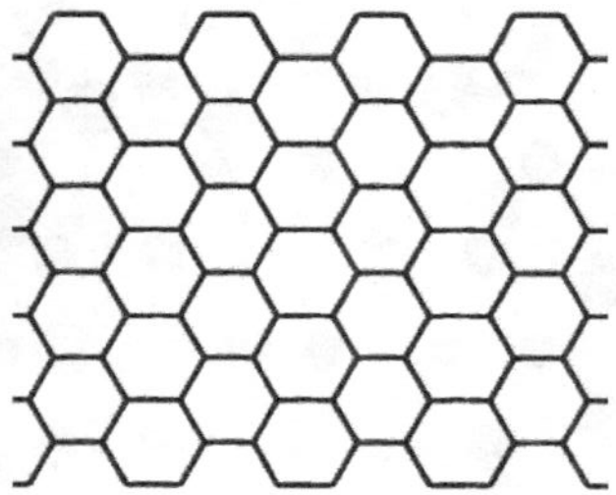	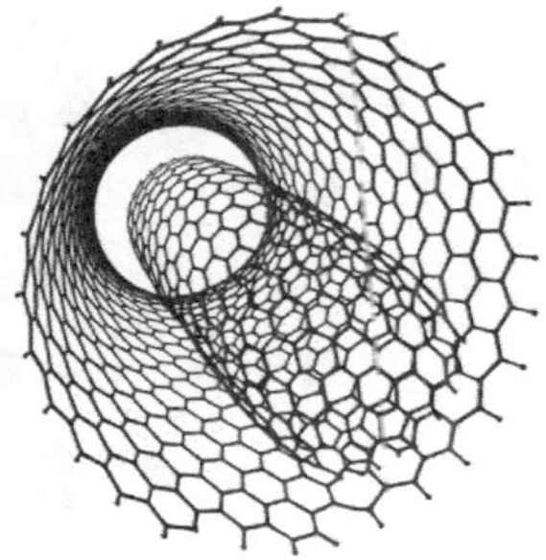		
A graphene layer seamlessly rolled into a tube	A single sheet or monolayer of carbon atoms tightly bound in a hexagonal honeycomb lattice	A concentric arrangement of SWCNTs, i.e., multiple layers of graphene seamlessly rolled into a concentric tubular structure	Graphene flakes are small platelets of graphite; GO/RGO are few-layer nanoscale graphites, have oxygen groups or have undergone additional reduction	A large set of multiple graphene sheets
Industrial-scale production (hundreds of tonnes)	Not designed as an additive to materials (production in kgs)	Industrial-scale production (thousands of tonnes)	Industrial-scale production (thousands of tonnes)	Industrial-scale production (millions of tonnes)

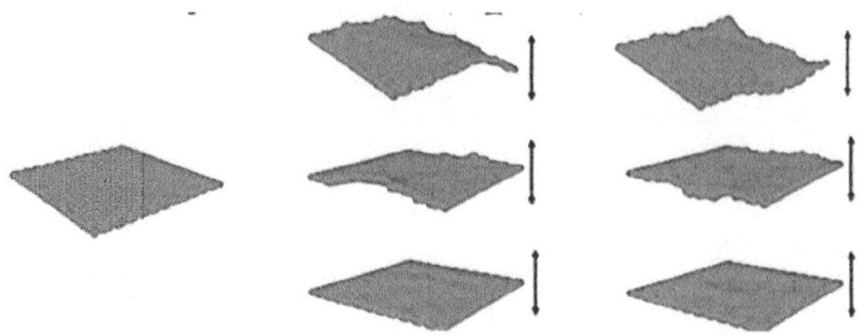

Figure 6.14 Graphical abstract to find effect of imperfections on CNT.

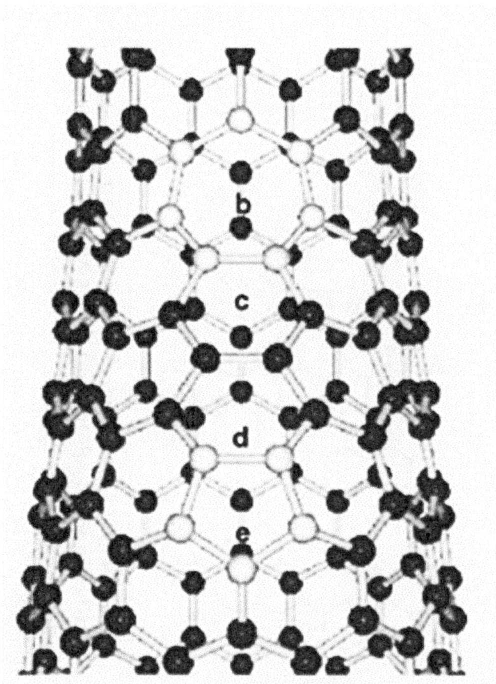

Figure 6.15 A heptagon-pentagon defect involves a change of the tube radius.

6.4 SYNTHESIS OF CARBON NANOTUBES

6.4.1 Introduction

A popular technique utilizes a hydrocarbon gas as the carbon source. A quartz tube filled with an inert gas and a catalyst-coated substrate are housed in a crucible. Next, a peak temperature is inserted to the tube. Subsequently, after entering the tube, the hydrocarbon gas is subjected to pyrolysis to transform into carbon atoms which form a bond with the substrate. This method produces both single- and multi-walled CNTs [12].

6.4.2 Arc Discharge

Using this method, pure graphite electrodes spaced 1 mm apart have voltage applied between 20 and 25 V while helium gas flows within the quartz chamber at a pressure of 500 torr (Figure 6.16). In certain situations, forcing the electrodes to collide results in the creation of an electric arc. The energy generated in the arc is transported to the anode, where it ionizes the carbon atoms in the pure graphite anode, resulting in the production of high-temperature atoms known as plasma and C^+ ions. As they approach the cathode, these positively charged carbon ions undergo reduction, deposition, and begin to produce carbon nanotubes (CNTs).

6.4.3 Laser Ablation

The graphite target is vaporized by a laser source using the Physical Vapor Deposition process (Figure 6.17). This technique involves placing the graphite target in the middle of a quartz chamber that has been filled with argon gas and heated to 1200°C. Either a continuous laser source or a pulsed laser source evaporates the graphite target.

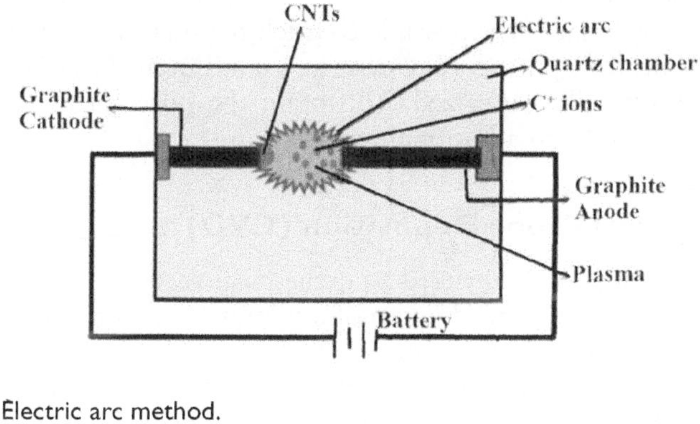

Figure 6.16 Electric arc method.

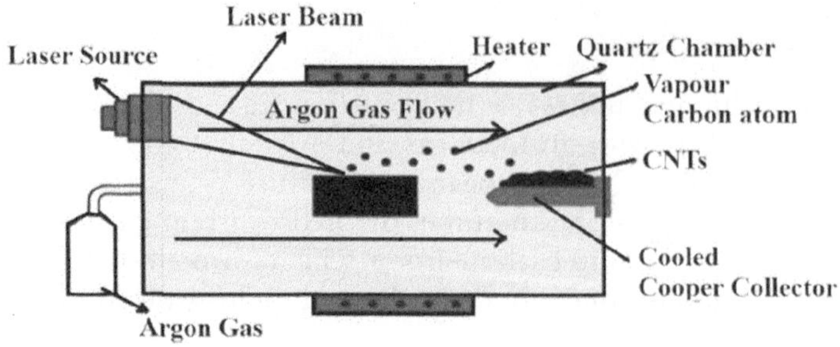

Figure 6.17 Laser ablation method.

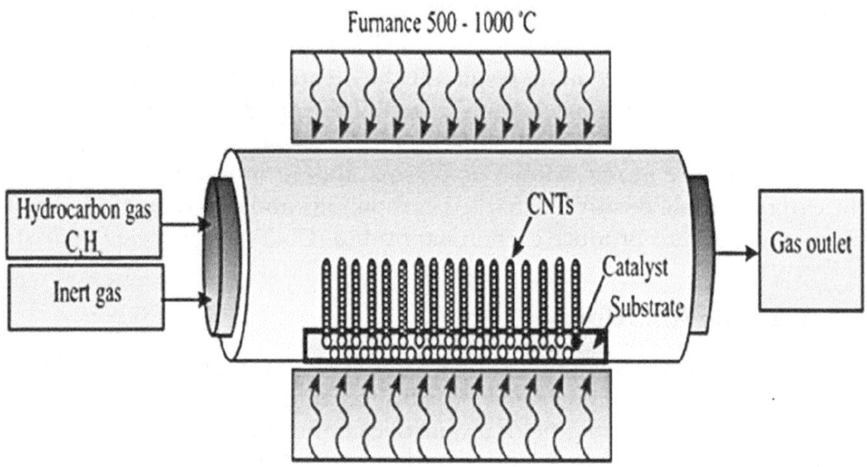

Figure 6.18 CVD method.

Because the amount of carbon atoms that evaporate with each shot is precisely proportionate, it is possible to track the quantity of CNTs created when a laser beam is pulsed. However, in a continuous laser beam, carbon atoms are continually vaporized. To obtain the pure form of CNTs, the resultant CNTs undergo additional purification.

6.4.4 Chemical Vapor Deposition (CVD)

Vaporized reactants are subjected to a chemical reaction in the CVD process, which results in the production of a nanomaterial product that is subsequently deposited onto the substrate (Figure 6.18).

This method involves placing a quartz tube within a furnace that has been heated by an RF heater to a high temperature (500–900°C). A quartz tube filled with an inert gas, such as argon gas, is positioned inside a crucible that holds the substrate coated with catalyst nanoparticles.

6.4.4.1 *Plasma Enhanced Chemical Vapor Deposition*

Plasma enhanced chemical vapor deposition (PECVD) is a frequently used technique for coating production. The main advantage of PECVD is its capacity to coat temperature-sensitive substrates, including polymers, by bringing the temperature down from a range of 400–2000°C to room temperature.

From low frequency (kHz region) to high frequency (MHz) and microwaves in the GHz range, electric fields are frequently used in a broad frequency range. To create the plasma in a practical manner, a number of modifications of the method have been devised [13].

6.4.4.2 *Alcohol Catalytic Chemical Vapor Deposition*

Several additional techniques were developed shortly thereafter, i.e., oxygen-aided, water-assisted, microwave plasma, and molecular beam synthesis [14]. Probably the simplest of these is the alcohol catalytic chemical vapor deposition (ACCVD) process, which is unique in that it permits the catalyst to be applied using a range of methods, such as combinatorial sputtering deposition and the original dip-coating technique.

6.4.4.3 *Laser-Assisted Thermal Chemical Vapor Deposition*

By locally heating a substrate, the CVD technique known as "laser-assisted chemical vapor deposition" (LCVD) uses a laser to produce thin films [15]. A laser beam heats a substrate inside a reaction chamber that is loaded with precursor gases. The gas precursor dissociates as a result, leaving behind a thin layer on the substrate. Figure 6.19 depicts the schematic framework of the approach [16].

- Applications
  In the semiconductor, MEMS, and composites industries, LCVD is utilized to produce high-purity films, fibers, and mechanical devices.
- Advantages
  Compared to traditional CVD, LCVD provides a number of benefits.
  Lower deposition temperatures: LCVD can be used to reduce deposition temperatures.
  Enhanced film purity: LCVD has the ability to enhance film purity.
  Complex thin-film patterns: LCVD allows for the direct writing of complex thin-film patterns.

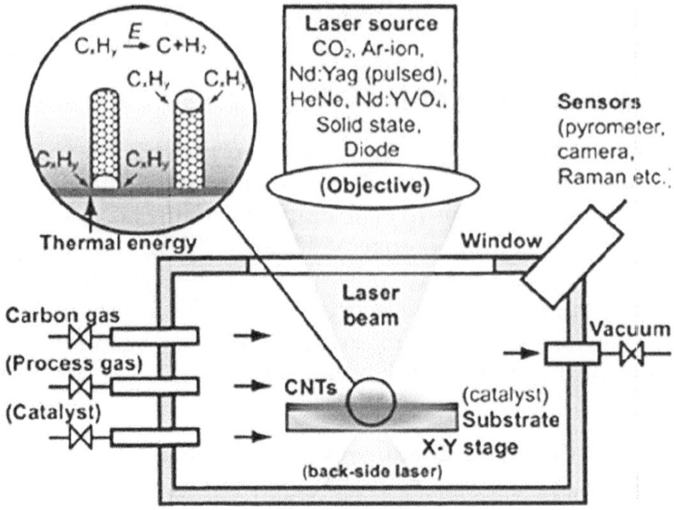

Figure 6.19 Laser-assisted thermal chemical vapor deposition.

6.4.4.4 Aero Gel Supported CVD

This can be used to modify aerogels in order to improve their properties and increase their use, as shown in Figure 6.20.

- CVD with support for templates
  This approach was used to create ultralight boron nitride aerogels by covering carbon aerogels with atomically thin boron nitride layers.
- Carbon black aerogel
  Using a catalyst-free CVD process, carbon black aerogel with low density, high compressive strength, and super-hydrophobic properties was created.
- Based on bacterial cellulose aerogel
  After freeze-drying, CVD silylation was performed to create multifunctional aerogels with high porosity, low heat conductivity, and high hydrophobicity [17].
- Da-β-CD/CMCS aerogel CVD was utilized to manufacture hydrophobic aerogels with good absorption capabilities for oils [18].

6.4.5 GROWTH MECHANISM OF CARBON NANOTUBES

This can be summed up using the "dissolution diffusion precipitation" model. After hydrocarbons are broken down, carbon first comes into contact [19].

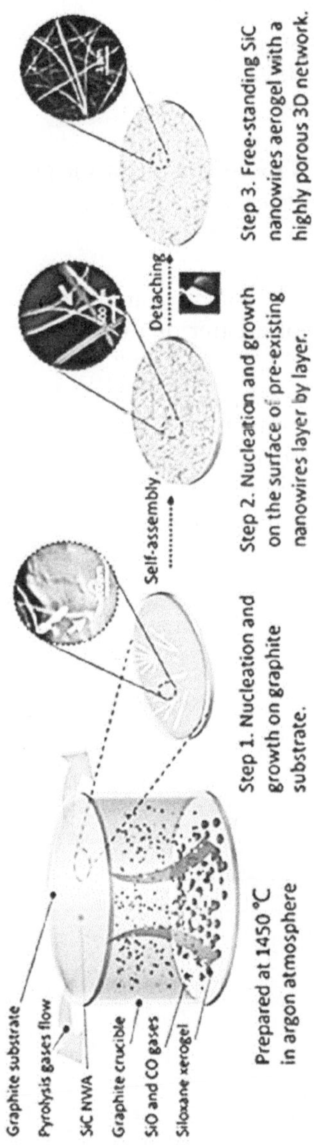

Figure 6.20 Graphical abstract of formation of carbon black aerogel.

- Growth mode
  The growth mode is based on the size of the catalyst item. Even for smaller particles, "base-growth", and for larger particles, "tip-growth" are the growth modes.
- CNT length
  CNT development stops when the catalytic nanoparticles are fully enclosed or poisoned. CNTs can reach lengths of up to 10 μm at their longest.
- CNT diameter
  The size of a SWCNT's seeding catalyst particle is determined by its diameter. The larger the SWCNT diameter, the larger the catalyst particle size.
- Growth models
  With the use of growth models, you can manage the growth conditions and produce CNTs of greater quality while also learning more about the growth process.

The metal has an acute contact angle with the substrate when the catalyst-substrate interaction is moderate, as Figure 6.21a illustrates. On the other hand, the metal has an obtuse contact angle with the substrate when the interaction is high, as seen in Figure 6.21b.

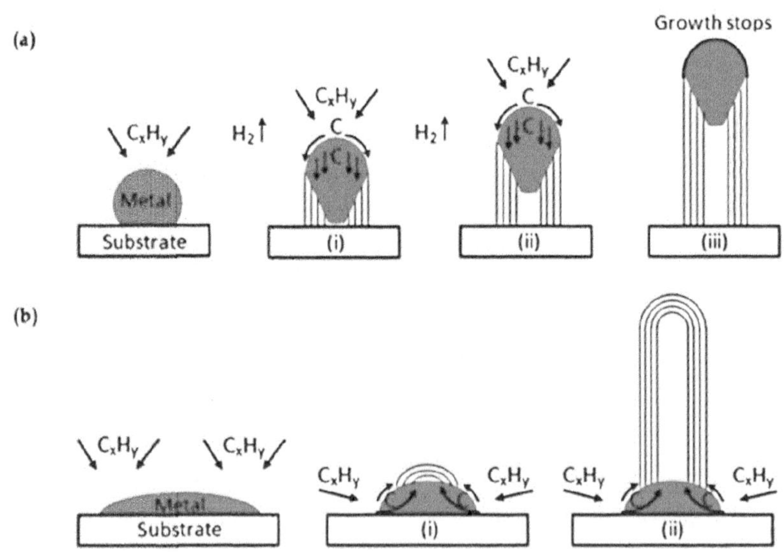

Figure 6.21 Formation of CNTs: (a) tip-growth model; (b) base-growth model.

6.5 FUNCTIONALIZATION OF CNT

6.5.1 Introduction

This technique can be carried out using covalent or non-covalent methods [20]. The functionalization procedure improves the CNTs' solubility and dispersion in solvents and polymer [21], as well as their interaction with matrices. The many modes of CNT functionalization are depicted in Figure 6.22.

6.5.2 Types of Functionalization of CNT

6.5.2.1 Endohedral Functionalization

Endohedral functionalization is the process of endowing endohedral metallofullerenes (EMFs) with new characteristics and converting them into useful materials. EMFs are fullerene cages containing metallic species (Figure 6.23).

- Applications
  Applications for endohedral functionalization include compound sensing, chemical reaction catalysis, and chemical reaction mediation.
- Structural features
  Knowledge of the structural properties of EMFs and their derivatives can help in comprehending their reactivity and regioselectivity.

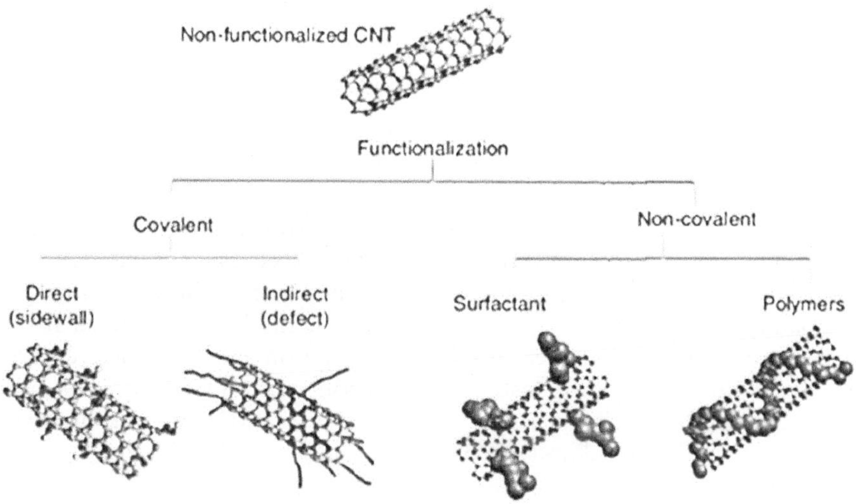

Figure 6.22 Types of functionalization of CNT.

- Classification

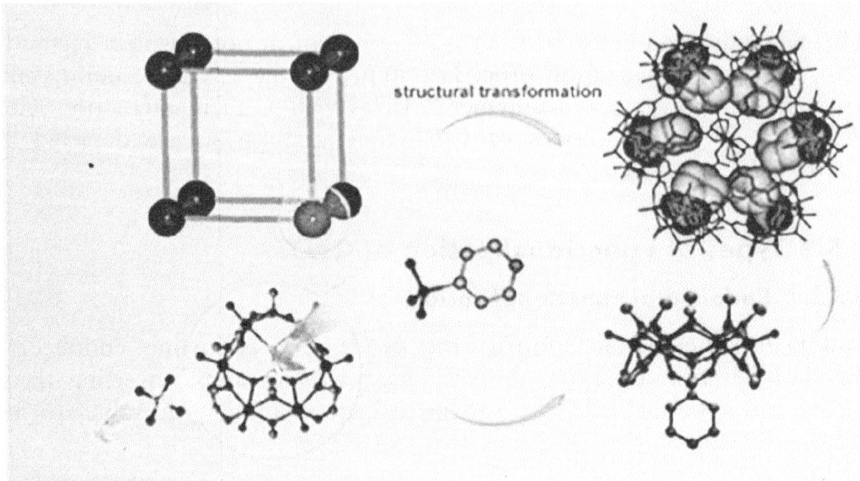

Figure 6.23 Endohedral functionalization.

Eleven classes can be distinguished amongst electromagnetic fields (EMFs) based on the composition of their intrinsic metallic species.

- Recent developments
  Two recent developments in EMF chemistry are new addition processes and advances in regioselective multi-additions.

6.5.2.2 Exohedral Chemical Functionalization

In contrast to earlier studies that primarily examined the optical, magnetic, geometric, and electrical structures of EMFs, there has been a notable increase in the functionalization of these materials in recent years. There are a lot of other unrelated processes that are also involved, like the coordination reaction, hydroxylation, dimerization [22], Diels–Alder reaction, Bingel–Hirsch reaction, carbene addition, benzyne addition, 1,3-dipolar cycloaddition, and free radical reaction. Thus, the basis of chemical oxidative purification is the idea of selective oxidation etching, where carbonaceous pollutants oxidize faster than carbon nanotubes [23]. The most common method of chemical purification involves oxidizing as-synthesized CNTs in both the liquid and gas phases.

6.5.3 Purification and Oxidization

6.5.3.1 Liquid Phase Oxidation

While there are many advantages to gas phase oxidation, one drawback is that metal particles cannot be removed. This limitation led to the development of liquid phase purification. It reliably gets rid of the amorphous carbon and metal catalyst simultaneously [24]. The oxidants that are used for oxidation are HNO_3, H_2O_2, and $KMnO_4$, or a combination of H_2O_2 and HCl, or a combination of H_2SO_4 and HNO_3.

6.5.3.2 Gas Phase Purification

The gas phase oxidative purification method purifies CNTs by oxidizing carbonaceous contaminants in an oxidizing atmosphere at temperatures between 225 and 760°C [25]. However, following the previously mentioned purification, only a tiny fraction of pure MWCNTs (1–2 weight percent) were left, which can be attributed to two factors. One is the limited capacity of carbon nanotubes to oxidize carbonaceous contaminants in a selective manner; the other is the uneven air exposure of CNTs throughout the oxidation process [26].

6.5.3.3 Electrochemical Oxidation

It follows logically that less defective carbon nanotubes (CNTs) should be more resistant to electrochemical oxidation than more defective carbon. Some researchers nvestigated the electrochemical cyclic voltammetric (CV) oxidation behavior of a sample of SWCNTs generated from an arc discharge in KOH solution [27]. After the amorphous carbon was removed, metal nanoparticles were visible, which made it easier to remove the metal contaminants with a second HCl wash.

6.5.4 COVALENT FUNCTIONALIZED CNTS

By using covalent bonds to affix chemical moieties to their tubular structure, carbon nanotubes (CNTs) can be functionalized. This process can:

- Improve dispersibility: Covalent functionalization can improve the dispersion and processability of CNTs.
- Improve reactivity: Covalent functionalization is a way to boost CNT reactivity.
- Improve adhesion: CNTs' ability to stick together can be improved via covalent functionalization.

- Improve stability: There is a chance that covalent functionalization will improve the stability of CNT dispersion.
- Fine-tune properties: Covalent functionalization can improve the physicochemical properties of CNTs for certain applications.
- Enable interactions with biological macromolecules: The functional groups on covalently functionalized carbon nanotubes (CNTs) can interact with biological macromolecules like proteins and polymers.

6.5.5 Non-Covalent Functionalized CNTs

Non-covalent functionalization is the method of enhancing the characteristics of CNTs by including polymers, aromatic compounds, biomacromolecules, and surfactants. Figure 6.24 shows a representation of the planned connective tissue, which is based on the non-covalent functionalization of the CNTs. Numerous interactions can be used to carry out this procedure [28], such as:

- π-π interactions: For these interactions, aromatic chemicals or polymers can be used.
- Electrostatic interactions: These interactions can be used to functionalize CNTs.
- *CH*-π interactions: Through these interactions, CNTs and biomolecules can functionalize one another.
- Improving biocompatibility: Non-covalent functionalization can improve the biocompatibility of carbon nanotubes (CNTs), making them more appropriate for DNA assays, bacterial detection, and enzyme biosensors.

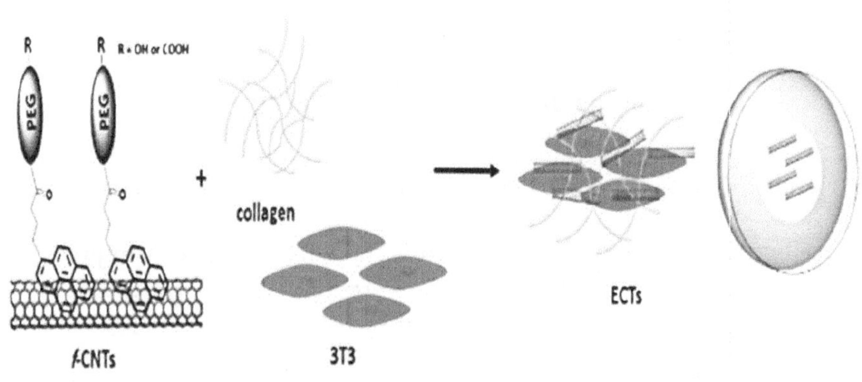

Figure 6.24 Non-covalent functionalization of the CNTs.

6.6 CARBON NANOTUBES APPLICATIONS

6.6.1 Introduction

Single-wall CNTs feature outstanding material properties like toughness, and extreme thermal and conductivity, which are a result of the unique nature of carbon and the molecular perfection of CNTs (Figure 6.25). With a strength comparable to that of a carbon–carbon bond [29], its extensive network of linkages across the periodic table makes it the only element possessing this unique property. Furthermore, the intrinsic heat conductivity of the high-frequency carbon-carbon bond vibrations is greater than that of diamond.

6.6.2 Nanocomposites

6.6.2.1 Polymer Matrix Composites

Numerous methods have been reported for the synthesis of polymer nanocomposites, including those containing layered materials and carbon nanotubes [30]. The following are the most noteworthy: i) in-situ polymerization; ii) template synthesis; iii) sol-gel technology; iv) direct mixture of polymer and particulates (Figure 6.26). Other papers describe various ways to include nanodispersoids into conducting polymers; the most well-known method is probably adding inorganic building blocks to organic polymers.

Figure 6.25 Basic structure of CNT.

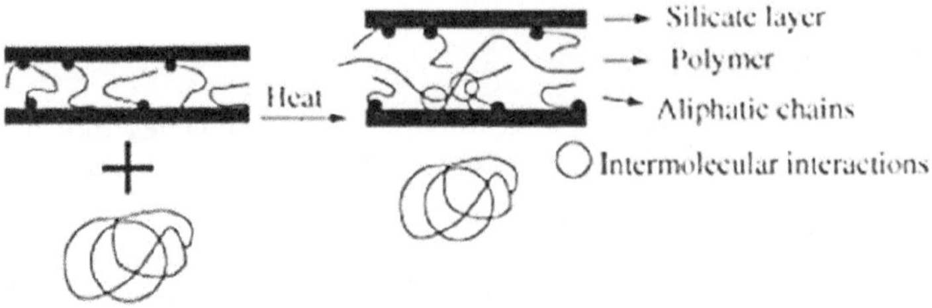

Figure 6.26 Melt intercalation synthesis of polymer/clay nanocomposites.

Table 6.5 Metal based nanocomposite systems [32]

Method	*Component*	*Process*
Spray Pyrolysis	Fe/MgO, W/Cu	To obtain the liquid source, dissolve the inorganic precursors in an appropriate solvent.
Liquid Infiltration	Pb/Cu, Pb/Fe, W/Cu, Nb/Cu	Combining the matrix metal materials with the reinforcing particles
Rapid Solidafication Process (RSP)	Al/Pb, Al/X/Zr	The metal components melt together. Maintaining the melt above the miscibility's critical line
RSP with ultrasonics	Al/Mo, Cu/W	Utilizing chemical processes to create material vapors and then consolidating them is known as sputtering or evaporating various components to create a vapor-phase CVD.

6.6.2.2 Metal Matrix Composites

Table 6.5 provides a list of systems created using these techniques. For instance, there are just two studies on Fe-based nanocomposites made using solidification processes [31]. The metallic glass made from a Fe-based alloy was quenched, and the glass precursor was devitrified by heat treatment above its crystallization temperature. The ultimate result of this was a material with a crystalline multi-phase microstructure.

6.6.3 Heat Sink Materials

Fins up to 300 μm in height and an aspect ratio of 30 are generated. The thermal characterization of the heat sink is achieved by means of an embedded microheater that functions as a temperature sensor. To do this, the low doped silicon bulk is employed as an electrical resistor [33, 34].

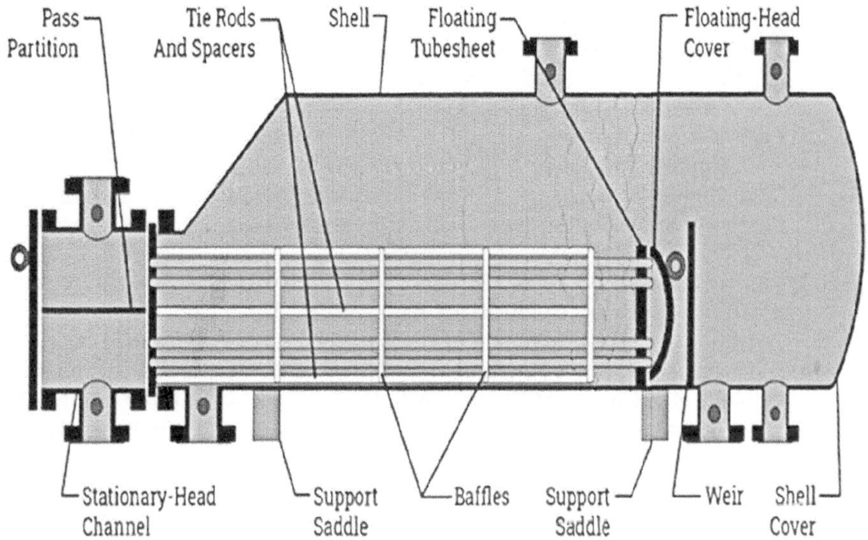

Figure 6.27 Front and bottom of the heat exchanger device.

The 0.6 Ω/K sensitivity of the sensor is shown. The CNT-based heat-sink's heat dissipation is evaluated using thermal characterization [35]. The results show that power usage can be reduced by up to 18% using the recommended CNT-based heat-sink configuration. A conceptual design for a heat-sink is shown in Figure 6.27.

6.7 CONCLUSION

This chapter highlights the special electrical, mechanical, and thermal properties of CNTs and discusses their production, properties, functionalization, characterization, and wide range of applications. There is discussion of several synthesis methods that affect the structure and quality of CNTs, such as chemical vapor deposition, laser ablation, and arc discharge. Because of their exceptional conductivity, strength, and flexibility, CNTs are used in a variety of industries, including electronics, energy storage, and biomedicine. However, issues with cost-effective production, purity, and scalability continue to be important research topics.

REFERENCES

[1] P.K. Sonkar, V. Ganesan, S.A. John, D.K. Yadav, R. Gupta, Non-enzymatic electrochemical sensing platform based on metal complex immobilized carbon nanotubes for glucose determination, *RSC Adv.* 6 (2016) 107094e107103.

[2] P.K. Sonkar, K. Prakash, M. Yadav, V. Ganesan, M. Sankar, R. Gupta, D.K. Yadav, Co(II)-porphyrindecorated carbon nanotubes as catalysts for oxygen reduction reactions: An approach for fuel cell improvement, *J. Mater. Chem. A* 5 (2017) 6263e6276.

[3] N. Yadav, M. Tyagi, S. Wadhwa, A. Mathur, J. Narang, Few biomedical applications of carbon nanotubes, *Methods Enzymol.* 630 (2020) 347..

[4] A. Tonkikh, V. Tsebro, E. Obraztsova, K. Suenaga, H. Kataura, A. Nasibulin, E. Kauppinen, E. Obraztsova, Metallization of single-wall carbon nanotube thin films induced by gas phase iodination, *Carbon* 94 (2015) 768e774.

[5] K. Ke˛dzierski, K. Rytel, B. Barszcz, A. Gronostaj, Ł. Majchrzycki, D. Wro´bel, Unusual conductivity temperature dependence of multiwalled carbon nanotube thin film, *Chem. Phys. Lett.* 712 (2018) 144e148.

[6] S. Sweeney, S. Hu, P. Ruenraroengsak, S. Chen, A. Gow, S. Schwander, J.J. Zhang, K.F. Chung, M.P. Ryan, A.E. Porter, Carboxylation of multiwalled carbon nanotubes reduces their toxicity in primary human alveolar macrophages, *Environ. Sci. Nano* 3 (2016) 1340e1350.

[7] P.K. Sonkar, V. Ganesan, D.K. Yadav, R. Gupta, Dual electrocatalytic behavior of oxovanadium(IV) salen immobilized carbon materials towards cysteine oxidation and cystine reduction: Graphene versus single walled carbon nanotubes, *ChemistrySelect* 1 (2016) 6726e6734.

[8] Y. Murakami, S. Chiashi, Y. Miyauchi, M.H. Hu, M. Ogura, T. Okubo, S. Maruyama, *Chem. Phys. Lett.* 385 (2004) 298.

[9] S. Maruyama, R. Kojima, Y. Miyauchi, S. Chiashi, M. Kohno, *Chem. Phys. Lett.* 360 (2002) 229.

[10] K. Hata, D.N. Futaba, K. Mizuno, T. Namai, M. Yumura, S. Iijima, *Science* 306 (2004) 1362.

[11] G.Y. Zhang, D. Mann, L. Zhang, A. Javey, Y.M. Li, E. Yenilmez, Q. Wang, J.P. McVittie, Y. Nishi, J. Gibbons, H.J. Dai, *Proc. Natl. Acad. Sci. U.S.A.* 102 (2005) 16141.

[12] G.F. Zhong, T. Iwasaki, K. Honda, Y. Furukawa, I. Ohdomari, H. Kawarada, *Jpn. J. Appl. Phys.* 44 (2005) 1558.

[13] G. Eres, A.A. Kinkhabwala, H. Cui, D.B. Geohegan, A.A. Puretzky, D.H. Lowndes, *J. Phys. Chem. B* 109 (2005) 16684.

[14] A. Jorio, M.A. Pimenta, A.G.S. Filho, Carbon nanotube photophysics, *J. Appl. Phys.* 101, no. 8 (Apr. 2007) 081721. doi: 10.1063/1.2719308

[15] M.S. Dresselhaus, G. Dresselhaus, P. Avouris, *Carbon Nanotubes: Synthesis, Structure, Properties, and Applications*. Berlin, Germany: Springer, 2001.

[16] Y. Murakami, Y. Miyauchi, S. Chiashi, S. Maruyama, *Chem. Phys. Lett.* 377, (2003) 49.

[17] S. Iijima, Helical microtubules of graphitic carbon, *Nature* 354, no. 6348 (Nov. 1991)56–58. doi: 10.1038/354056a0

[18] S. Noda, H. Sugime, T. Osawa, Y. Tsuji, S. Chiashi, Y. Murakami, S. Maruyama, *Carbon* 44 (2006) 1414.

[19] A. Jorio, G. Dresselhaus, M.S. Dresselhaus, *Carbon Nanotubes: Advanced Topics in the Synthesis, Structure, Properties, and Applications*. Berlin, Germany: Springer, 2008.

[20] M.H. Hu, Y. Murakami, M. Ogura, S. Maruyama, T. Okubo, *J. Catal.* 225 (2004) 230.

[21] H. Dai, Carbon nanotubes: synthesis, integration, and properties, *Acc. Chem. Res.* 35, no. 12 (Dec. 2002) 1035–1044. doi: 10.1021/ar0101641

[22] S.J. Tans, M.H. Devoret, H. Dai, C. Dekker, Individual single-wall carbon nanotubes as quantum wires, *Nature* 386, no. 6624 (Apr. 1997) 474–477. doi: 10.1038/386474a0

[23] K. Hernadi, A. Siska, L. Thien-Nga, L. Forro, I. Kiricsi, *Solid State Ionics* (2001) 141.

[24] M. Kumar, Y. Ando, Controlling the diameter distribution of carbon nanotubes grown from camphor on a zeolite support. *Carbon*, 43(3) (2005) 533–540..

[25] R. Saito, G. Dresselhaus, M.S. Dresselhaus, *Physical Properties of Carbon Nanotubes*. London, U.K.: Imperial College Press, 1998.

[26] H. Hu, B. Zhao, M.E. Itkis, R.C.J. Haddon, *Phys. Chem. B.* 107(50) (2003) 13838.

[27] F. Ikazaki, S. Ohshima, K. Uchida, Y. Kuriki, H. Hayakawa, M. Yumura, et al. *Carbon* 32 (8) (1994) 1539.

[28] P.M. Ajayan, O.Z. Zhou, Applications of carbon nanotubes, *Topics in Appl. Phys.* 80 (2001) 391–425. doi: 10.1007/3-540-39947-X_14

[29] V. N. Popov, Carbon nanotubes: Properties and application, *Mater. Sci. Eng. R: Rep.* 43, no. 3, (Dec. 2004) 61–102. doi: 10.1016/j.mser.2003.10.001

[30] P.J.F. Harris, *Carbon Nanotube Science: Synthesis, Properties and Applications*. Cambridge, U.K.: Cambridge University Press, 2009.

[31] C. Dekker, Carbon nanotubes as molecular quantum wires, *Physics Today* 52, no. 5 (May 1999) 22–28. doi: 10.1063/1.882658

[32] A.K. Geim, K.S. Novoselov, The rise of graphene, *Nat. Mater.* 6, no. 3 (Mar. 2007) 183–191. doi: 10.1038/nmat1849

[33] D.B. Tuckerman, R.F.W. Pease, "High-performance heat sinking for VLSI", *IEEE Electron. Dev. Lett.* (1981).

[34] Y. Fu, , T. Wang, J. Liu, Carbon nanotubes as cooling fins in microelectronic systems, *IEEE-NANO* (2009).

[35] X. Wang, H. Liu, J. Wang, W. Zhang, Z. Li, Package heat dissipation with integrated carbon nanotube micro heat sink, *International Conference on Electronic Packaging Technology & High Density Packaging*, IEEE (2009).

Chapter 7

Organic Field-Effect Transistor

Introduction, Modeling, and Process

Mahesh Kumar Singh and Shriti Suman

7.1 INTRODUCTION

Organic field-effect transistors (OFETs) are a captivating and rapidly advancing area of research within organic electronics. These devices, which use organic semiconductors as their active materials, have attracted considerable interest from both industry and academia due to their distinctive properties and potential applications. OFETs merge the fundamental operating principles of conventional field-effect transistors with the flexibility and processability of organic materials, paving the way for innovative, flexible, lightweight, and potentially low-cost electronic devices.

The development of OFETs began in the late 1980s, building on earlier research in organic semiconductors and the discovery of conducting polymers. Since then, significant strides have been made in enhancing the performance, stability, and overall understanding of these devices [1]. These days, OFETs are being intensively investigated for several applications [2], such as flexible displays, chemical and biological sensors, RFID tags, and large-area electronics [3]. They are no longer an experimental curiosity.

At the core of OFETs are several essential components: an organic semiconductor layer serving as the active channel material; the electrodes of the drain and source that collect and inject charge carriers; a gate electrode that modulates the channel's conductivity; and a dielectric layer that insulates the gate from the channel. The distinct properties of organic semiconductors—such as their capacity for solution processing, mechanical flexibility, and tunability via chemical synthesis—differentiate OFETs from their inorganic counterparts.

This comprehensive exploration of OFETs will cover their fundamental principles, the materials involved in their construction, fabrication techniques, characterization methods, and both current and potential applications. Additionally, we will address the challenges facing OFET technology and the ongoing research aimed at overcoming these obstacles.

 DOI: 10.1201/9781003533535-7

7.2 FUNDAMENTAL PRINCIPLES OF OFETs

To understand the operation of OFETs, it's essential to first grasp the basics of organic semiconductor materials and how these materials differ from traditional inorganic semiconductors like silicon.

Organic semiconductors: Materials based on carbon with semiconducting qualities are known as organic semiconductors. Organic semiconductors are bound by lower van der Waals forces than inorganic semiconductors, which have covalent crystal structures. This leads to distinct electronic properties and unique charge transport mechanisms.

The two main classes of organic semiconductor materials used in OFETs are:

1. **Small molecules:** These are low molecular weight compounds, often based on aromatic systems like pentacene, rubrene, or oligothiophenes.
   a. **Pentacene:**
   Pentacene is an organic compound devised of five benzene rings integrated together in a linear arrangement, placing it within the polycyclic aromatic hydrocarbon (PAH) family (as shown in Figure 7.1). Its chemical formula is $C_{22}H_{14}$.
   b. **Rubrene:**
   Rubrene (shown in Figure 7.2) is an organic compound recognized for its outstanding semiconducting properties, especially in organic electronic applications. With the chemical formula $C_{42}H_{28}$, rubrene is a member of the polycyclic aromatic hydrocarbons (PAHs) but features a more complex structure compared to simpler PAHs like pentacene.
   c. **Oligothiophenes:**
   Oligothiophenes (shown in Figure 7.3) are a class of organic compounds composed of a small number (oligo-) of thiophene rings. Thiophene is a five-membered heterocyclic compound with the formula C_4H_4S, in which sulfur replaces one of the carbon atoms in the ring. Oligothiophenes are extensively researched in the field of organic electronics because of their semiconducting properties.

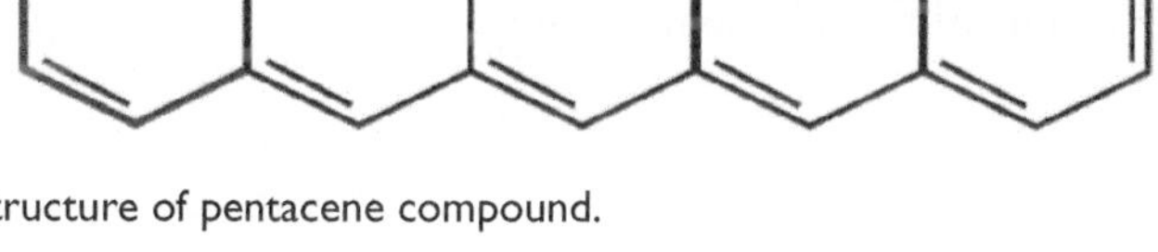

Figure 7.1 Structure of pentacene compound.

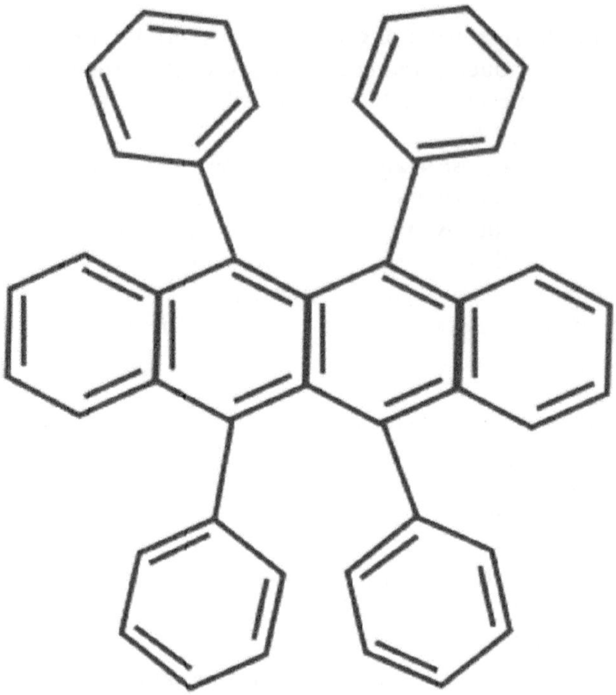

Figure 7.2 Structure of rubrene compound.

2. **Polymers:** These are long-chain molecules composed of repeating units, such as polythiophenes or polyfluorenes.
   a. **Polythiophenes**
   Polythiophenes (shown in Figure 7.4) are polymers made up of repeating thiophene units. Thiophene is a five-membered aromatic ring surrounding four carbon atoms and one sulfur atom. Polythiophenes are important in organic electronics because of their conductive and optical properties.
   b. **Polyfluorene**
   Polyfluorene (shown in Figure 7.5) is a conjugated polymer commonly used in organic electronics, especially in light-emitting diodes (LEDs), due to its outstanding electronic and optical properties. The polymer is composed of repeating fluorene units, which are aromatic compounds formed by two benzene rings integrated to a central five-elemented ring [4].

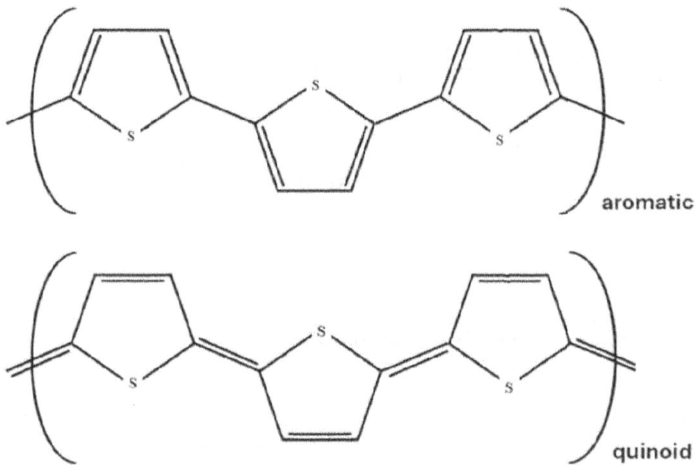

Figure 7.3 Structure of oligothiophenes compound: (a) aromatic, (b) quinoid

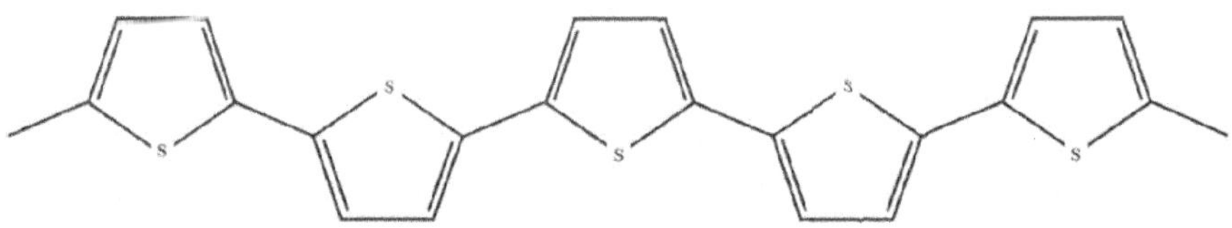

Figure 7.4 Structure of polythiophenes.

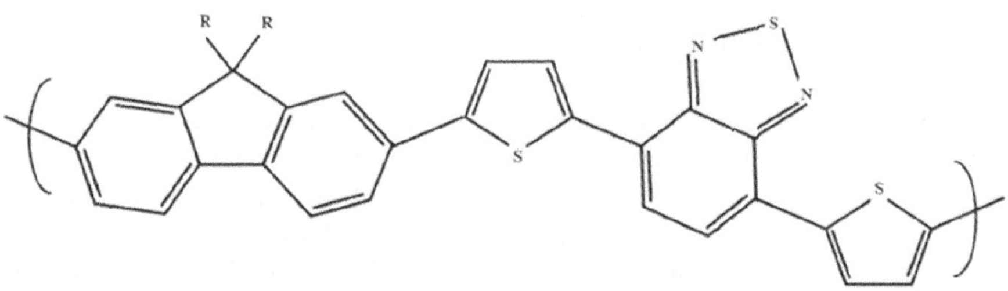

Figure 7.5 Structure of polyfluorene.

Both classes can form ordered structures through various intermolecular interactions, which is crucial for charge transport.

Charge transport in the organic semiconductors: In organic semiconductors, the movement of charge mainly happens through a hopping mechanism between localized states, as opposed to the band-like transport characteristic of crystalline inorganic semiconductors. This hopping process is typically slower and more susceptible to disorder, which helps account for

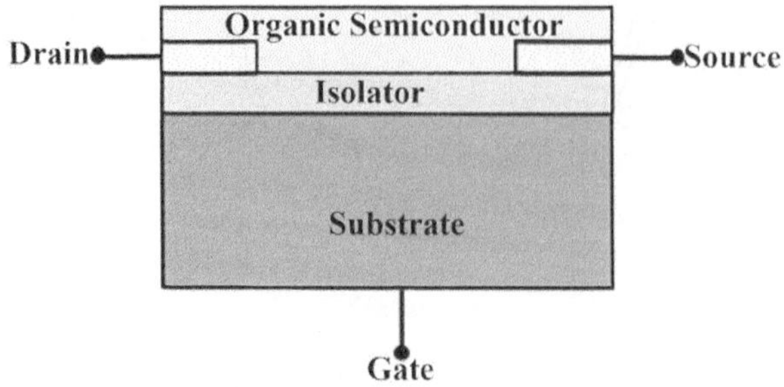

Figure 7.6 Structure of organic field effect transistor.

the lower mobilities of charge carriers observed in organic semiconductors when compared to inorganic ones.

An OFET typically comprises the following layers (as shown in Figure 7.6):

1. **Substrate:** Often made of a rigid material like silicon or glass, though flexible materials are sometimes used for specific applications.
2. **Gate Electrode:** A conductive layer responsible for controlling the electric field within the device.
3. **Dielectric Layer:** An insulating band that isolates the gate electrode from the semiconductor.
4. **Organic Semiconductor Layer:** The active layer where charge transport takes place.
5. **Drain and Source Electrodes:** Conductive contacts that facilitate charge injection and collection.

An electric field produced at the gate electrode modifies the conductivity of the semiconductor channel, which is how an OFET operates. An electric field that passes through the semiconductor and dielectric layers is created when a voltage is supplied to the gate. At the semiconductor-dielectric interface, this field causes charge carriers (either electrons or holes) to accumulate, forming a conductive channel [5].

A field-effect transistor (FET) regulates the current flowing between the drain and source terminals by using an electric field generated by a voltage supplied to the gate terminal (as shown in Figure 7.7). A positive gate voltage in an n-channel FET drags electrons to the channel, enhancing its conductivity and allowing current to flow from the drain to the source. The gate voltage adjusts the channel's width, thereby controlling the current. The channel is "off", meaning that no current may pass across it, when the gate voltage falls below a predetermined threshold [6]. FETs' high input impedance and low power consumption make them a popular choice for switching and amplification in electronic circuits.

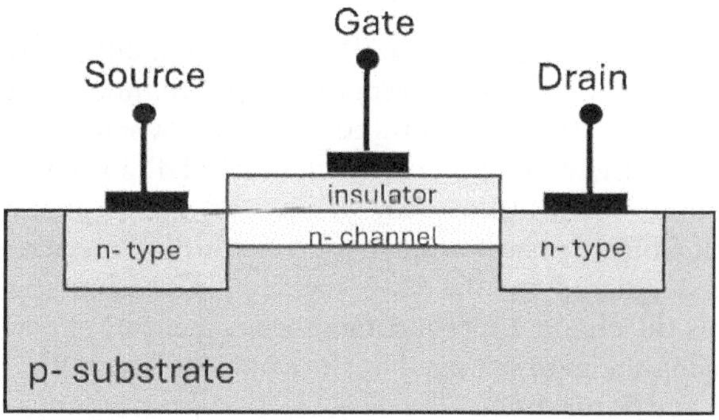

Figure 7.7 Structure of field effect transistor.

The nature of charge carriers is decided by the characteristics of the organic semiconductor:

- In n-type materials, electrons are the primary carriers.
- In p-type materials, holes are the primary carriers.
- Certain materials can function in both modes, and these are referred to as ambipolar semiconductors.
- The accumulated charge carriers can move if a voltage is placed between the drain and source electrodes, producing a current. By adjusting the gate voltage, the OFET may function as a voltage-controlled current-source, much like its inorganic counterparts, and the quantity of this current can be controlled [7].

7.3 OPERATING PRINCIPLE

1. **Gate Voltage Regulation:** The voltage exerted to the gate (V_G) generates an electric field which influences the channel's conductivity. In a JFET, the gate is reverse-biased, whereas in a MOSFET, an insulating layer (typically silicon dioxide) isolates the gate from the channel.
2. **Channel Formation:**
    - **For n-Channel FET:** Applying a positive gate voltage attracts electrons into the channel, enhancing its conductivity.
    - **For p-Channel FET:** Applying a negative gate voltage attracts holes, increasing conductivity.
3. **Current Flow:**
    - **With No Gate Voltage (V_G = 0):** In a depletion-mode FET, the channel is inherently conductive, permitting current to outflow from the source to the drain. Conversely, in an enhancement-mode

FET, the channel remains non-conductive without an applied gate voltage.
- **With Gate Voltage ($V_G \neq 0$):** For an n-channel FET, applying a positive gate voltage strengthens the channel, enabling current (ID) to flow from the drain to the source when a voltage is exerted between them (VDS). For a p-channel FET, a negative gate voltage facilitates current flow [8].

4. **Current Control:** The amount of current outflowing across the channel (ID) is regulated by the gate voltage. Increasing the gate voltage expands the channel, permitting more current to pass through, while decreasing the gate voltage has the opposite effect [9].
5. **Cutoff and Saturation:**
   - **Cutoff Region:** If the gate voltage is below a specific threshold ($V_{GS} < V_{th}$), the channel is "off," and no current flows between the drain and source.
   - **Saturation Region:** If the gate voltage is significantly above the threshold, the channel is fully open, and the current flow remains relatively constant and independent of V_{DS}.

7.4 KEY PARAMETERS OF OFETS

Several parameters are used to characterize the performance of OFETs:

1. **Charge Carrier Mobility:** This quantifies the speed at which charge carriers travel through the semiconductor in response to an exerted electric field. In an OFET, carrier mobility (μ) indicates how easily charge carriers (electrons or holes) flow across the organic semiconductor material when an electric field is applied. It is defined by the equation:

$$\mu = \frac{L^2}{V_{DS} \cdot C_i} \cdot \frac{\partial I_{DS}}{\partial V_{GS}}$$

   where L is the channel length, V_{DS} is the drain-source voltage, C_i is the gate capacitance per unit area, I_{DS} is the drain current, and $\partial I_{DS}/\partial V_{GS}$ is the slope of the transfer characteristic [10]. High mobility indicates efficient charge transport, critical for OFET performance. Higher mobility is generally desirable for faster device operation.
2. **On/Off Current Ratio:** The ratio of current between the "on" state (high gate voltage) and the "off" state (low or zero gate voltage). In an OFET, the current ratio of on/off (I_{on}/I_{off}) assesses the device's capability to transition between its conducting ("on") and non-conducting ("off") states [11]. It is defined as:

$$On - Off\ Ratio = \frac{I_{on}}{I_{off}}$$

where I_{on} represents the drain current when the OFET is in the "on" state (with a high gate voltage V_{GS}), and I_{off} denotes the drain current when the OFET is in the "off" state (with a low or zero gate voltage) [12]. A high current ratio of on/off is essential for digital applications, as it signifies minimal leakage current and effective switching in electronic devices.

3. **Threshold Voltage:** The minimal gate voltage needed to establish a conductive channel. In an OFET, the threshold voltage (Vth) is the gate voltage necessary to create a conducting channel between the source and drain electrodes [13]. Mathematically, it is often expressed as:

$$V_{th} = V_{GS} - V_{th0}$$

where V_{GS} and V_{th0} denote the gate-to-source voltage and the threshold voltage at zero gate-source voltage, respectively [14]. In an OFET, the threshold voltage (V_{th}) can also be affected by factors such as the properties of the gate insulator and the characteristics of the organic semiconductor. The threshold voltage denotes the point at which the OFET begins to conduct and is essential for the device's operation. A lower threshold voltage is often preferred for low-power applications.

4. **Subthreshold Swing:** This parameter reflects how quickly the device switches on as the gate voltage increases. In an OFET, subthreshold swing (S) gauges the efficiency of transitioning between the off and on states. It measures how rapidly the drain current (I_D) rises as the gate voltage (V_{GS}) exceeds the threshold voltage (V_{th}). Mathematically, it is given by:

$$S = \frac{dV_{GS}}{d(\log I_D)}$$

In the subthreshold region, I_D follows:

$$I \alpha e^{(V_{GS} - V_{th})/nV_{th}}$$

where n is the subthreshold swing coefficient and V_{th} is the thermal voltage. A lower subthreshold swing (S) signifies a steeper transition and improved performance. A smaller subthreshold swing facilitates more efficient switching.

5. **Contact resistance:** The resistance at the junction between the electrodes and the organic semiconductor affects the device's overall performance, especially in low-voltage applications. Mathematically, the total resistance R_{total} of an OFET is the summation of the contact resistance R_c and the channel resistance R_{ch}:

$$R_{total} = R_c + R_{ch}$$

The contact resistance (R_c) is typically determined using techniques such as the transmission line model (TLM) or the transfer length method [15]. An increased R_c can restrict charge injection, thereby diminishing the efficiency of the OFET. Reducing contact resistance is crucial for enhancing device performance.

Understanding these fundamental principles and parameters is crucial for designing, fabricating, and optimizing OFETs for various applications. In the next section, we'll explore the materials commonly used in OFET construction and how their properties influence device performance.

7.5 MATERIALS USED IN OFETS

The performance and attributes of OFETs are significantly influenced by the materials used in their construction. Each element of the OFET structure—namely the organic semiconductor, dielectric layer, electrodes, and substrate—plays a vital role in shaping the device's overall performance. Let's explore each of these components in detail:

1. Organic Semiconductors

The organic semiconductor is central to the OFET, as it is responsible for charge transport. The selection of semiconductor material greatly affects important device parameters like charge carrier mobility, stability, and ease of processing. Organic semiconductors used in OFETs can generally be categorized into two main parts:

a. **Small Molecules:** Small molecule semiconductors are low molecular weight compounds, typically with a well-defined structure and high purity. Some prominent examples include:
   - Pentacene and its derivatives: Known for high hole mobility, but susceptible to oxidation.
   - Rubrene: Exhibits high hole mobility in single-crystal form.
   - Oligothiophenes: Offer good stability and tunable properties through chemical modification.
   - Fullerenes (e.g., C60): Excellent electron transport materials.

Small molecules [16] typically form polycrystalline films upon deposition, with charge transport predominantly occurring along the π-π stacking orientation of the molecules [17].

b. **Polymers:** Semiconducting polymers are elongated molecules that provide excellent solution processability and mechanical flexibility. Examples include:
    - Poly(3-hexylthiophene) (P3HT): A extensively researched p-type polymer.
    - Poly(9,9-dioctylfluorene-co-bithiophene) (F8T2): A type of light-emitting semiconductor.
    - Naphthalene diimide (NDI)-based polymers: Recognized for their strong n-type performance.

Polymers typically form semicrystalline or amorphous films, with charge transport occurring both along the polymer backbone and between adjacent chains.

Recent advancements in organic semiconductor design have focused on:

- Improving air stability, particularly for n-type materials.
- Enhancing charge carrier mobility through better molecular packing.
- Developing ambipolar materials that can have both carriers, i.e., holes and electrons.
- Creating solution-processable small molecules that combine the advantages of both material classes.

2. Dielectric Materials

The dielectric layer in an OFET acts as an insulator between the port of gate electrode and the semiconductor channel, playing a vital role in defining the electric field at the juncture between the dielectric and semiconductor [18]. Key considerations for dielectric materials include:

- Dielectric constant: A high dielectric constant allows for lower operating voltages.
- Breakdown strength: The capacity to endure high electric fields without experiencing failure. Surface properties: The dielectric surface can significantly impact the growth and ordering of the semiconductor layer.

Common dielectric materials include:

a. **Inorganic Dielectrics:**
    - Silicon dioxide (SiO_2): Widely used due to its compatibility with silicon-based fabrication.

   - Aluminum oxide (Al_2O_3): Offers a higher dielectric constant than SiO_2.
   - Hafnium oxide (HfO_2): Provides an even higher dielectric constant, allowing for low-voltage operation.

b. **Organic Dielectrics:**
   - Polymers such as poly(methyl methacrylate) (PMMA) or polyvinyl alcohol (PVA): Provide solution processability and flexibility.
   - Self-assembled monolayers (SAMs): Can alter the dielectric surface to enhance semiconductor growth and minimize charge trapping [19].

c. **Hybrid Dielectrics:** These are combinations of organic and inorganic materials designed to take advantage of the benefits offered by both, such as through multilayer structures or nanocomposites.

3. Electrode Materials

The choice of electrode materials affects charge injection/collection and overall device performance. Key factors include:

- Work function: Should align with the HOMO levels (for p-type) or LUMO levels (for n-type) of the organic semiconductor to ensure effective charge injection.
- Chemical stability: Resistance to degradation when in contact with the organic materials.
- Processability: Compatibility with the chosen fabrication methods.

Common electrode materials include:

- Gold (Au): Widely used for p-type OFETs due to its high work function.
- Silver (Ag): Often used in printed electronics due to its good conductivity and processability.
- Aluminum (Al): Suitable for n-type OFETs due to its lower work function.
- Indium tin oxide (ITO): Transparent conductor, useful for applications requiring optical transparency [20].
- Conductive polymers (e.g., PEDOT:PSS): Offer flexibility and solution processability.

4. Substrate Materials

The substrate offers mechanical support for the OFET and can affect the formation of the layers that are applied afterward. Substrate choices depend on the intended application and fabrication method:

- Silicon wafers: Common for research devices, often with a thermally grown oxide layer serving as the dielectric.
- Glass: Provides a smooth surface and optical transparency.

- Flexible plastics (e.g., polyethylene terephthalate, PET): Enable flexible and conformable devices.
- Paper: Explored for ultra-low-cost, disposable electronics.

7.6 MATERIAL INTERFACES AND INTERACTIONS

The interfaces between different materials in an OFET are crucial to device performance:

- Semiconductor-dielectric interface: This is where the conductive channel forms. The quality of this interface significantly affects charge transport.
- Electrode-semiconductor interface: Computes the efficiency of injection and extraction charges. Modifying this interface with self-assembled monolayers or dopants can improve device performance.
- Substrate-semiconductor interface: Influences the growth and morphology of the semiconductor layer, particularly for bottom-gate device architectures.

Understanding and optimizing these interfaces is an active area of research in OFET development.

7.7 EMERGING MATERIALS

Research in OFET materials continues to evolve, with new classes of materials being explored:

- 2D materials: Comprising graphene and transition metal dichalcogenides (TMDs), which provide distinctive electronic properties.
- Perovskites: Known for their excellent optoelectronic properties in solar cells, now being investigated for transistor applications.
- Biodegradable and biocompatible materials: For environmentally friendly and biomedical applications.
- Nanocomposites: Combining organic semiconductors with inorganic nanoparticles to enhance specific properties.

The choice of materials for OFETs involves balancing various factors including performance, processability, stability, and cost. As the field progresses, new materials and innovative combinations continue to push the boundaries of OFET performance and applicability.

7.8 FABRICATION TECHNIQUES FOR OFETS

The fabrication of OFETs entails depositing and patterning various material layers to form the desired device structure. The choice of fabrication method is influenced by factors such as the materials involved, the performance requirements, cost considerations, and the intended application. OFET fabrication techniques can generally be divided into two primary categories: vacuum-based methods and solution-based methods.

1. Vacuum-Based Fabrication Techniques

Vacuum-based techniques are often used for small molecule semiconductors and offer precise regulation over film thickness and purity [21].

a. **Thermal Evaporation:**
   Thermal evaporation is a crucial method in the fabrication of OFETs for applying thin films of organic semiconductors and metal electrodes. In this process, the material to be accumulated, usually an organic semiconductor or metal, is placed in a vacuum chamber and heated until it evaporates [22]. The evaporated atoms or molecules then travel through the vacuum and accumulate onto a substrate, creating a thin film.

   This technique is commonly employed because it provides explicit control over the thickness and uniformity of the deposited layer, which is vital for the performance of OFETs. The process is executed in a high vacuum to avoid contamination and ensure the production of a high-quality film [23]. Thermal evaporation is especially beneficial for depositing organic semiconductors that are sensitive to air and moisture, as the vacuum environment safeguards the material during deposition. The technique's versatility makes it a key method in OFET fabrication.

   Key takeaways:
   - **Process:** The source material is heated within a vacuum chamber, causing it to evaporate and then deposit onto the substrate.
   - Advantages: High purity films, good control over thickness.
   - Limitations: Limited to thermally stable materials, can be expensive for large-area production.

b. **Organic Molecular Beam Deposition (OMBD):**
   Organic Molecular Beam Deposition (OMBD) is a specialized technique used in the fabrication of OFETs to deposit ultra-thin, high-quality organic semiconductor layers. In OMBD, organic molecules are evaporated from heated crucibles in a high-vacuum chamber, forming a molecular beam. This beam then travels towards the substrate, where the molecules condense and form a uniform thin film.
   OMBD offers precise control over the deposition rate and film thickness, which is crucial for tailoring the electrical properties of OFETs. The high vacuum environment minimizes impurities and defects, ensuring the deposited organic layer has excellent

crystalline quality and purity. This technique is particularly effective for depositing complex organic molecules with minimal thermal degradation, preserving their functional properties.

OMBD is valued for its ability to produce highly ordered molecular films, which enhances charge transport in OFETs, leading to better device performance and stability.

Key takeaways:

- Process: Similar to thermal evaporation, but offers greater precision in controlling the deposition rate and substrate temperature.
- Advantages: Enables the formation of highly ordered films and even single crystals.
- Limitations: Characterized by slow deposition rates and the need for costly equipment.

c. **Sputtering:**

Sputtering is a technique to deposit vapor physically. The process is also known as the physical vapor deposition (PVD) technique, and it is employed in the fabrication of OFETs to apply thin films of materials such as metal electrodes, dielectric layers, or specific organic substances. In this process, a target material (like a metal or insulator) is bombarded with high-energy ions within a vacuum chamber, causing atoms to be evicted from the target and deposited onto a substrate [24].

Sputtering provides explicit control over film thickness, uniformity, and composition, making it ideal for producing highly uniform and dense layers essential for optimal OFET performance. It is especially effective for depositing metal electrodes with excellent adhesion and conductivity. Moreover, sputtering is suitable for depositing dielectric layers with high purity and controlled stoichiometry, crucial for OFET gate insulators. The process is compatible with various substrates, including flexible materials, and can be scaled up for large-area deposition, making it a versatile tool in OFET fabrication.

Key takeaways:

- Process: Atoms are removed from a target material by being bombarded with high-energy particles and then deposited onto the substrate.
- Advantages: Good for depositing metal electrodes and some dielectric materials.
- Limitations: Can potentially damage organic layers, not suitable for most organic semiconductors.

2. Solution-Based Fabrication Techniques

Solution-based methods are particularly attractive for organic electronics due to their potential for low-cost implementation and its large-area fabrication.

a. **Spin Coating:**

Spin coating is a commonly used solution-based fabrication method in OFET production, ideal for applying uniform thin films of organic semiconductors, dielectrics, and other functional layers. During this process, a small quantity of solution containing the desired material is placed at the center of a substrate, which is then rapidly spun at high speeds. The centrifugal force evenly distributes the solution across the surface, creating a thin film.

The film thickness can be regulated by modifying the spin speed, duration, and solution concentration. Spin coating is especially beneficial for producing highly uniform and smooth films, which are essential for efficient OFET performance. This method is straightforward and reproducible, making it ideal for both research and small-scale production.

Spin coating is compatible with a broad spectrum of organic materials and plays a crucial role in forming the active semiconductor layers in OFETs, ensuring uniform electrical properties throughout the device.

Key takeaways:

- Process: A material solution is applied to a substrate, which is then rapidly rotated to spread the solution and evaporate the solvent.
- Advantages: Simple, reproducible, good for uniform thin films.
- Limitations: Material waste, difficulty with patterning, mainly suitable for small substrates.

b. **Drop Casting:**

Drop casting is a straightforward solution-based fabrication method used in OFET manufacturing to accumulate thin films of organic semiconductors and other functional materials. In this process, a solution containing the desired material is gently dropped onto a substrate, where the solvent evaporates, separating behind a thin film of the material.

This technique is simple and doesn't require complex equipment, making it suitable for lab-scale experiments. However, the resulting film's thickness and uniformity are less controllable compared to other methods like spin coating. The morphology of the film in drop casting is heavily influenced by factors such as the solvent evaporation rate, solution concentration, and substrate properties [25].

Despite its simplicity, drop casting can yield high-quality films with good crystallinity, especially for organic semiconductors. This makes it valuable for studying the fundamental attributes of materials in OFETs or for creating devices where film uniformity is less crucial.

Key takeaways:

- Process: A solution of the material is simply dropped onto the substrate and allowed to dry.
- Advantages: Very simple, can produce thicker films.
- Limitations: Less control over film uniformity, slow drying can lead to non-uniform crystallization.

c. **Dip Coating:**

Dip coating is a flexible solution-based method used in the fabrication of OFETs to accumulate thin films of organic semiconductors, dielectrics, and other materials [26]. In this process, a substrate is immersed in a solution containing the desired material. As the substrate is slowly withdrawn, a thin, uniform film forms as the solution drains off.

The thickness of the deposited film can be regulated by adjusting the dipping speed, solution concentration, and drying conditions. Dip coating is advantageous for creating smooth, uniform films over large surfaces and is especially effective for coating flexible or irregularly shaped substrates. Additionally, it is a relatively simple and cost-effective technique.

In OFET fabrication, dip coating ensures good coverage and consistency, which are crucial for achieving uniform electrical characteristics across the device. The technique is suitable for both lab-scale experiments and large-scale production.

Key takeaways:

- Process: The substrate is immersed in a solution of the material and then slowly withdrawn at a controlled rate [27].
- Advantages: Can produce very uniform films, good for large area coating.
- Limitations: Requires careful control of withdrawal speed and solution properties.

d. **Inkjet Printing:**

Inkjet printing is a versatile solution-based method used in OFET fabrication to deposit precise patterns of organic semiconductors, dielectrics, and conductors [28]. In this process, an inkjet printer ejects droplets of a solution containing the desired material onto a substrate. The printer's nozzle regulates the size and positioning of the droplets, enabling high-resolution patterning.

Inkjet printing provides many advantages, including its ability to deposit materials selectively, reducing waste and allowing the formation of complex device structures. It is compatible with various substrates, including flexible and large-area ones, and can be used for both research and commercial production. The technique allows explicit control over film thickness and pattern resolution, which is essential for optimizing OFET performance. However, achieving consistent results requires careful management of ink properties and printing conditions.

Key takeaways:

- Process: Droplets of material solution are precisely deposited on the substrate using a printhead.
- Advantages: Allows for direct patterning, minimal material waste, scalable to large areas.
- Limitations: Resolution limited by droplet size, potential for non-uniform drying effects.

e. **Screen Printing:**

Screen printing is an additive manufacturing method used in OFET fabrication to deposit various functional layers, including organic semiconductors, dielectrics, and conductive electrodes. In this process, a patterned stencil (or screen) is positioned over a substrate, and ink containing the desired material is spread across the screen with a squeegee. The ink flows through the open areas of the stencil, transferring the pattern onto the substrate.

Screen printing offers several benefits for OFET fabrication, including simplicity, low cost, and scalability, making it ideal for large-area electronics and flexible devices. The technique enables the deposition of thick, uniform layers with precise control over pattern geometry. It is adaptable with a wide range of materials, including conductive polymers and organic semiconductors [29]. However, the resolution is generally lower than other methods, which can limit the minimum feature size. Despite this, it remains a popular choice for cost-effective, high-throughput production of OFETs.

Key takeaways:

- Process: Ink is forced through a patterned mesh screen onto the substrate.
- Advantages: Simple, fast, suitable for thick films and large areas.
- Limitations: Lower resolution compared to other techniques, limited to certain ink formulations.

f. **Gravure Printing:**

Gravure printing is a high-speed roll-to-roll method used in OFET fabrication to deposit thin, precise layers of materials such as organic semiconductors, dielectrics, and conductive inks. In this process, an engraved cylinder, known as the gravure cylinder, transfers ink from an ink reservoir onto a substrate. The cylinder's surface features recessed cells that hold the ink, and as the cylinder rotates, the ink is transferred to the substrate, creating a continuous pattern.

Gravure printing is ideal for manufacturing large-area, flexible electronic devices because it can deposit thin, uniform layers at high throughput. The process provides excellent regulation over the thickness and uniformity of the printed layers, which is crucial for OFET performance. Its high resolution enables the creation of fine features, making it an efficient method for mass production. However, the initial setup costs can be high due to the requirement for engraved cylinders.

Key takeaways:

- Process: Ink is transferred from engraved cells on a printing cylinder to the substrate.
- Advantages: High-speed, large-area production, good control over film thickness.
- Limitations: High initial cost for cylinder production, best suited for large production runs.

g. **Blade Coating/Doctor Blading:**

Blade coating, or doctor blading, is a smooth and versatile technique used in OFET fabrication to accumulate thin films of organic semiconductors, dielectrics, and other functional materials. In this process, a solution containing the desired material is applied to a substrate. A blade, usually made of metal or a similar rigid material, is then drawn across the substrate at a controlled height, spreading the solution into a uniform thin film.

The thickness of the film can be precisely regulated by adjusting the gap between the blade and the substrate, as well as the speed at which the blade moves. Blade coating is preferred for its simplicity, scalability, and compatibility with a broad range of materials, including those used in OFETs. It is especially useful for fabricating large-area devices and can be applied in both laboratory and industrial settings. This method also enables the creation of films with controlled morphology, which is crucial for optimizing OFET performance.

Key takeaways:

- Process: A sharp blade is used to spread a solution of material across the substrate.
- Advantages: Simple, scalable to large areas, good for creating uniform films.

- Limitations: Patterning requires additional steps, sensitive to blade geometry and coating speed.

3. Patterning Techniques

Patterning techniques are essential in the fabrication of OFETs because they determine the geometry of the device's active layers, including the organic semiconductor, dielectric, and electrodes. These methods provide precise control over the layout, which impacts the device's performance and reliability.

Photolithography is a widely used patterning technique where a photoresist is employed to a substrate, unveiled to light through a mask, and then evolved to generate the desired pattern. Other methods include inkjet printing, which directly prints patterns onto the substrate, and soft lithography, which employs elastomeric stamps to transfer patterns onto the substrate.

Advanced methods like nanoimprint lithography and laser ablation offer a high resolution, suitable for creating intricate features required in miniaturized OFETs. Each technique balances resolution, throughput, and material compatibility, allowing for tailored approaches depending on the specific OFET design and application. These patterning methods are essential for integrating OFETs into circuits and optimizing their electrical characteristics.

a. **Photolithography:**

   Photolithography is a commonly employed patterning technique in the fabrication of OFETs, used to define complex patterns of organic semiconductors, dielectrics, and metal electrodes. In this process, a photosensitive material known as the photoresist is employed to the substrate and exposed to ultraviolet (UV) light through a patterned mask. The UV light induces chemical changes in the photoresist, which are subsequently developed to reveal the desired pattern.

   The process consists of several steps: coating the substrate with photoresist, aligning and exposing the photoresist through a mask, developing the exposed photoresist, and then etching or depositing materials according to the remaining resist pattern. Photolithography offers a high resolution and precision, allowing for the creation of fine features and complex device geometries that are crucial for high-performance OFETs. While it is very effective for rigid substrates, adaptations are required for flexible or irregular surfaces. The technique is appreciated for its scalability and repeatability in device fabrication.

Key takeaways:

- Process: A photoresist is exposed to light through a mask and then developed to form patterns.
- Advantages: High resolution, well-established in the semiconductor industry.
- Limitations: Multiple steps involved, potential for chemical incompatibility with organic materials.

b. **Shadow Masking:**

Shadow masking is a patterning technique used in OFET fabrication to create precise patterns of organic semiconductors, dielectrics, and metal electrodes. In this method, a thin, patterned mask is positioned directly over a substrate, which is then exposed to deposition processes such as thermal evaporation or sputtering. The mask features openings that delineate the required pattern, permitting material deposition only in selected areas while protecting the remaining parts of the substrate.

The key advantage of shadow masking is its simplicity and ability to produce well-defined patterns without the need for complex photoresist processing. It is particularly useful for depositing materials onto small or irregularly shaped substrates and for creating large-area devices with straightforward pattern requirements. However, achieving high resolution and aligning multiple layers can be challenging. Shadow masking is valued for its cost-effectiveness and ease of use in both research and production settings.

Key takeaways:

- Process: A physical mask is placed on the substrate during material deposition to create patterns.
- Advantages: Simple, compatible with vacuum deposition techniques.
- Limitations: Lower resolution compared to photolithography, alignment can be challenging.

c. **Soft Lithography:**

Soft lithography is a versatile and cost-effective patterning technique used in the fabrication of OFETs to create detailed patterns of organic semiconductors, dielectrics, and electrodes. In this method, an elastomeric stamp, usually made from polydimethylsiloxane (PDMS), is employed to transfer a pattern onto a substrate. The stamp is first molded from a master pattern, then inked with the desired material and pressed onto the substrate to transfer the pattern.

Soft lithography allows for the precise control of pattern features with high resolution and can be applied to flexible and irregularly shaped substrates. It is particularly useful for creating micro- and nanoscale patterns without the need for expensive photolithography equipment. This technique is ideal for producing high-quality, reproducible patterns in OFETs, making it suitable for both research and low-cost manufacturing of flexible and large-area electronic devices.

Key takeaways:

- Process: Uses elastomeric stamps to transfer patterns, often in combination with self-assembled monolayers.
- Advantages: Can achieve high resolution, potentially less damaging to organic materials.
- Limitations: Stamp deformation can affect pattern fidelity.

d. **Laser Ablation:**

Laser ablation is a precise patterning technique used in the fabrication of OFETs to create intricate patterns on substrates. In this method, a focused laser beam is directed at the material's surface, causing it to quickly evaporate or be removed through photothermal or photochemical processes. The intensity and wavelength of the laser can be adjusted to target specific materials and achieve high-resolution patterns.

Laser ablation is particularly advantageous for its ability to pattern both organic and inorganic materials with high precision and minimal chemical contamination. It allows for direct writing of intricate designs without the need for masks or photoresists, making it suitable for prototyping and custom applications. Additionally, it can be used on various substrates, including flexible and irregular surfaces. The technique provides excellent control over pattern dimensions and feature sizes, crucial for optimizing OFET performance and integrating complex device structures.

Key takeaways:

- Process: A laser is employed to selectively remove material from the substrate.
- Advantages: Direct writing capability, no masks required.
- Limitations: Serial process (slow for large areas), potential for thermal damage.

4. Self-Aligned Fabrication Techniques

Self-aligned fabrication techniques are advanced methods used in the production of the OFETs to streamline the patterning process and enhance device performance. These techniques take advantage of the natural alignment of different layers during deposition or processing, thereby minimizing the need for the precise alignment steps typically necessary in traditional lithography.

One common self-aligned approach involves the use of materials or processes that inherently align with previously deposited layers. For example, in self-aligned gate structures, the gate electrode can be deposited in such a way that it automatically aligns with pre-existing source and drain contacts. This can be achieved using techniques like self-assembled monolayers or patterned deposition methods that ensure proper alignment without additional masking steps.

Self-aligned techniques offer several advantages, including reduced fabrication complexity, fewer alignment errors, and the ability to create high-resolution features. They are particularly useful for scaling down OFETs and improving their electrical characteristics and reliability.

a. **Self-Aligned Inkjet Printing:**

   Self-aligned inkjet printing techniques are becoming increasingly popular in the fabrication of OFETs due to their precision and efficiency. This method enables the direct deposition of functional materials onto specific areas of the substrate, eliminating the need for traditional photolithography and reducing material waste. By precisely aligning the inkjet-printed layers, the process achieves minimal overlap and accurate patterning, which is essential for the performance and miniaturization of OFETs. Additionally, self-aligned inkjet printing facilitates the creation of multilayer structures, allowing for the sequential and controlled integration of gate electrodes, dielectric layers, and organic semiconductors. This technique not only simplifies the manufacturing process but also supports scalable production, making it a promising approach for developing low-cost, flexible electronics [30].

   Key takeaways:

   - Process: Uses differences in surface energy to confine printed materials to specific areas.
   - Advantages: Can achieve smaller channel lengths, reduces alignment requirements.

b. **Vertical OFETs:**

   Vertical OFETs using self-aligned fabrication techniques represent a significant advancement in organic electronics. These methods simplify the manufacturing process by precisely aligning the gate electrode with the drain and source, which reduces parasitic capacitance and improves device performance. In self-aligned vertical OFETs, the channel length is generally determined by the thickness of the organic semiconductor layer, allowing for ultra-short channels that facilitate high-speed operation. The fabrication process often involves deposition techniques such as vapor phase deposition or inkjet printing, which are compatible with flexible substrates. This alignment precision not only enhances electrical characteristics like current density and switching speed but also reduces leakage currents. Self-aligned vertical OFETs are particularly attractive for applications in flexible displays, sensors, and low-power electronics due to their simplified fabrication, potential for large-area production, and compatibility with roll-to-roll processes.

Key takeaways:

- Process: Layers are deposited sequentially to create a vertical channel structure.
- Advantages: Can achieve very short channel lengths, potentially higher current densities.

5. Emerging Fabrication Techniques

Several new approaches are being explored to improve OFET fabrication:

a. **Roll-to-Roll Processing:**

Roll-to-roll (R2R) processing is an innovative fabrication technique for OFET production that provides a scalable and cost-effective solution for manufacturing flexible electronics. This method uses continuous rolls of flexible substrates, enabling the sequential deposition of organic semiconductors, dielectric layers, and electrodes. R2R processing is ideal for large-scale production as it reduces material waste and production time. It accommodates various deposition techniques like inkjet printing, slot-die coating, and gravure printing, which ensure precise layer formation with minimal defects. The flexibility of R2R processing allows for the integration of OFETs into diverse applications, comprising flexible displays, wearable electronics, and smart packaging. This technique not only increases the throughput of OFET fabrication but also lowers production costs, making it a pivotal technology in advancing organic electronics and developing next-generation flexible devices.

Key takeaways:

- Continuous fabrication on flexible substrates, promising for large-scale, low-cost production.

b. **3D Printing:**

3D printing is an emerging technique in OFET fabrication, offering innovative possibilities for designing and producing flexible and complex electronic devices. Unlike traditional methods, 3D printing enables the direct, additive construction of OFET layers—such as organic semiconductors, dielectrics, and electrodes—in a single, continuous process. This approach facilitates the creation of intricate geometries and multi-layered structures with precise control over material placement and thickness. By using a range of printable inks, including conductive polymers and organic materials, 3D printing can produce OFETs on various substrates, such as flexible plastics or textiles. The technique is highly versatile, supporting rapid prototyping and making it ideal for customized electronics and small-scale production. Additionally, 3D printing reduces material waste and manufacturing costs, speeding up the development of innovative applications in flexible displays, wearable electronics, and smart sensing technologies.

Key takeaways:

- Exploring the use of 3D printing technologies for direct fabrication of complete OFET structures.

c. **Meniscus-Guided Coating:**

Meniscus-guided coating is an advanced technique in OFET fabrication that provides precise control over the deposition of organic semiconductor layers. This method utilizes a meniscus, which forms between a coating blade and the substrate, to direct the deposition of a solution-based organic material. As the substrate moves relative to the blade, a thin film of the semiconductor is applied, with its thickness and uniformity regulated by the substrate's speed, the concentration of the solution, and the shape of the meniscus.

Meniscus-guided coating offers significant advantages for producing highly uniform films with controlled molecular alignment, which improves charge mobility in OFETs. This technique is compatible with large-area substrates and can be fused into roll-to-roll processes, making it ideal for mass production. Its capacity to generate high-quality films with minimal material waste and precise control over layer properties makes it a valuable tool for advancing flexible and cost-effective OFETs.

Key takeaways:

- Uses surface tension effects to guide material deposition, can produce highly aligned films.

d. **Aerosol Jet Printing:**

Aerosol jet printing is an innovative fabrication technique for OFETs that enables precise deposition of electronic materials onto various substrates. This method utilizes an aerosolized ink composed of organic semiconductors, dielectrics, or conductive materials, which is focused into a fine jet and directed onto the substrate. The process allows for high-resolution patterning, with feature sizes down to tens of micrometers, making it ideal for creating intricate OFET structures.

A key advantage of aerosol jet printing is its ability to deposit materials on both flat and three-dimensional surfaces, enabling the production of flexible and conformal electronic devices [31]. The technique is also compatible with a wide range of substrates, comprising flexible plastics, textiles, and paper. Moreover, aerosol jet printing is a non-contact method that minimizes the risk of damage to delicate materials, enhancing the performance and reliability of OFETs. Its versatility makes it a promising solution for scalable, cost-effective, and high-performance organic electronics.

Key takeaways:

- Allows for finer feature sizes compared to traditional inkjet printing.

e. **Organic Vapor Jet Printing:**
Organic vapor jet printing (OVJP) is an advanced technique in OFET fabrication that provides a solvent-free method for depositing organic semiconductors with high precision. In OVJP, organic materials are vaporized and carried through a nozzle by a carrier gas, where they then condense onto a cooler substrate to create thin films. This process offers explicit regulation over the film's thickness, morphology, and crystallinity, which are essential for optimizing OFET performance.

OVJP is notable for its capability to produce uniform, high-purity films without the use of liquid solvents, which minimizes the risk of contamination and improves device stability. This technique is scalable and suitable for both small-area patterning and large-area deposition, making it adaptable with flexible substrates and roll-to-roll processes. Its precision and versatility make OVJP an appealing choice for manufacturing high-performance OFETs, particularly for applications that require flexible, lightweight, and large-area electronics.

Key takeaways:

- Combines aspects of vapor deposition and jet printing for improved resolution and material utilization.

7.9 CHALLENGES AND CONSIDERATIONS IN OFET FABRICATION

1. Material Compatibility: Ensuring that solvents and processes used for one layer don't damage previously deposited layers.
2. Interfacial Engineering: Controlling the properties of interfaces between different layers to optimize device performance.
3. Scalability: Developing techniques that can be scaled up for large-area or high-volume production.
4. Environmental Control: Many organic materials are sensitive to oxygen and moisture, requiring careful control of fabrication environments.

5. Reproducibility: Achieving consistent device performance across multiple fabrication runs.

The field of OFET fabrication continues to evolve, with ongoing efforts to develop techniques that balance performance, cost, and scalability. The choice of fabrication method often depends on the specific materials, device architecture, and intended application of the OFETs.

7.10 CHARACTERIZATION METHODS FOR OFETS

Accurate characterization of OFETs is crucial for understanding device performance, optimizing fabrication processes, and developing new materials. Characterization methods can be broadly categorized into electrical characterization, which focuses on device performance, and materials characterization, which examines the properties of individual components and interfaces.

1. Electrical Characterization

Electrical characterization is crucial for assessing the performance of the OFETs and determining key device parameters. This process usually involves measuring the current-voltage (I-V) characteristics to extract important metrics such as mobility, on/off ratio, and threshold voltage.

a. **Current-Voltage (I-V) Characteristics:**
   - Output characteristics: The drain current (I_D) is measured against the drain-source voltage (V_{DS}) at different gate voltages (V_G).
   - Transfer characteristics: The drain current (I_D) is measured against the gate voltage (V_G) at a constant drain-source voltage (V_{DS}).
   - From these measurements, several key parameters can be determined:
     - Charge carrier mobility (μ)
     - Threshold voltage (V_T)
     - On/off current ratio
     - Subthreshold swing (SS)
     - Contact resistance.

The current in an OFET can be modeled using the following mathematical expressions:

- **Linear Regime:** When the drain-source voltage (V_{DS}) is low, the OFET functions in the linear regime, and the drain current (I_{DS}) is described by:

$$I_{DS} = \mu C_i \frac{W}{L}\left(\left(V_{GS} - V_{th}\right)V_{DS} - \frac{V_{DS}^2}{2}\right)$$

where μ represents the charge carrier mobility, Ci is the capacitance per unit area of the gate dielectric, W is the channel width, L is the channel length, VGS is the gate-source voltage, and Vth is the threshold voltage [32].

- **Saturation Regime:** When V_{DS} exceeds $V_{GS}-V_{th}$, the OFET enters the saturation regime, and I_{DS} becomes:

$$I_{DS} = \mu \frac{C_i}{2} \frac{W}{L} \left(V_{GS} - V_{th}\right)^2$$

These equations are employed to determine the mobility (μ) and threshold voltage (Vth) from experimental data [33]. The I_{on}/I_{off} ratio, which represents the ratio of the maximum to the minimum drain current, is also a key parameter that reflects the switching efficiency of the OFET. High I_{on}/I_{off} ratios and mobility indicate high-performance OFETs, making them useful for applications in flexible and organic electronics.

2. Capacitance-Voltage (C-V) Measurements:

Capacitance-voltage (C-V) measurements are essential for analyzing the electrical properties of the OFETs, especially for understanding the behavior of the gate dielectric and the accumulation of charge carriers in the channel. The capacitance in an OFET, primarily related to the gate dielectric, varies with the gate-source voltage (V_{GS}) [34]. The total gate capacitance (C_G) is expressed as:

$$C_G = \frac{\partial Q_G}{\partial V_{GS}}$$

where Q_G represents the charge induced in the channel by the gate voltage. As V_{GS} increases, charges accumulate at the dielectric-semiconductor interface, resulting in a change in capacitance. In the accumulation region, where V_{GS} is positive (for p-type OFETs), the capacitance approaches its maximum value, which is near the geometrical capacitance of the dielectric (C_i), given by:

$$C_i = \frac{\epsilon_i \epsilon_0 A}{d}$$

where ϵ_i is the dielectric constant, ϵ_0 is the permittivity of free space, A is the gate area, and d is the thickness of the dielectric layer.

The C-V curve offers valuable information about the threshold voltage (V_{th}), interface trap densities, and dielectric properties. The point at which accumulation begins helps determine V_{th}, while deviations from ideal behavior in the C-V curve can indicate the presence of traps or defects. These measurements are crucial for optimizing dielectric materials and enhancing OFET performance, particularly in flexible and low-power electronics.

- Offers insights into charge accumulation in the channel and the quality of the semiconductor-dielectric interface.
- Can be employed to estimate the density of trap states.

3. Impedance Spectroscopy:

Impedance spectroscopy is a valuable technique for examining the frequency-dependent electrical properties of the OFETs. By employing an AC signal to the OFET and measuring the resulting impedance across different frequencies, this method offers detailed insights into charge transport, interface states, and dielectric properties.

The impedance $Z(\omega)$ of an OFET can be expressed as:

$$Z(\omega) = \frac{V(\omega)}{I(\omega)}$$

where $V(\omega)$ and $I(\omega)$ are the voltage and current at angular frequency ω, respectively. The impedance is generally complex and can be represented as:

$$Z(\omega) = Z'(\omega) + jZ''(\omega)$$

where $Z'(\omega)$ represents the real part (resistive component) and $Z''(\omega)$ denotes the imaginary part (reactive component, associated with capacitance and inductance).

The OFET impedance can be modeled using an equivalent circuit, often comprising resistive elements (channel resistance, R) and capacitive elements (gate dielectric capacitance, C_i). The frequency dependence of these components allows for the separation of different physical processes, such as carrier injection, transport, and trapping.

At low frequencies, the impedance is primarily influenced by the gate capacitance (Ci), whereas at higher frequencies, resistive elements become more prominent. Analyzing the impedance spectrum can provide insights into charge carrier mobility, interface trap densities, and the quality of the gate dielectric. This makes impedance spectroscopy an essential tool for optimizing OFET performance, particularly in complex, flexible, and organic electronic systems.

- Analyzes the frequency-dependent response of the device.
- Useful for understanding charge transport dynamics and interfacial properties.

4. Noise Measurements:

Noise measurements in OFETs offer valuable information about device performance, stability, and the quality of the organic materials employed [35]. The main types of noise observed in OFETs are $\frac{1}{f}$ noise (also known as flicker noise) and thermal noise.

a. **Flicker Noise:** Flicker noise, often dominant at low frequencies, can be modeled by:

$$S_v(f) = \frac{K}{f^{\alpha}}$$

where $S_v(f)$ is the power spectral density of the voltage noise, K is a constant related to the material and device quality, f is the frequency, and α typically ranges from 0.7 to 1.0. This type of noise is linked to charge carrier fluctuations and defects at the dielectric-semiconductor interface.

b. **Thermal Noise:** Thermal noise (Johnson-Nyquist noise) is due to the random thermal motion of carriers and is given by:

$$S_v(f) = 4\kappa_B TR$$

where $S_v(f)$ is the noise power spectral density, κ_B is Boltzmann's constant, T is the absolute temperature in Kelvin, and R is the resistance of the channel.

c. **Shot Noise:** Shot noise arises from the discrete nature of charge carriers and is more significant in devices with high current densities. Its power spectral density $S_I(f)$ is given by:

$$S_I(f) = 2qI_D\Delta f$$

where q is the charge of an electron and I_D is the drain current.

Noise measurements involve analyzing the power spectral density of the noise signal to assess device quality, stability, and performance. These measurements help in understanding the impact of material defects, interface traps, and other imperfections on the OFET's operation, ultimately guiding improvements in device design and fabrication.

- Characterizes the noise properties of the device, which is important for sensor applications.
- Can provide insights into charge trapping and de-trapping processes.

5. Transient Measurements:

Transient measurements in the OFETs are used to study the dynamic response of the device to time-varying signals, providing insights into charge carrier dynamics, switching characteristics, and overall device performance. These measurements typically involve applying a step or pulse voltage to the gate and observing the corresponding changes in drain current over time.

a. **Switching Transients:** The transient response of an OFET when switching between on and off states can be examined by analyzing the drain current (I_{DS}) as a function of time (t). When a step voltage is applied to the gate, the drain current can be expressed as:

$$I_{DS}(t) = I_{DS,\max}\left(1 - e^{-\frac{t}{\tau}}\right)$$

where $I_{DS,\max}$ represents the maximum drain current in the steady state, and τ denotes the time constant associated with the accumulation or depletion of charge carriers.

b. **Charge Carrier Dynamics:** The charge carrier mobility μ and relaxation time τ_r can be extracted from the transient response. For a pulse input, the drain current I_{DS} can be analyzed using:

$$I_{DS}(t) = \frac{V_{GS}(t) - V_{th}}{R_{ch}} . e^{-\frac{t}{\tau_r}}$$

where $V_{GS}(t)$ is the gate-source voltage over time, V_{th} is the threshold voltage, and R_{ch} is the channel resistance.

Transient measurements help in understanding the switching speed, charge transport efficiency, and stability of OFETs, which are crucial for optimizing device performance in applications like flexible displays and high-speed electronics.

- Examines the time-dependent response of the device.

- Useful for understanding switching speed and charge carrier dynamics.

6. Bias Stress Measurements:
Bias stress measurements are used to evaluate the stability and reliability of the OFETs under prolonged electrical stress. This technique involves employing a constant voltage bias to the OFET and monitoring changes in its electrical characteristics over time to assess the impact of stress on device performance.

a. **Threshold Voltage Shift:** A common metric is the shift in threshold voltage (Vth) resulting from bias stress. The variation in Vth can be modeled as:

$$\Delta V_{th}(t) = V_{th}(t) - V_{th}(0)$$

where V_{th} (t) is the threshold voltage at time t under bias stress, and V_{th} (0) is the initial threshold voltage. This shift is indicative of charge trapping or other degradation mechanisms.

b. **Time-Dependent Dielectric Breakdown:** For assessing the degradation of the dielectric layer, the time-dependent dielectric breakdown (TDDB) can be described using:

$$t_{BD} = A.\left(\frac{V_{BD}}{E_{BD}}\right)^{-n}$$

where t_{BD} is the time to breakdown, V_{BD} is the breakdown voltage, E_{BD} is the electric field, A is a pre-exponential constant, and n is an empirical fitting parameter.

c. **Drain Current Degradation:** The drain current I_{DS} degradation can be expressed as:

$$\Delta I_{DS}(t) = I_{DS}(t) - I_{DS}(0)$$

where I_{DS} (t) is the drain current at time t under stress, and I_{DS} (0) is the initial drain current [36].

Bias stress measurements provide crucial information about the long-term stability and reliability of OFETs, helping to identify potential failure mechanisms and optimize device design for enhanced performance in real-world applications.

- Evaluates device stability under prolonged electrical stress.
- Important for understanding long-term reliability.

7.10.1 Materials Characterization

Materials characterization is essential for precise modeling and optimization of the OFETs. Gaining insight into the properties of the materials used in OFETs—such as the organic semiconductor, dielectric, and electrode materials—has a direct effect on device performance and reliability. Important elements of materials characterization include:

1. Morphological Characterization:
Morphological characterization in OFETs entails examining the physical structure and surface properties of the active layers and interfaces. This analysis is vital for understanding how material morphology influences device performance, including charge transport and overall reliability.

a. **Film Thickness and Uniformity:** Film thickness d is a key parameter, typically measured using ellipsometry or atomic force microscopy (AFM). For an OFET, the thickness of the organic semiconductor layer can be computed by:

$$d = \frac{n\lambda}{2\delta}$$

where n is the refractive index of the film, λ is the wavelength of light used in ellipsometry [37], and δ is the phase shift measured. Uniformity is assessed by mapping thickness variations across the substrate.

b. **Surface Roughness:** Surface roughness R_q quantifies the mean deviation from the mean surface height and is measured using AFM. The root mean square roughness is given by:

$$R_q = \sqrt{\frac{1}{N}(\sum_{i=1}^{N}\left(h_i - \bar{h}\right)^2}$$

where h_i is the height of the ith point, $\bar{h}$ is the mean height, and N is the number of sampled points.

c. **Grain Size and Morphology:** Grain size affects charge transport and is often analyzed using AFM or scanning electron microscopy (SEM). The average grain size D can be estimated from the SEM images by:

$$D = \frac{1}{N}\sum_{i=1}^{N} D_i$$

where D_i is the diameter of the ith grain and N is the total number of grains measured.

These morphological characteristics influence the electronic properties of OFETs, including charge carrier mobility, threshold voltage, and overall device performance. Proper morphological characterization ensures that the organic layers are well-suited for efficient device operation.

2. Structural Characterization:

Structural characterization of OFETs involves examining the arrangement and quality of materials at the atomic and molecular scales. Techniques like grazing incidence X-ray diffraction (GIXD) and X-ray diffraction (XRD) offer insights into the crystallinity and molecular orientation of the organic semiconductor. Atomic force microscopy (AFM) and scanning electron microscopy (SEM) determine the surface morphology and film uniformity. These methods highlight how structural properties affect charge transport and device performance, influencing factors like charge carrier mobility and threshold voltage. Understanding these structural details is essential for enhancing the efficiency and stability of OFETs.

3. Spectroscopic Techniques:

Spectroscopic techniques for OFETs include ultraviolet-visible (UV-Vis) spectroscopy, which provides information on absorption and electronic transitions in the organic materials. Photoluminescence (PL) spectroscopy reveals emission properties and energy levels. Infrared (IR) spectroscopy identifies functional groups and molecular interactions. Raman spectroscopy offers insights into vibrational modes and molecular structure. These techniques are instrumental in understanding the electronic properties, material composition, and interactions within the OFET, which supports the optimization of device performance and stability. Each spectroscopic method offers valuable insights into the behavior and quality of the organic semiconductor and dielectric layers.

a. **UV-Visible Spectroscopy:**

UV-Visible spectroscopy is an essential technique for analyzing OFETs, as it provides insights into the electronic structure and optical characteristics of organic semiconductor materials. By measuring the absorption of ultraviolet and visible light by these materials, this technique reveals information about their electronic transitions and energy levels.

- **Absorption Spectrum:** The absorption spectrum of an organic semiconductor is typically plotted as absorbance $A(\lambda)$ versus wavelength λ. The absorbance can be related to the absorbance spectrum using the Beer-Lambert law:

$$A(\lambda) = \log_{10}\left(\frac{I_0}{I(\lambda)}\right)$$

where I_0 denotes the intensity of the incident light and I(λ) implies the intensity of the transmitted light at wavelength λ.

- **Optical Band Gap:** The optical band gap Eg of the organic material can be determined from the absorption spectrum. The absorption coefficient α(λ) can be obtained from:

$$\alpha(\lambda) = 2.303^{*}\frac{A(\lambda)}{d}$$

where d is the thickness of the sample. The optical band gap is then estimated by plotting (α·ħω)1/n versus photon energy ħω, where n is the transition type (usually n=2 for direct allowed transitions) and extrapolating to zero absorption.

- **Electronic Transitions:** UV-Vis spectroscopy helps identify electronic transitions between different energy levels (e.g., π-π* transitions), which are crucial for understanding charge transport and device operation in OFETs.

By analyzing these spectra, one can optimize the material properties for improved performance in OFET applications, such as adjusting the bandgap and maximizing absorption.

b. Photoluminescence Spectroscopy:

Photoluminescence (PL) spectroscopy is a valuable method for examining the optical properties and electronic structure of organic semiconductors in the OFETs. This technique measures the light emitted by a material following photon absorption, offering insights into electronic transitions, energy levels, and the overall quality of the material.

i. **Photoluminescence Emission:** In PL spectroscopy, a sample is excited using a light source (typically UV or visible light), and the emitted light is then measured. The emission spectrum IPL(λ) as a function of wavelength λ [38] displays the intensity of photoluminescence. The emission spectrum can be described as:

$$I_{PL}(\lambda) = \int_0^{\infty} \phi(\lambda,t)\,dt$$

where ϕ(λ,t) is the time-resolved PL emission at wavelength λ and time t.

ii. **Photoluminescence Quantum Yield (PLQY):** PLQY measures the efficiency of light emission and is defined as:

$$PLQY = \frac{Number\ of\ photons\ emitted}{Number\ of\ photons\ absorbed}$$

It is calculated from the ratio of emitted light intensity to the absorbed light intensity. A high PLQY indicates efficient radiative recombination, which is beneficial for device performance.

iii. **Energy Levels and Defects:** PL spectroscopy helps identify the energy levels of the organic materials by analyzing emission peaks. Defects or impurities in the material can cause additional emission peaks or broadening of the spectrum.

By studying PL spectra, one can optimize the electronic properties and minimize defects in OFET materials, leading to improved device performance and stability.

c. Raman Spectroscopy:

Raman spectroscopy is a valuable tool for characterizing OFETs, as it provides insights into the vibrational modes and molecular structure of organic materials. This technique measures the inelastic scattering of monochromatic light, revealing information about molecular vibrations and the properties of the material.

i. **Raman Scattering and Shift:** In Raman spectroscopy, the analysis focuses on the intensity of scattered light I(ω) as a function of the Raman shift Δω. The Raman shift is given by:

$$\Delta\omega = \omega_{\text{incident}} - \omega_{\text{scattered}}$$

where ω_{incident} is the frequency of the incident light and $\omega_{\text{scattered}}$ is the frequency of the scattered light, representing the energy difference associated with the vibrational modes of the molecules [39].

ii. **Raman Spectrum:** The Raman spectrum provides peaks corresponding to different vibrational modes of the organic material. These peaks are described by:

$$I(\Delta\omega) = \sum_i A_i \cdot \exp\left(-\frac{(\Delta\omega - \Delta\omega_i)^2}{2\sigma_i^2}\right)$$

where A_i is the amplitude of the peak i, $\Delta\omega_i$ is the peak position, and σ_i is the width of the peak [40].

iii. **Structural Analysis:** Raman spectroscopy helps identify molecular bonds, conjugation, and structural changes. Peaks related to the C-C stretching and C-H bending modes are particularly relevant for analyzing the conjugated systems in OFETs.

By analyzing Raman spectra, one can assess material quality, monitor changes during processing, and understand how molecular structure affects the electrical performance of OFETs, aiding in material optimization and device design.

d. X-ray Photoelectron Spectroscopy (XPS):

X-ray photoelectron spectroscopy (XPS) is an essential method for analyzing the surface chemistry and elemental composition of OFETs. XPS delivers comprehensive information about the chemical states of elements within the organic semiconductor and dielectric layers.

i. **Photoelectron Emission:** In XPS, the sample is exposed to X rays, which eject photoelectrons from the surface. The kinetic energy (Ek) of these photoelectrons is measured, and the binding energy (Eb) is calculated as follows:

$$E_b = E_{X-ray} - E_K - \phi$$

where $E_{X\text{-ray}}$ represents the energy of the incident X-rays, E_k is the kinetic energy of the emitted photoelectron, and ϕ is the work function of the spectrometer.

ii. **XPS Spectrum:** The XPS spectrum plots the intensity of photoelectrons as a function of binding energy. The spectrum can be expressed as:

$$I(E_B) = \sum_i A_i \cdot \exp\left(-\frac{(E_b - E_{bi})^2}{2\sigma_i^2}\right)$$

where A_i is the peak area corresponding to element i, E_{bi} is the binding energy of the i-th element, and σ_i is the peak width.

iii. **Chemical State Analysis:** XPS gives for the determination of chemical states by analyzing the binding energy shifts of core-level electrons. For instance, shifts in the carbon 1s peak can indicate different bonding environments (e.g., C-C vs. C=O).

By providing detailed chemical and elemental analysis, XPS helps in understanding the surface composition, contamination, and chemical states of materials in OFETs, facilitating material optimization and improving device performance and stability.

4. Thermal Analysis:

Thermal analysis of the OFETs is crucial for understanding the impact of temperature fluctuations on the performance, stability, and reliability of these devices. Common techniques for thermal analysis are thermogravimetric analysis (TGA), Differential scanning calorimetry (DSC), and thermal imaging.

i. **Thermogravimetric Analysis (TGA):** TGA determines the change in mass of a material as a function of temperature. The mass loss Δm is given by:

$$\Delta m = m_0 - m(T)$$

where m_0 is the initial mass and m(T) is the mass at temperature T. TGA helps in determining thermal stability and decomposition temperatures of organic materials in OFETs.

ii. **Differential Scanning Calorimetry (DSC):** DSC determines the difference in heat flow between the sample and a reference material as a function of temperature. The heat flow ΔH during a temperature ramp can be described by:

$$\Delta H = \frac{dQ}{dT}$$

where dQ is the heat absorbed or released, and dT is the temperature change. DSC provides information on phase transitions, such as melting and glass transitions, which are critical for understanding material behavior under operational conditions.

iii. **Thermal Imaging:** Thermal imaging helps visualize temperature distribution and hotspots on OFETs during operation. By capturing thermal images, one can assess thermal management and identify potential areas of overheating.

Thermal analysis is crucial for optimizing the performance and longevity of OFETs by ensuring materials and device structures can withstand the operational temperature ranges without degradation.

5. Surface Energy Measurements:

Surface energy measurements are vital in the OFETs for understanding wettability, adhesion, and interface properties, which affect the device's performance and reliability. Surface energy quantifies the energetic cost associated with creating a new surface and influences interactions between the OFET layers and substrates.

i. **Contact Angle Measurements:** A common method for determining surface energy involves measuring the contact angle θ of a liquid droplet on the OFET surface. The surface energy γ_s can then be calculated using Young's equation:

$$\gamma_s = \gamma_{SL} - \gamma_{LV}\ cos\theta$$

where γ_{SL}, γ_{LV}, and θ denote the solid-liquid interfacial energy, liquid-vapor interfacial energy, and contact angle of the liquid droplet, respectively [41].

ii. **Owens-Wendt-Rabel-Kaelble (OWRK) Method:** To compute the dispersive and polar components of surface energy, the OWRK method [42] can be used. The total surface energy γs is divided into dispersive γsd and polar γsp components:

$$\gamma_s = \gamma_s^d + \gamma_s^p$$

where γ_s^d and γ_s^p are obtained from contact angle measurements with different probe liquids, using the relationship:

$$\gamma_{LV}\ cos\theta = \gamma_s^d + \gamma_s^p$$

These measurements provide insights into the compatibility of different materials in the OFET structure, affecting deposition, adhesion, and charge transport. By optimizing surface energy, one can enhance the device's efficiency and stability.

7.10.2 Advanced Characterization Techniques

Advanced characterization techniques for OFETs include time-resolved spectroscopy, which offers insights into charge carrier dynamics and recombination processes. Synchrotron-based methods, such as X-ray absorption spectroscopy (XAS) and high-resolution X-ray photoelectron spectroscopy (XPS), provide detailed analyses of electronic structures and chemical states. Atomic force microscopy (AFM) and scanning tunneling microscopy (STM) allow for high-resolution imaging of surface morphology and molecular arrangements. Kelvin probe force microscopy (KPFM) measures surface work functions and local electronic properties. These advanced techniques are crucial for optimizing material properties and enhancing the performance and reliability of OFETs.

1. In-situ Characterization

In-situ characterization techniques are essential for observing real-time changes in the OFETs throughout fabrication, operation, and exposure to different environments. These methods offer valuable insights into the dynamic behavior and stability of OFETs under varying conditions.

a. **In-situ Spectroscopy:**
    - **In-situ UV-Vis Spectroscopy:** Measures optical properties during processing or operation. The absorbance A(λ) at wavelength λ can be tracked in real-time using the Beer-Lambert law:

$$A(\lambda) = \log_{10}\left(\frac{I_0}{I(\lambda)}\right)$$

where I_0 and I(λ) represent the incident light intensity and the transmitted light intensity, respectively.
    - **In-situ X-ray Photoelectron Spectroscopy (XPS):** Monitors chemical changes during device operation. The binding energy E_b [43] of core electrons is given by:

$$E_b = E_{X-ray} - E_\kappa - \phi$$

where E_{X-ray} is the X-ray energy, E_k is the kinetic energy of the ejected electrons, and ϕ is the work function of the spectrometer.

b. **In-situ Electrical Measurements:**
    - **In-situ I-V Measurements:** Evaluates real-time changes in electrical characteristics. The drain current I_{DS} can be monitored as a function of drain-source voltage V_{DS} and gate-source voltage V_{GS}, using:

$$I_{DS} = \frac{W}{L}\cdot\mu\cdot C_{ox}\left((V_{GS} - V_{th})V_{DS} - \frac{V_{DS}^2}{2}\right)$$

where W, L, C_{ox}, μ and V_{th} denote the channel width, channel length, gate oxide capacitance, carrier mobility, and threshold voltage, respectively.

c. **In-situ Atomic Force Microscopy (AFM):**
    - **AFM:** Monitors surface morphology and changes during device operation. The surface roughness R_q is calculated as:

$$R_q = \sqrt{\frac{1}{N}(\sum_{i=1}^{N}(h_i - \bar{h})^2}$$

where h_i, $\bar{h}$, and N are the height of the ith point, the mean height, and number of sampled points, respectively.

In-situ techniques enable comprehensive monitoring of OFETs, providing insights into their behavior and stability throughout different stages of operation and processing.

2. Synchrotron-based Techniques

Synchrotron-based techniques are advanced methods used to analyze the OFETs with high precision, providing detailed information on material properties and electronic structures.

a. **X-ray Absorption Spectroscopy (XAS):** XAS computes the absorption of X-rays as a function of energy to investigate the local electronic structure around specific atoms. The absorption μ(E) as a function of photon energy E is given by:

$$\mu(E) = \frac{I_0 - I(E)}{I_0}$$

where I_0 is the incident X-ray intensity and I(E) is the transmitted X-ray intensity at energy E. XAS provides information on oxidation states, coordination environments, and electronic transitions.

b. **X-ray Photoelectron Spectroscopy (XPS):** XPS, when conducted at synchrotron sources, offers high-resolution data on core-level electron binding energies. The binding energy E_b is calculated as:

$$E_b = E_{X-ray} - E_\kappa - \phi$$

where $E_{X\text{-ray}}$, E_k, and ϕ denote the X-ray photon energy, the kinetic energy of emitted electrons and the work function of the spectrometer, respectively. XPS reveals chemical states and surface composition.

c. **Grazing Incidence X-ray Diffraction (GIXD):** GIXD analyzes the crystallinity and orientation of thin films by measuring diffracted X-rays at a shallow angle. The diffraction pattern intensity I(q) as a function of scattering vector qqq can be expressed as:

$$I(q) = \frac{d\sigma}{d\Omega} \cdot \exp\left(-\frac{q^2}{2\sigma^2}\right)$$

where $\frac{d\sigma}{d\Omega}$ is the differential scattering cross-section and σ is the Gaussian broadening parameter. GIXD provides insights into molecular packing and layer alignment.

3. Scanning Probe Microscopy Techniques

Scanning probe microscopy (SPM) techniques are essential for characterizing OFETs at the nanoscale. AFM provides high-resolution mapping of surface topography and roughness, calculating root mean square roughness RqR_qRq from height variations. STM delivers atomic-scale images by measuring tunneling current III as a function of the distance between the tip and the sample. KPFM measures the contact potential difference (CPD) to determine surface work functions. These techniques offer detailed insights into surface features, electronic properties, and material interactions, which are crucial for optimizing OFETs.

4. Ultrafast Spectroscopy

Ultrafast spectroscopy techniques are essential for studying the rapid dynamics of charge carriers and excitons in OFETs. These techniques provide insights into the fundamental processes that govern device performance and efficiency.

a. **Time-Resolved Photoluminescence (TRPL):** TRPL measures the emission of light from excited states over time with high temporal resolution, typically in the range of picoseconds to nanoseconds. It helps in understanding the carrier recombination dynamics and the effects of trapping and quenching. The decay of photoluminescence I(t) can be fitted to an exponential function:

$$I(t) = I_0 \exp\left(-\frac{t}{\tau}\right)$$

where I_0 is the initial intensity, t is time, and τ is the photoluminescence lifetime.

b. **Time-Resolved Absorption Spectroscopy (TRAS):** TRAS monitors changes in optical absorption as a function of time after excitation. This technique provides information on transient states and carrier dynamics. The transient absorption signal $\Delta A(t)$ is expressed as:

$$\Delta A(t) = A(t) - A_0$$

where A(t) is the absorption at time t and A_0 is the baseline absorption before excitation.

c. **Ultrafast Pump-Probe Spectroscopy:** This technique uses a pump pulse to excite the sample and a delayed probe pulse to measure the changes in optical properties. It provides real-time information on charge and energy transfer processes.

These ultrafast spectroscopy techniques are crucial for optimizing OFET materials and device architectures by providing detailed insights into the fast dynamics of charge carriers and excitons.

5. Neutron Scattering

Neutron scattering techniques are valuable for studying OFETs due to their sensitivity to light elements and their ability to probe structural and dynamic properties at the atomic level.

a. **Neutron Reflectometry (NR):** NR measures the reflectivity of neutrons as a function of incident angle, providing detailed information about the thickness, density, and layer structure of thin films and interfaces in OFETs [44]. The reflectivity R(θ) is related to the neutron wavelength λ and the scattering vector qz by:

$$R(\theta) = \left| \frac{r_s - r_i}{r_s + r_i} \right|^2$$

where r_s and r_i are the scattering amplitudes of the sample and incident medium, respectively. NR helps in characterizing layer thicknesses and interfacial roughness.

b. **Small-Angle Neutron Scattering (SANS):** SANS is used to investigate nanostructures and morphological features on length scales from nanometers to micrometers [45]. The scattering intensity I(q) as a function of the scattering vector q is given by:

$$I(q) = \frac{d\sigma}{d\Omega} \propto form\ factor \times structure\ factor$$

where $\frac{d\sigma}{d\Omega}$ is the differential cross-section. SANS provides information on the size, shape, and distribution of nanoparticles and domains within OFET materials.

c. **Neutron Diffraction:** Neutron diffraction reveals crystallographic information, including lattice parameters and phase transitions. The diffraction pattern intensity I(q) depends on the atomic arrangement and can be described by Bragg's law:

$$n\lambda = 2dsin\theta$$

where n is the order of diffraction, λ is the neutron wavelength, d is the interplanar spacing, and θ is the diffraction angle.

These neutron scattering techniques are essential for understanding the structural and dynamic properties of OFETs, aiding in material optimization and device performance enhancement.

6. Electron Paramagnetic Resonance (EPR)

Electron paramagnetic resonance (EPR) techniques are crucial for probing the electronic and magnetic properties of materials used in OFETs. EPR detects unpaired electron spins in materials, providing insights into charge carrier dynamics, defect states, and interactions.

a. **EPR Spectroscopy:** EPR measures the resonance of unpaired electrons in a magnetic field. The EPR signal is characterized by the g-factor, which reflects the electronic environment of the paramagnetic centers. The EPR absorption A is given by:

$$A = \frac{P_0}{\sqrt{1 + \left(\frac{B - B_0}{\Delta B}\right)^2}}$$

where P_0 is the maximum absorption, B is the applied magnetic field, B_0 is the resonance field, and ΔB is the line width. This technique helps in identifying different paramagnetic species and their concentration.

b. **Time-Resolved EPR:** Time-resolved EPR provides information on the dynamics of charge carriers and transient states. By measuring changes in the EPR signal over time, one can study charge recombination, trapping, and transport processes. The time-dependent EPR signal S(t) can be analyzed to obtain:

$$S(t) = S_0 exp\left(-\frac{t}{\tau}\right)$$

where S_0 is the initial signal intensity and τ is the relaxation time.

c. **EPR Imaging:** EPR imaging maps the spatial distribution of paramagnetic centers within an OFET, providing insights into local defects and electronic structures. This technique is used to visualize inhomogeneities and optimize device performance.

EPR techniques are essential for understanding charge transport mechanisms, defect states, and material properties in OFETs, facilitating improvements in device design and functionality.

7.10.3 Device-Specific Characterization

Device-specific characterization for the OFETs is tailored to their intended applications to ensure optimal performance and functionality. Here's an elaboration of the necessary characterization for various applications:

1. For Display Applications

a. **Luminance and Color Measurements:** For light-emitting OFETs, it's crucial to measure luminance (brightness) and color accuracy. Luminance can be measured using a photometer, while color accuracy is assessed through colorimeter devices to ensure that the OFETs produce the desired color range and intensity. This is essential for high-quality display performance.

b. **Pixel Addressing and Crosstalk Evaluation:** In display arrays, pixel addressing involves ensuring that each pixel can be individually controlled without interference from adjacent pixels. Crosstalk evaluation measures the extent to which signals from one pixel affect neighboring pixels, which is crucial for maintaining image clarity and color accuracy in high-resolution displays.

2. For Sensor Applications

a. **Sensitivity and Selectivity Measurements:** These tests determine how effectively the OFET sensor detects and responds to specific analytes. Sensitivity measures the minimal concentration of an analyte that can be estimated, while selectivity assesses the sensor's ability to distinguish between the target analyte and other substances, ensuring accurate readings.

b. **Response Time and Recovery Characteristics:** The response time measures how quickly the OFET reacts to the presence of an analyte, while recovery characteristics assess how fast the sensor returns to its baseline state once the analyte is removed. Both metrics are critical for evaluating the performance of sensors in dynamic environments.

3. For Flexible Electronics

a. **Mechanical Stress Testing:** This involves subjecting the OFETs to mechanical stresses such as bending and stretching to simulate real-world conditions. Testing ensures that the OFETs maintain performance and reliability under mechanical deformation, which is vital for flexible and wearable electronics.

b. **Performance Evaluation Under Deformation:** This includes assessing how mechanical deformation (e.g., bending, twisting) affects the electrical characteristics of the OFETs. Performance evaluation involves measuring changes in electrical properties like mobility, threshold voltage, and current, which can impact the functionality of flexible electronic devices.

4. For Printed Electronics

a. **Print Quality Assessment:** This involves evaluating the resolution and uniformity of printed OFETs. Resolution ensures that the fine features of the printed circuits are accurately reproduced, while uniformity checks for consistent layer thickness and device performance across the print area.
b. **Adhesion and Durability Testing:** Adhesion testing evaluates how well the printed layers adhere to the substrate, which affects device reliability and longevity. Durability testing assesses the stability of printed electronics under various conditions, such as environmental exposure and mechanical wear, ensuring long-term performance.

7.10.4 Reliability and Stability Testing

Long-term performance evaluation is essential for ensuring that OFETs function reliably in practical applications. This involves comprehensive testing to address various factors that could impact device longevity and performance. Here's a detailed look at the key aspects:

1. **Environmental Stability**
   a. **Performance Testing Under Various Temperature and Humidity Conditions:**
      - **Temperature Testing:** OFETs are exposed to a range of temperatures, from extremely low to high, to evaluate how temperature fluctuations affect their electrical characteristics, such as carrier mobility, threshold voltage, and current conduction. This testing ensures that the devices maintain consistent performance across different operating temperatures and do not suffer from thermal degradation or instability.
      - **Humidity Testing:** Devices are tested under different humidity levels to assess their susceptibility to moisture. High humidity can lead to water absorption, which may affect the organic materials' properties, cause swelling, or lead to electrical short-circuits. Evaluating how OFETs perform in varying humidity conditions helps in understanding their resilience and suitability for environments with fluctuating moisture levels.

b. **Exposure to Light, Oxygen, and Other Potential Degradation Factors:**
    - **Light Exposure:** Prolonged exposure to light, especially UV light, can cause photooxidation and degradation of the organic semiconductor materials. Testing OFETs under continuous light exposure helps in assessing the impact of light-induced degradation on device performance, ensuring that they can operate reliably under typical lighting conditions.
    - **Oxygen Exposure:** Exposure to oxygen can lead to oxidative degradation of organic materials, affecting the device's electrical properties. Testing OFETs in oxygen-rich environments helps to evaluate their resistance to such degradation and ensures that they maintain performance over time.

2. **Operational Stability**
    a. **Continuous Operation Tests:**
        - These tests involve running OFETs continuously for extended periods, often under normal operating conditions, to simulate long-term usage. Key parameters like drain current, threshold voltage, and mobility are monitored to detect any gradual degradation. Continuous operation tests help to identify potential issues related to material fatigue, thermal effects, and electrical drift over time, providing insights into the device's long-term reliability.

    b. **Cyclic Stress Tests:**
        - Cyclic stress tests simulate real-world usage by applying repetitive mechanical and electrical stresses, such as bending, stretching, and cycling through different voltage levels. This testing helps to evaluate how the OFETs perform under conditions that mimic everyday use, assessing their durability and stability. Cyclic stress tests reveal any potential issues related to mechanical fatigue, performance degradation due to repeated stress, and long-term reliability.

7.10.5 Modeling and Simulation

Modeling and simulation are critical tools for designing and optimizing OFETs. Although they are not direct characterization methods, they complement experimental techniques by offering theoretical insights and predictions regarding device behavior and material properties. Here is a detailed explanation of the key aspects of modeling and simulation used alongside experimental characterization:

1. Device Physics Modeling

Device physics modeling involves creating mathematical and computational models to understand and predict the behavior of OFETs based on fundamental physical principles. These models help elucidate the underlying mechanisms of charge transport and device operation.

- **Charge Transport Mechanisms:** Models such as Drift-Diffusion models and the Marcus-Hush theory are used to describe how charge carriers move through the organic semiconductor material. For instance, the Drift-Diffusion model accounts for both drift due to electric fields and diffusion due to concentration gradients [46]. The equations governing charge transport in the device are:

$$J = \mu n\left(E - \frac{\kappa T}{e}\nabla n\right)$$

  where J is the current density, μ is the carrier mobility, n is the charge carrier concentration, E is the electric field, k is the Boltzmann constant, T is the temperature, and e is the elementary charge.
- **Potential Distribution:** The Poisson-Boltzmann equation is often used to model the electrostatic potential distribution within the OFET. This helps in understanding how the gate voltage influences the carrier concentration and the resulting device performance.

2. Circuit-Level Simulations

Circuit-level simulations predict how OFETs will behave when integrated into complex electronic circuits. These simulations are crucial for designing and optimizing OFET-based systems.

- **Spice Simulations:** OFETs are often modeled using SPICE (Simulation Program with Integrated Circuit Emphasis) simulations. The OFETs are represented as equivalent circuit elements, and the simulations predict their behavior in various circuit configurations. Parameters such as threshold voltage, transconductance, and output conductance are incorporated into the circuit models to assess overall circuit performance.
- **Behavioral Modeling:** This involves creating simplified models of OFETs to predict their behavior in different operating conditions and circuit designs. These models can help in optimizing device performance and power consumption in applications like displays and sensors.

3. Molecular Dynamics Simulations

Molecular dynamics (MD) simulations determine a detailed view of material properties at the atomic level, offering insights into the structural and dynamic aspects of organic semiconductors.

- **Atomic Scale Studies:** MD simulations involve modeling the motion of atoms and molecules in the organic semiconductor material over time. This helps in understanding how molecular interactions, crystallinity, and defects affect material properties and device performance. Key aspects studied include:

$$F_i = m_i \frac{d^2 r_i}{dt^2}$$

where F_i, m_i, and r^i represent the force on atom i, the mass of atom i, and the position of atom I, respectively.

- **Material Properties:** MD simulations can reveal information about the packing density, intermolecular interactions, and stability of organic semiconductors. This information is crucial for optimizing material properties such as charge transport and mechanical flexibility.

4. Integration of Modeling and Simulation with Experimental Data

Modeling and simulation are used in tandem with experimental characterization to validate theoretical predictions and refine models based on experimental observations. This iterative process helps in improving the accuracy of models, optimizing device design, and guiding material selection. For example, discrepancies between simulated and experimental results can lead to adjustments in model parameters or material formulations, leading to more reliable and efficient OFETs.

7.11 CHALLENGES IN OFET CHARACTERIZATION

1. **Sample Preparation:** Sample preparation is a critical challenge in OFET characterization due to the sensitivity of organic materials. Organic semiconductors can be easily altered by the preparation process, which can impact their intrinsic properties. Ensuring that characterization methods do not introduce artifacts or alter the material's characteristics involves:
    - **Minimizing Contamination:** Organic materials are prone to contamination from environmental sources, which can affect performance. Cleanroom environments and rigorous handling procedures are necessary to avoid contamination.
    - **Maintaining Structural Integrity:** Techniques like spin-coating, thermal evaporation, and solution processing must be carefully controlled to avoid introducing stress or defects that could alter the material's properties.
    - **Reproducibility:** Variations in sample preparation can lead to inconsistencies in characterization results. Developing and

adhering to precise and reproducible preparation protocols is crucial for obtaining reliable data.

2. **Standardization:** Developing standardized protocols is essential for ensuring consistent and comparable results across different research groups. Challenges in standardization include:
   - **Defining Benchmarks:** Establishing universal benchmarks for device performance and material properties helps in comparing results from different studies.
   - **Protocol Consistency:** Standardizing testing conditions, such as temperature, humidity, and measurement techniques, ensures that comparisons are meaningful.
   - **Calibration and Validation:** Regular calibration of measurement instruments and validation of methods against known standards are necessary to maintain consistency and accuracy in results.
3. **In-situ and Operando Measurements:** Studying OFETs under actual operating conditions presents several challenges:
   - **Real-time Measurement:** Techniques must be developed to capture real-time data while the device is in operation. This includes integrating sensors and measurement systems into the device setup.
   - **Environmental Control:** Maintaining consistent operating conditions (e.g., temperature, humidity, electrical bias) while performing measurements is challenging but essential for accurate in-situ characterization.
   - **Non-intrusiveness:** Measurement techniques must be non-intrusive to avoid affecting the device's performance. Developing advanced techniques that minimize the impact on device operation is critical.
4. **Big Data Analysis:** The increasing complexity and volume of data from combinatorial studies and high-throughput characterization create challenges in managing and interpreting results:
   - **Data Management:** Efficiently storing, organizing, and retrieving large datasets requires robust data management systems. This includes using databases and data storage solutions capable of handling large volumes of information.
   - **Data Analysis:** Analyzing complex datasets involves sophisticated algorithms and computational tools. Machine learning and statistical methods are often employed to extract meaningful patterns and correlations from the data.
   - **Integration of Results:** Combining data from multiple sources (e.g., different characterization methods) requires methods for integrating and reconciling results to provide a comprehensive understanding of device behavior.

Evolving Techniques and Comprehensive Understanding

The field of OFET characterization is rapidly evolving, with enhancements aimed at overcoming these challenges. New techniques and methodologies are continuously being developed to provide deeper insights into device physics and material properties. Combining multiple characterization methods—such as electrical, optical, and structural techniques—often offers the most comprehensive understanding of OFET performance and behavior. This multi-faceted approach helps in addressing the limitations of individual methods and provides a more complete picture of device operation and material interactions.

7.12 APPLICATIONS OF OFETS

The OFETs have attracted considerable attention due to their distinctive characteristics, such as mechanical flexibility, the ability to be processed over large areas, and potential for low-cost production. These features make OFETs ideal for various applications, some of which are already on the market, while others are still undergoing research and development.

1. Flexible and Printed Electronics

One of the most promising areas for OFET application is in flexible and printed electronics.

a. Flexible Displays:
  - OFETs can serve as pixel-driving elements in flexible organic light-emitting diode (OLED) displays.
  - Advantages include the potential for rollable or foldable screens, lightweight designs, and conformable displays that can be integrated into non-flat surfaces.

b. E-paper and E-readers:
  - OFETs can be used as backplane driving circuits for electronic paper displays.
  - The low power consumption of OFETs is particularly advantageous for these applications.

c. Smart Labels and Packaging:
  - Integration of OFETs into product packaging for interactive displays or sensors.
  - Applications include temperature-sensitive labels for perishable goods or tamper-evident packaging.

2. Sensors and Biosensors

The high surface-to-volume ratio and chemical tunability of organic semiconductors make OFETs excellent candidates for sensing applications.

a. Chemical Sensors:
    - Detection of gases, vapors, or dissolved analytes through changes in OFET electrical characteristics.
    - Applications include environmental monitoring, food quality control, and industrial safety.
b. Biosensors:
    - Detection of biomolecules such as proteins, DNA, or small metabolites.
    - Potential applications in point-of-care diagnostics, wearable health monitors, and lab-on-a-chip devices.
c. Pressure and Strain Sensors:
    - Utilizing the mechanical flexibility of OFETs to create sensors that respond to physical deformation.
    - Applications in electronic skin, wearable electronics, and structural health monitoring.

3. Radio-Frequency Identification (RFID) Tags
OFETs can be used in the creation of low-cost, printable RFID tags.

- Advantages include the potential for large-area manufacturing and integration into flexible substrates.
- Applications in inventory tracking, supply chain management, and smart packaging.

4. Large-Area Electronics
The ability to fabricate OFETs over large areas opens up unique application possibilities.

a. Solar Cell Backplanes:
    - OFETs can be used as switching elements in large-area photovoltaic panels, particularly for building-integrated photovoltaics.
b. X-ray Imaging Arrays:
    - Large-area OFET arrays for flat-panel X-ray detectors in medical imaging or security screening [47].
c. Structural Health Monitoring:
    - Integration of OFET-based sensor networks into large structures like buildings or bridges for real-time monitoring.

5. Wearable and Textile Electronics
The mechanical flexibility and potential for low-temperature processing make OFETs suitable for integration into textiles and wearable devices.

a. Smart Textiles:
    - Integration of OFETs into fabrics for creating interactive or responsive clothing.

    - Applications in sportswear, military gear, or healthcare monitoring garments.

b. Wearable Displays:
    - Flexible OFET-driven displays that can be integrated into clothing or accessories.

c. Skin-Mounted Electronics:
    - Ultra-thin, conformable OFET circuits for on-skin electronics and medical monitoring devices.

6. Neuromorphic Computing

OFETs are being explored for use in neuromorphic computing systems, which aim to mimic the function of biological neural networks.

- Synaptic transistors using OFETs to emulate the behavior of biological synapses.
- Potential applications in artificial intelligence, machine learning, and brain-computer interfaces.

7. Memories and Data Storage

While not as prevalent as in other areas, OFETs are being investigated for memory applications.

- Organic ferroelectric field-effect transistors for non-volatile memory.
- Potential for creating flexible, printable memory elements.

8. Active Matrix Backplanes

OFETs can serve as switching and amplifying elements in active matrix backplanes for various devices.

- Organic light-emitting diode (OLED) displays.
- Large-area sensor arrays.
- Flexible or rollable displays.

9. Logic Circuits

While not competitive with silicon for high-performance computing, OFETs can be used to create simple logic circuits for specific applications.

- Low-cost, disposable electronic circuits.
- Integrated circuits for smart packaging or RFID tags.

10. Energy Harvesting and Management

OFETs are being explored for use in energy harvesting and management systems.

- Integration with organic photovoltaics for solar energy harvesting.
- Low-power circuit elements for energy-efficient systems.

Challenges and Future Directions:
While OFETs show promise in many applications, several challenges need to be addressed:

1. Performance Improvement: Enhancing charge carrier mobility, stability, and uniformity of OFET performance.
2. Scalability: Developing manufacturing processes that can reliably produce OFETs at industrial scales.
3. Encapsulation and Packaging: Improving device longevity through better protection from environmental factors.
4. Integration: Developing techniques to effectively integrate OFETs with other electronic components and systems.
5. Standardization: Establishing industry standards for OFET characterization and performance metrics.

Future research directions include:

- Development of new organic semiconductor materials with improved performance and stability.
- Exploration of hybrid organic-inorganic systems to accumulate the advantages of both material classes.
- Investigation of novel device architectures to enhance performance or enable new functionalities.
- Advancement in printing and patterning technologies for high-resolution, large-area fabrication.

The field of OFET applications continues to evolve, with new possibilities emerging as the technology advances [48]. The unique properties of OFETs, particularly their mechanical flexibility and potential for low-cost, large-area fabrication, position them as enabling components for next-generation electronics in areas where traditional silicon-based technologies may be less suitable.

7.13 CHALLENGES AND FUTURE PROSPECTS FOR OFET TECHNOLOGY

While OFETs have shown significant promise in various applications, several challenges require to be addressed to fully realize their potential. At the same time, ongoing research and development are opening up exciting new prospects for the technology. Let's explore both the challenges and future directions:

Challenges:

1. **Performance and Stability:**
   a. Charge Carrier Mobility:
      - Challenge: Improving mobility to levels comparable with inorganic semiconductors.
      - Current efforts: Developing new materials and optimizing molecular packing.

   b. Operational Stability:
      - Challenge: Enhancing long-term stability under continuous operation and environmental stresses.
      - Current efforts: Developing more robust organic semiconductors and better encapsulation techniques.

   c. Threshold Voltage Control:
      - Challenge: Achieving precise control and minimizing threshold voltage shifts over time.
      - Current efforts: Interface engineering and development of new dielectric materials.

2. **Manufacturing and Scalability:**
   a. Large-Area Uniformity:
      - Challenge: Ensuring consistent performance across large substrate areas.
      - Current efforts: Improving deposition techniques and developing self-assembling materials.

   b. High-Resolution Patterning:
      - Challenge: Achieving fine feature sizes comparable to silicon electronics.
      - Current efforts: Adapting advanced lithography techniques and developing new patterning methods.

   c. Yield and Reproducibility:
      - Challenge: Improving manufacturing yield and device-to-device consistency.
      - Current efforts: Enhancing process control and developing more robust device architectures.

3. **Material Development:**
   a. n-Type Semiconductors:
      - Challenge: Developing high-performance, air-stable n-type organic semiconductors.
      - Current efforts: Exploring new molecular designs and doping strategies.

   b. Dielectric Materials:
      - Challenge: Creating high-k organic dielectrics for low-voltage operation.
      - Current efforts: Investigating nanocomposites and hybrid organic-inorganic materials.

c. Contact Materials:
  - Challenge: Optimizing electrode materials for efficient charge injection/collection.
  - Current efforts: Developing new electrode materials and interface modification techniques.

4. **Device Physics Understanding:**

a. Charge Transport Mechanisms:
  - Challenge: Fully understanding and modeling charge transport in disordered organic systems.
  - Current efforts: Combining experimental studies with advanced theoretical modeling.

b. Interface Phenomena:
  - Challenge: Elucidating the complex processes occurring at material interfaces.
  - Current efforts: Developing in-situ characterization techniques and interface-sensitive measurements.

5. **Integration and System-Level Challenges:**

a. Circuit Design:
  - Challenge: Developing circuit design methodologies tailored for OFET characteristics.
  - Current efforts: Creating OFET-specific design rules and simulation tools.

b. Hybrid Integration:
  - Challenge: Integrating OFETs with other electronic components (e.g., silicon ICs).
  - Current efforts: Exploring heterogeneous integration techniques and flexible interconnect technologies.

Future Prospects:

1. **Advanced Materials:**

a. 2D Organic Semiconductors:
  - Prospect: Developing atomically thin organic semiconductors for ultra-high performance.
  - Potential impact: Dramatic improvements in mobility and subthreshold swing.

b. Self-Assembling Materials:
  - Prospect: Creating materials that spontaneously organize into optimal structures.
  - Potential impact: Simplified manufacturing and improved device uniformity.

c. Biocompatible and Biodegradable Materials:
  - Prospect: Developing OFETs that are safe for in-vivo use or environmentally degradable.
  - Potential impact: New applications in bioelectronics and eco-friendly disposable electronics.

2. **Novel Device Architectures:**
    a. Vertical OFETs:
        - Prospect: Developing transistors with vertical channel orientation.
        - Potential impact: Higher current densities and easier integration into 3D circuits.
    b. Dual-Gate and Multi-Gate Structures:
        - Prospect: Creating OFETs with enhanced gate control.
        - Potential impact: Improved switching characteristics and multi-functional devices.
    c. Steep Subthreshold Swing Devices:
        - Prospect: Developing OFETs with subthreshold swing below the thermal limit.
        - Potential impact: Ultra-low power consumption for IoT applications.
3. **Advanced Manufacturing Techniques:**
    a. Roll-to-Roll Printing:
        - Prospect: Continuous, high-speed fabrication of OFETs on flexible substrates.
        - Potential impact: Dramatic cost reduction and scalability for large-area electronics.
    b. 3D Printing of Complete Devices:
        - Prospect: Additive manufacturing of entire OFET-based circuits.
        - Potential impact: Customizable, on-demand electronic systems.
    c. Self-Aligned Fabrication Methods:
        - Prospect: Developing techniques that automatically align critical device features.
        - Potential impact: Improved performance and reduced manufacturing complexity.
4. **Emerging Applications:**
    a. Neuromorphic Computing:
        - Prospect: Using OFETs to emulate biological neural networks.
        - Potential impact: Energy-efficient AI hardware and brain-computer interfaces.
    b. Quantum Computing:
        - Prospect: Exploring organic materials for quantum bits (qubits).
        - Potential impact: New paradigms in quantum information processing.
    c. Human-Machine Interfaces:
        - Prospect: Developing conformable, skin-like electronic systems based on OFETs.
        - Potential impact: Revolutionary advancements in prosthetics and human augmentation.

5. **Sustainability and Circular Economy:**
    a. Green Electronics:
        - Prospect: Developing fully sustainable OFET-based electronic systems.
        - Potential impact: Reduced electronic waste and environmental impact.
    b. Energy Harvesting Integration:
        - Prospect: Combining OFETs with organic photovoltaics and thermoelectrics.
        - Potential impact: Self-powered electronic systems for IoT and remote sensing.
6. **Artificial Intelligence and Machine Learning:**
    a. Materials Discovery:
        - Prospect: Using AI to accelerate the discovery of new organic semiconductor materials.
        - Potential impact: Rapid advancement in OFET performance and functionality.
    b. Automated Design and Optimization:
        - Prospect: AI-driven design of OFET devices and circuits.
        - Potential impact: Faster development cycles and optimized system performance.

The field of OFET technology is at an exciting juncture [49]. While significant challenges remain, the ongoing research and emerging prospects suggest a bright future. The unique properties of OFETs, combined with advancements in materials science, manufacturing techniques, and device engineering, are likely to enable new applications and potentially disruptive technologies in the coming years. As the technology matures, we can expect to see OFETs playing an increasingly important role in next-generation electronics, particularly in areas where flexibility, large-area coverage, and low-cost manufacturing are crucial [50].

7.14 CONCLUSION

Organic field-effect transistors (OFETs) have attracted significant attention in recent years and have been extensively studied for applications such as smart tags, displays, and sensors. OFETs are structurally similar to conventional metal-oxide-semiconductor field-effect transistors (MOSFETs); however, due to the inherently low conductivity of organic semiconductor materials compared to their inorganic counterparts, a thin-film transistor (TFT) architecture is typically employed. This chapter described the comprehensive study of distinct materials used in OFETs design, followed by a discussion of the distinct fabrication techniques and characterization approaches, and a description of the challenges and applications of OFETs.

Finally, we outlined several challenges, open issues, and future research directions.

REFERENCES

[1] R. W. Li and G. Liu, *Flexible and Stretchable Electronics – Materials, Designs, and Devices*. CRC Press, 2019.

[2] V. Greanya, *Bioinspired Photonics – Optical Structures and Systems Inspired by Nature*. CRC Press, 2019.

[3] H. Y. Tseng, "All inkjet-printed, fully self-aligned transistors for low-cost circuit applications," *Org. Electron.*, vol. 12, no. 2, pp. 101–108, 2011.

[4] J. R. Reynolds, B. C. Thompson, and T. A. Skotheim, *Conjugated Polymers – Perspective, Theory, and New Materials*. CRC Press, 2019.

[5] M. Aliofkhazraei, N. Ali, W. I. Milne, C. S. Ozkan, S. Mitura, and J. L. Gervasoni, *Graphene Science Handbook – Size-Dependent Properties*. CRC Press, 2019.

[6] Y. Huang and K.-N. Tu, *Silicon and Silicide Nanowires – Applications, Fabrication, and Properties*. Pan Stanford, 2019.

[7] D. Nirmal and J. Ajayan," *Handbook for III-V High Electron Mobility Transistor Technologies*. CRC Press, 2019.

[8] C. K. Sarkar, *Technology Computer Aided Design – Simulation for VLSI MOSFET*. CRC Press, 2019.

[9] S. Zhang, *Nanostructured Thin Films and Coatings – Functional Properties*. CRC Press, 2019.

[10] J. Jin and J. Grote, *Materials Science of DNA*. CRC Press, 2019.

[11] Z. Bao and J. Locklin, *Organic Field Effect Transistors*. CRC Press, 2018.

[12] C. S. Suchand Sangeeth, P. Stadler, S. Schaur, N. S. Sariciftci, and R. Menon, "Interfaces and traps in pentacene field-effect transistor," *J. Appl. Phys.*, vol. 107, no. 6, pp. 064509, 2010.

[13] Y. Zhao, W. Wang, Z. He, B. Peng, C.-A. Di, and H. Li, "High-performance and multifunctional organic field-effect transistors," *Chinese Chemical Letters*, vol. 34, no. 5, pp. 1–7, 2023.

[14] W. K. Chen, *Memory, Microprocessor, and ASIC*. CRC Press, 2019.

[15] E. T. Yu and M. O. Manasreh, *III-V Nitride Semiconductors: Applications & Devices*. CRC Press, 2022.

[16] S.-S. Sun and L. R. Dalton, *Introduction to Organic Electronic and Optoelectronic Materials and Devices*. CRC Press, 2016.

[17] J. Youn, M.-C. Chen, Y.-J. Liang, H. Huang, et al., "Novel semiconductors based on functionalized benzo dithiophenes and the effects of thin film growth conditions on organic field effect transistor performance," *Chem. Mater.*, vol. 22, no. 19, pp. 5517–5526, 2010.

[18] F. M. Li, A. Nathan, Y. Wu, and B. S. Ong, *Organic Thin Film Transistor Integration*. Wiley, 2011.

[19] J. Locklin, M. E. Roberts, S. C. B. Mannsfeld, and Z. Bao, "Optimizing the thin film morphology of organic field-effect transistors: The influence of molecular structure and vacuum deposition parameters on device performance," *J. Macromol. Sci. C Polym. Rev.*, vol. 46, no. 2, pp. 135–169, 2006.

[20] F. Abraham, W. E. Ford, F. Scholz, G. Nelles, G. Sandford, and F. von Wrochem, "Surface energy and work function control of AlOx/Al surfaces by Fluorinated Benzylphosphonic Acids," *ACS Appl. Mater. Interfaces*, vol. 14, no. 3, pp. 3521–3530, 2022.

[21] K. Ravikumar and M. S. Dangate, "Advancements in stretchable organic optoelectronic devices and flexible transparent conducting electrodes: Current progress and future prospects," *Heliyon*, 2024.

[22] G. Harsányi, *Polymer Films in Sensor Applications – Technology, Materials, Devices and Their Characteristics*. CRC Press, 2017.

[23] Z. R. Li, *Organic Light-Emitting Materials and Devices*. CRC Press, 2017.

[24] S. A. Kadam, K. P. Kadam, and N. R. Pradhan, "Advancements in 2D MXene-based supercapacitor electrodes: Synthesis, mechanisms, electronic structure engineering, flexible wearable energy storage for real-world applications, and future prospects," *J. Mater. Chem. A*, vol. 12, no. 3, pp. 1342–1365, 2024.

[25] T. A. Skotheim, J. Reynolds, and T. A. Skotheim, *Conjugated Polymers – Processing and Applications*. CRC Press, 2006

[26] I. Inamuddin, M. I. Ahamed, R. Boddula, and T. Altalhi, *Polymers in Energy Conversion and Storage*. CRC Press, 2022.

[27] W. J. van Ooij and H. R. Anderson, *First International Congress on Adhesion Science and Technology*. CRC Press, 2023.

[28] M. Jamshidi and J. M. Gardner, "Copper(I) Iodide thin films: Deposition methods and hole-transporting performance," *Molecules*, vol. 29, no. 2, pp. 488, 2024.

[29] W. O. Soboyejo and T. S. Srivatsan, *Advanced Structural Materials – Properties, Design Optimization, and Applications*. CRC Press, 2019.

[30] C. Zhou, M. Zhang, and C. Y. Yang, *Nanocarbon Electronics*. Jenny Stanford Publishing, 2020.

[31] K. Iniewski, *Wireless Technologies – Circuits, Systems, and Devices.*, CRC Press, 2017.

[32] G. Nisato, D. Lupo, and S. Ganz, *Organic and Printed Electronics – Fundamentals and Applications*. Pan Stanford, 2019.

[33] S. Lakshmi Priya, T. W. Huang, K. Agrahari, and Y. W. Wang, "The comprehensive study of hybrid dielectric layer adopted organic thin film transistors for low voltage operation," *J. Mol. Liq.*, vol. 365, pp. 131827, 2024.

[34] T. Ndjountche, *Amplifiers, Comparators, Multipliers, Filters, and Oscillators*. CRC Press, 2019.

[35] C.-A. Di, Y. Liu, G. Yu, and D. Zhu, "Interface engineering: An effective approach toward high-performance organic field-effect transistors," *Acc. Chem. Res.*, vol. 42, no. 3, pp. 379–387, 2009.

[36] N. V. V. Subbarao, M. Gedda, P. K. Iyer, and D. K. Goswami, "Organic field-effect transistors as high performance humidity sensors with rapid response, recovery time and remarkable ambient stability," *Org. Electron.*, vol. 31, pp. 298–305, 2016.

[37] S.-S. Sun and L. R. Dalton, *Introduction to Organic Electronic and Optoelectronic Materials and Devices*. CRC Press, 2019.

[38] T. Vo-Dinh, *Biomedical Photonics Handbook – Biomedical Diagnostics*. CRC Press, 2019.

[39] A. Giwa, A. O. Yusuf, H. A. Balogun, A. Balaraman, et al., *Production of Chemicals and Energy*. Elsevier BV, 2024.

[40] M. Razeghi, "*The MOCVD Challenge – Volume 2: A Survey of GaInAsP-GaAs for Photonic and Electronic Device Applications*". CRC Press, 2019.

[41] C. Chappey, K. Fatyeyeva, E. Rynkowska, W. Kujawski, L. Karpenko-Jereb, A.-M. Kelterer, and S. Marais, "Sulfonic membrane sorption and permeation properties: Complementary approaches to select a membrane for pervaporation," *J. Phys. Chem. B*, vol. 121, no. 12, pp. 2938–2951, 2017.

[42] Y. Wang, C. J. Hansen, C.-C. Wu, E. J. Robinette, and A. M. Peterson, "Effect of surface wettability on the interfacial adhesion of a thermosetting elastomer on glass," *RSC Adv.*, vol. 11, no. 50, pp. 31665–31674, 2021.

[43] E. DiMasi and L. B. Gower, *Biomineralization Sourcebook – Characterization of Biominerals and Biomimetic Materials*. CRC Press, 2014.

[44] B. I. Kharisov, O. V. Kharissova, and U. Ortiz-Mendez, *CRC Concise Encyclopedia of Nanotechnology*. CRC Press, 2019.

[45] S. Mahamuni, D. Sidhaye, and S. Kulkarni, *Foundations of Experimental Physics*. CRC Press, 2020.

[46] P. Lall, M. G. Pecht, and E. B. Hakim, *Influence of Temperature on Microelectronics and System Reliability*. CRC Press, 2020.

[47] Metal-Halide, *Perovskite Semiconductors*. Springer Science and Business Media LLC.

[48] M. Piksa, C. Lian, I. C. Samuel, K. J. Pawlik, I. D. W. Samuel, and K. Matczyszyn, "The role of the light source in antimicrobial photodynamic therapy," *Chem. Soc. Rev.*, vol. 52, no. 15, pp. 5439–5463, 2023.

[49] Y. Sun, Y. Liu, and D. Zhu, "Advances in organic field-effect transistors," *J. Mater. Chem.*, vol. 15, no. 12, pp. 1168–1182, 2005.

[50] Sivashankar Krishnamoorthy, and Krzysztof (Kris) Iniewski, *Advances in Fabrication and Investigation of Nanomaterials for Industrial Applications*. Springer Science and Business Media LLC, 2024.

Chapter 8

Contact Resistance in an Organic Field Effect Transistor

Samayun Saikh and Ayash Kanto Mukherjee

8.1 INTRODUCTION

In the midst of rapidly advancing technology, Organic Field Effect Transistors (OFETs) are gaining popularity in the realm of flexible and portable electronic devices due to ease of fabrication over large areas. Despite recent advancement in device performance and various advantages that OFETs offer over conventional silicon-based devices, they have failed to make a niche in the market. This is because of several bottlenecks that are crippling OFETs with regards to practical use. One of the severely degrading parameters is contact resistance at the charge injecting contacts. This chapter explores different aspects of contact resistance in relation to OFETs and discusses various methods for minimizing it.

Metals are typically used for electrodes in organic semiconducting devices. Whenever a metal comes into physical atomic level contact with an organic semiconductor (OSC), the interface thus formed gives rise to contact resistance if certain conditions are met. This parasitic resistance presents a hindrance to the movement of charge carriers, which affects the efficiency and performance of an OFET. As a result, it also leads to a usage-based degradation of the device's overall performance. Therefore, thorough understanding of the origin and working of contact resistance between the metal-OSC contact is very much needed. Additionally, it is also essential to minimize the influence of this parasitic resistance in OFETs and other organic electronic devices. Addressing the causes and impact of contact resistance is crucial for diagnosing this issue and advancing the technology towards industrialization.

In the very beginning, this chapter systematically explains the physics of contact resistance in OFETs and how it affects device performance. It is followed by discussion on the impact that contact resistance has over key factors like mobility and threshold voltage. Next, the chapter provides a detailed analysis of different methods for measuring contact resistance. It explores various techniques to mitigate contact resistance in OFETs. The chapter concludes with a discussion on models used to characterize

 DOI: 10.1201/9781003533535-8

and quantify this resistance. Additionally, the chapter also discusses the limitations of various model and current measurement techniques. Challenges in minimizing contact resistance are also explored at the end.

The purpose of this chapter is to present an in-depth analysis of contact resistance and its impact. It is designed to help researchers and engineers overcome one of the major obstacles in organic electronics. By achieving necessary control over contact resistance, OFET technology can set forth a new era of electronic devices in the commercial market by unlocking its full potential.

8.2 PHYSICS BEHIND CONTACT RESISTANCE

The existence of contact resistance (R_C) in an OFET creates a *traffic jam*-like situation. It impedes the smooth flow of the charge carriers in and out of the device. R_C is characterized by the relationship between the potential difference across the contact region and the current passing through it. A high R_C reduces the efficiency of charge carrier injection and negatively impacts the performance of the device in question. It is to be noted that proper understanding of the underlying physics of contact resistance is complex and depends on various factors, such as the barrier for the charge injection between the metal and organic semiconductor (OSC), the quality of the interface, the fabrication methods for metallic contacts, the nature of the OSC, and the design of the OFET. This is essential for reducing R_C, thus improving injection, and enhancing the overall performance of OFETs.

Contact resistance is a combination of two essential components [1–3]: (1) injection-based contact resistance, $R_{C,inj}$, which is related to the energy barrier present at metal-OSC interface; and (2) morphology-based contact resistance at the vicinity of interface, $R_{C,bulk}$, which is due to the resistance in the OSC material near the interface. In other words, $R_C = R_{C,inj} + R_{C,bulk}$. Figure 8.1 shows a schematic of an OFET in two configurations, one for top contact and other for bottom contact. The top and bottom configurations are also known as *staggered* and *planar*, respectively. There is a difference in the geometrical orientation of electrodes in these two configurations. In a staggered OFET geometry, the source and drain electrodes are positioned on the opposite side of the gate electrode, whereas in *planar* configuration, all three electrodes are located at one side, albeit a dielectric layer separates the drain and source electrodes from the gate electrode. In both the cases, it can be observed that the two components exist in a series combination. It is crucial to develop an understanding of the origin of $R_{C,inj}$ and $R_{C,bulk}$.

At the atomic contact, the difference in work functions (WF) between a metal electrode and an OSC causes charge transfer between them. The primary reason for charge transfer is to balance out the Fermi levels. Since most OSCs are naturally *p*-type, their WF is higher than that of the metal. This implies that the Fermi level of the metal is positioned higher than that

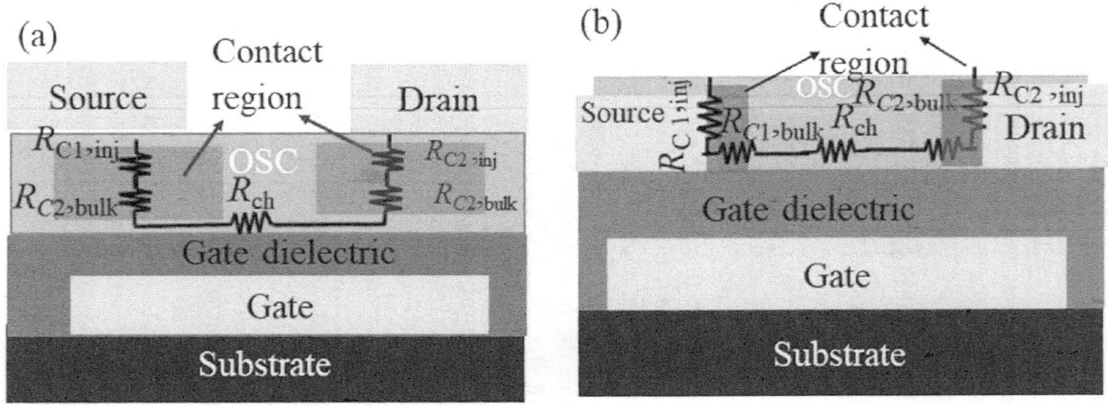

Figure 8.1 Schematics: (a) staggered OFET; (b) planar OFET. $R_{C,inj}$, and $R_{C,bulk}$ are resistance caused by the injection barrier and contact resistance due to bulk contribution. (Copyright 2024, Springer [3].)

of the OSC, as illustrated in Figure 8.2. The transferred charge accumulates across the interface, thus creating a charge double-layer. The positive head of double-layer is located to the metal-side while the negative tail is at the OSC molecule. Since an electric field exists across the charge double layer, the molecular orbitals, specifically the Highest Occupied Molecular Orbital (HOMO) and Lowest Unoccupied Molecular Orbital (LUMO), curve in the region occupied by the charge double layer. Further, the balancing action of the Fermi level also results in the formation of the hole-barrier $q\phi_{b0}$ positioned between the metal's Fermi level and the HOMO of the OSC, as depicted in Figure 8.2. This injection barrier is the primary cause of the high $R_{C,inj}$ at the interface between the electrode and OSC. The value of $R_{C,inj}$ depends exponentially on the height of the energy barrier [1–3]. Consequently, even a small change in the barrier height can lead to a significant variation in the injection resistance, $R_{C,inj}$. Typical values of $q\phi_{b0}$ range from 0.1 eV to 0.85 eV [4]. Next, it is necessary to address the origin of $R_{C,bulk}$.

The metal-OSC interface isn't uniform but exhibits an abrupt transition. Usually, the metal electrodes are fabricated by physical vapour deposition process. In some cases, radio frequency or DC sputtering is also used. In all these deposition processes, there occurs some degree of penetration of metal

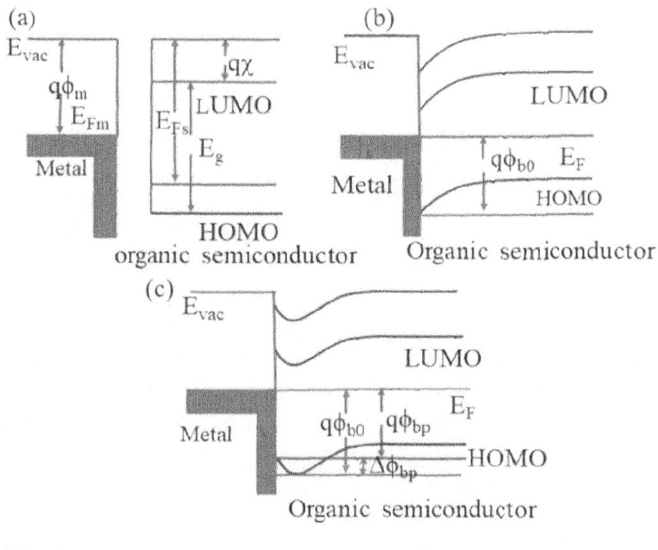

Figure 8.2 Schematics of energy band of metal-organic semiconductor contact: (a) before contact; (b) after contact. E_{vac}, $q\phi_m$, E_{FS}, E_g $q\chi$, $q\phi_{b0}$, and $\Delta\phi_{bp}$ are the vacuum level, work-function of metal, fermi level of OSC, HOMO-LUMO gap of OSC, electron affinity of OSC, injection barrier height at zero electric field and barrier lower due to image force, respectively. (Copyright 2024, Springer [3].)

atoms into the OSC [5,6]. This undesired penetration is limited to the region just below the metal-OSC interface for the source-drain electrode in the top-contact-configuration, as shown in Figure 8.1(a). For the bottom-contact-configuration, the OSC is spin-coated/drop-cast/thermally deposited over the pre-deposited metal electrodes. Since OSCs are big molecules, they carry significant amount of momentum. During deposition of these OSCs by any of the above mentioned methods, the OSC molecule loses its momentum and the momentum is transferred to the metal atoms of pre-deposited electrode. Some of these metal atoms could break out from the metallic bond and penetrate into the bulk of OSCs. Although the level of this undesired penetration is not high, it is adequate enough to create a region of different morphology and composition than the rest of bulk of OSC. This region is located near the immediate vicinity of the interface and it has dissimilar electrical properties as compared to the bulk of OSC. The presence of metal atoms at interstitial spaces of OSC molecules lead them to act as incoherent scattering centres thus enhancing the resistivity of this region. This lower conducting region is called the *contact region* [3]. The *contact region* typically has lower charge carrier mobility due to the increased disorder and local potential barriers that charge carriers encounter when moving from the electrodes into the OSC. Consequently, this results in higher contact resistance, denoted as $R_{C,bulk}$. Besides, $R_{C,inj}$ and $R_{C,bulk}$, there could be another additional factor that could contribute to R_C.

An insulating layer between the electrode and the OSC also increases R_C [7]. During contact deposition techniques—like point contact, sputtering, thermal evaporation, or electron beam deposition—an insulating or oxide layer might form in the region between the metal and OSC. This scenario is commonly encountered during deposition of aluminium and silver metals. Even though point contacts are commonly used, those are not perfect. Moreover, during deposition under high-vacuum, substrates or samples are exposed to the air before deposition, leading to the quick formation of an insulating layer [7]. This layer, whose thickness ranges between 10 Å to 20 Å [7], acts as an extra barrier to charge injection. This severely impedes the ability to achieve a good electrical contact.

The contributions of $R_{C,inj}$ and $R_{C,bulk}$ to the total contact resistance differ between *staggered* and *planar* OFET geometries [8]. In *planar* OFET devices, charge injection happens directly from the metal to the accumulation layer of OSC, so $R_{C,bulk}$ has a smaller impact [8]. In this case, most of the total R_C comes from $R_{C,inj}$. In contrast, for staggered OFETs, the charge carriers must traverse the OSC to reach the accumulation layer at the interface between the OSC and the gate dielectric, which means $R_{C,bulk}$ has a greater effect on the total R_C [8]. Despite this, staggered OFETs often have lower overall R_C than planar devices [9]. This is because metal atoms materials diffuse into the OSC in the *contact region* [5,6]. This may form a conductive pathway between the carrier-injecting-electrode and accumulation layer.

R_C is also influenced by the physical properties of the OSC. In crystalline OSC, bulk-mobility is generally higher than disordered systems. This allows for faster charge movement in the contact area and improved charge injection, thus leading to lower R_C. More detailed analysis on this is discussed in the section about modelling. Conversely, in disordered OSC, Lower mobility values hinder charge flow between the electrodes, leading to increased contact resistance at the *contact region*.

8.3 IMPACT OF CONTACT RESISTANCE

Contact resistance in OFET is an undesirable parasitic element. The impact of this parasitic parameter is profound. In an OFET, R_C reduces the magnitude of drain currents I_D. Alongside the presence of R_C, the characteristics of the OFET are found to deviate from standard metal oxide semiconductor field effect transistor (MOSFET) characteristics [10,11]. The presence of high R_C causes non-linearity in the transfer characteristics at low gate voltage (V_G) [10] and results in an *S*-shaped output characteristic under low bias conditions [11] as depicted in Figure 8.3(a) and 8.3(b). Additionally, the high R_C prevents the I_D from reaching saturation at higher values of V_D and V_G, as shown in Figure 8.3(c) [12]. Not only this, the presence of high R_C leads to inaccuracies in the characterization parameters such as charge carrier mobility (μ), as illustrated in Figure 8.3(d) [10]. Further, the presence of the R_C introduces an error in the extracted threshold voltage (V_T) values [13]. Therefore, it is important to understand how R_C affects μ and V_T.

The applied drain voltage (V_D) is distributed across three distinct regions: (1) the channel region where the voltage drop is V_{ch}; (2) across the region of the source contact, where the voltage drop is denoted as V_{sc}; and (3) across the drain contact region, where it is represented as V_{dc}, i.e., $V_D = V_{sc} + V_{ch} + V_{dc}$. So, the actual voltage drop across the channel is $V_{ch} = V_D - I_D (R_{C1} + R_{C2})$, Here, R_{C1} represents the resistance of source contact, R_{C2} represent resistance of drain contact, while I_D denotes the drain current. In the case of a MOSFET the resistances R_{C1} and R_{C2} are negligibly small and hence their impact is uncountable [14,15]. On the contrary, in OFETs, these resistances can be relatively high or even comparable to the channel resistance (R_{ch}). Therefore, their contributions are significant and cannot be overlooked [16–18]. μ and V_T are two key parameters of an OFET, typically extracted by fitting the classical MOSFET equations to the OFET characteristics [19]. This has remained the *modus operandi* for the determination of μ and V_T since the advent of OFET. However, utilizing MOSFET equations for this purpose introduces errors in the extracted values of μ [10, 19] and V_T [13], as these equations do not account for the effects of R_C [20]. Even minor inaccuracies in the extracted parameters can result in significant errors when designing new active materials, in developing new architectures, and creating new fabrication methods.

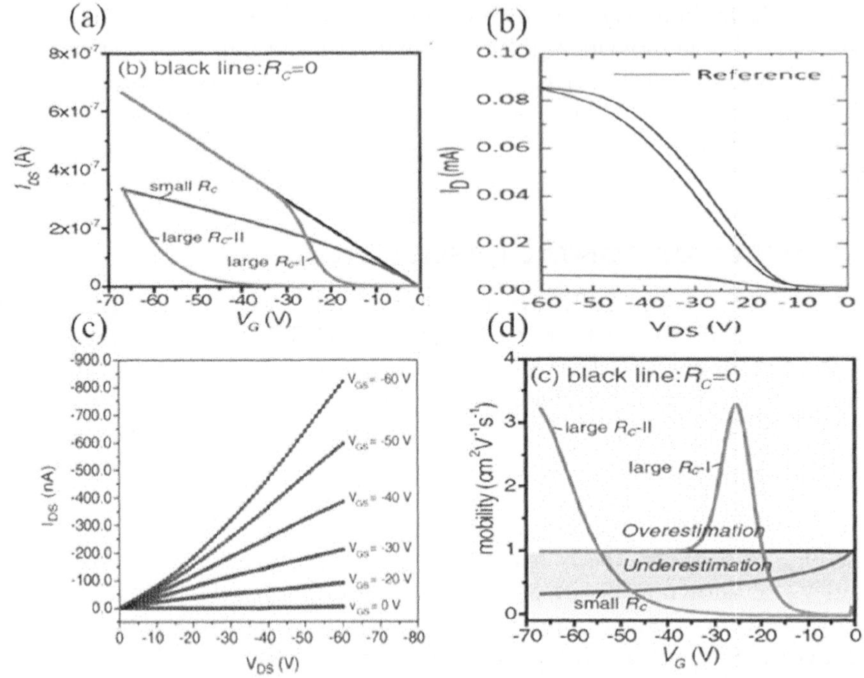

Figure 8.3 Various effects of contact resistance on the performance of an OFET: (a) non-linear transfer characteristic at low gate voltage; (b) S-shaped output characteristics at low bias; (c) non-saturation output characteristics at high voltage; (d) inaccuracies in the extracted values of charge carrier mobility. ((a,d) Copyright 2017, American Physical Society and SciPris [10]; (b) Copyright 2018, John Wiley & Sons [11]; (c) Copyright 2006, Elsevier [12].)

8.4 MEASUREMENTS TECHNIQUES

The impact of this detrimental parasitic resistance on the performance of an OFET is substantial. Therefore, accurate quantification of this parasitic element is crucial for optimizing device performance and advancing future designs. To extract this invasive parameter, several methods are commonly employed, including the Transfer Line Method, Four-Probe Method, and Scanning Kelvin Probe Microscopy. This section explores the theory behind these techniques, their limitations, and improvements.

a. Transfer Line Method

This is the most common method for the evaluation of total R_C in an OFETs, often known as the Gated Transfer Line Method (G-TLM) [21,22]. This method entails measuring the total resistance (R_{Tot}), which is the combined resistance of both R_C and R_{ch}. The total resistance (R_{Tot}) is expressed as: $R_{Tot} = R_C + R_{ch}$, where R_C denotes the sum of the R_{C1} and R_{C2}, *i.e.*, $R_C = R_{C1} + R_{C2}$. Therefore, R_{ch} is expressed as [21,22]:

$$R_{ch} = \frac{L}{W\mu_{ch} C_i (V_G - V_T)} \tag{8.1}$$

where L is the channel length, W is the channel width, μ_{ch} is the channel mobility, and C_i is the gate capacitance per unit area. The total normalized resistance in an OFET is:

$$R_{Tot} W = \frac{L}{W\mu_{ch} C_i (V_G - V_T)} + R_C W \tag{8.2}$$

By measuring R_{Tot} for OFETs with varying L and plotting R_{Tot} as a function of L, the total R_C can be obtained from the point where the plot intersects the y-axis, as shown in Figure 8.4(a). During this process, V_G is maintained at a constant value. The procedure relies on several assumptions:

1. μ_{ch}, V_T, and R_C are considered to be independent of L.
2. The OSC film is assumed to be uniform across all devices.
3. All devices are assumed to have identical electrical characteristics.
4. V_G is assumed to be evenly distributed across all resistors.

However, these assumptions may not always hold true. Variations in μ_{ch} and V_T with channel length can lead to inaccuracies in R_C measurements [23]. Further, this method requires several OFETs with varying channel lengths. Also, this measurement method is valid only for linear operational region of an OFET, which can be a practical limitation.

To improve the accuracy of R_C measurements obtained through the TLM, the Modified Transfer Line Method (M-TLM) has been developed [24]. This approach helps address some of the limitations posed by traditional TLM by making the results less sensitive to variations in channel length.

In M-TLM method, R_{Tot} is calibrated based on L using the following formula:

$$R_{Tot} \frac{W}{L} = \frac{1}{W\mu_{ch} C_i (V_G - V_T)} + R_C \frac{W}{L} \tag{8.3}$$

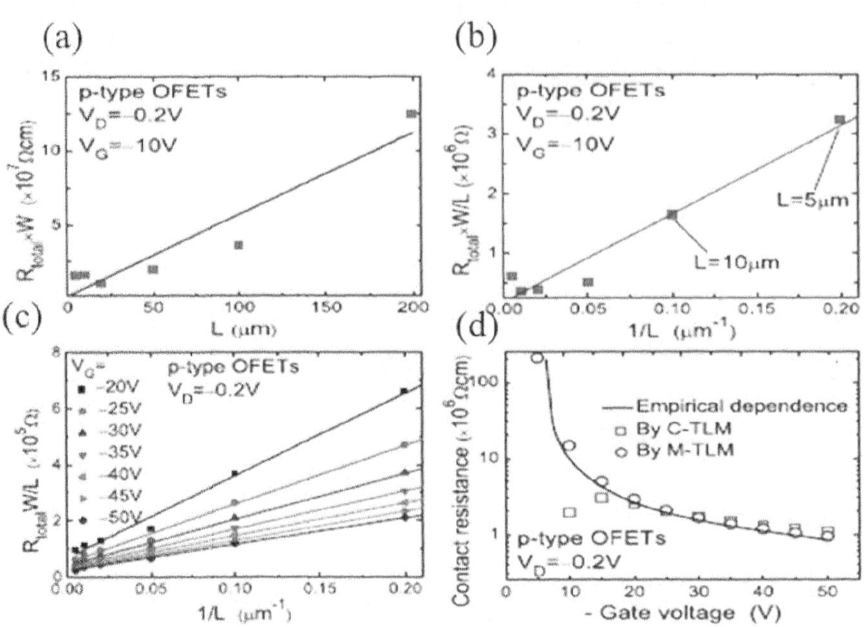

Figure 8.4 (a) Total resistance versus channel length graph. (b) Total resistance versus 1/channel length graph for single gate voltage. (c) Total resistance versus 1/channel length graph for different gate voltage. (d) Gate voltage versus total contact resistance extracted by conventional TLM and modified TLM. (Copyright 2010, American Institute of Physics [24].)

The term $R_{Tot}\frac{W}{L}$ is plotted against $1/L$ as depicted in Figure 8.4(b) and (c), and the slope of the resulting linear graph is used to determine the R_C. Figure 8.4(d) displays the variation of the extracted total R_C values with respect to V_G.

Another modification of TLM is done by introducing the channel length dependency of threshold voltage [25], which is expressed in the following relation:

$$V_T(L) = V_{T,\infty} + V_{T,0}\frac{1}{1+\frac{\rho_{ch}}{R_C}L} \tag{8.4}$$

where $V_{T,\infty}$, $\Delta V_{T,0}$, and ρ_{ch} are the threshold voltage for sufficiently high L shift of V_T shift for zero channel length and channel resistivity.

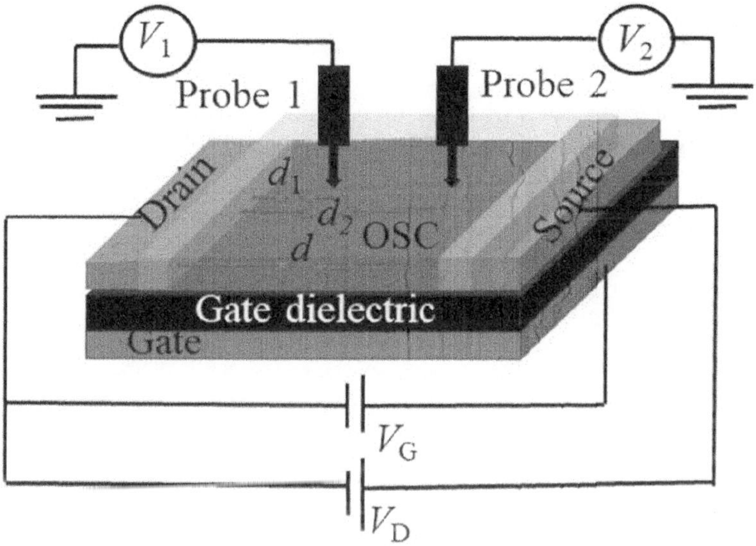

Figure 8.5 The schematic of a conventional four-probe method includes several key components: V_1 represents the measured potential at Probe 1, while V_2 indicates the measured potential at Probe 2. The distance from the source contact to Probe 1 is denoted as d_1, and the distance from the source contact to Probe 2 is labelled d_2. Additionally, d represents the distance between the source and drain electrode. (Copyright 2024, Institute of Physics [26].)

These modifications ensure more accurate measurement of the contact resistance than those obtained from traditional TLM methods. Additionally, the M-TLM approach results in a lower standard deviation, reflecting improved precision [24,25] in the R_C measurements.

b. Four-Probe Method

Another commonly used method for extracting contact resistance is the four-probe method [26–28], originally developed to measure semiconductor resistivity. In this setup, the outer two probes function as current sources, while the inner two probes serve as voltage sensors. The probes are arranged linearly, with the voltage-sensing probes separated from the current-supplying probes. The detailed schematics of the traditional four-probe method is presented in Figure 8.5.

The method involves positioning two voltage-sensing probes at separation distances of d_1 and d_2 from the source electrode, where $d_2 > d_1$. The measured voltages at these probes are V_1 and V_2 . The voltage drop across the channel (V_{ch}) is calculated as:

$$V_{ch} = \frac{V_2 - V_1}{d_2 - d_1} d \tag{8.5}$$

where d is the separation of the source-drain electrodes. Then, V_{sc} and V_{dc} are given by the following relations:

$$V_{sc} = V_1 - \frac{V_2 - V_1}{d_1 - d_1} d_1 \tag{8.5a}$$

and

$$V_{dc} = V_D - V_{sc} - \frac{V_2 - V_1}{d_2 - d_1} d \tag{8.5b}$$

The R_{C1} and R_{C2} are then calculated as:

$$R_{C1} = \frac{V_{sc}}{I_D} \text{ and } R_{C2} = \frac{V_{dc}}{I_D} \tag{8.6}$$

Although the four-probe method can separately determine R_{C1} and R_{C2}, it has limitations similar to the TLM. When transistors reach saturation, the measured R_C can be affected by errors due to pinch-off [12,29–30] near the drain contact, which causes a sudden increase in potential drop. This pinch-off effect renders the four-probe method less accurate in such conditions. Additionally, errors can arise from the voltage-sensing probes themselves, as the dimension of contact-patch can create a lateral shunting effect in the channel [31]. This can cause an increased drop in the channel region, leading to a strong local electric field that introduces significant errors in mobility measurements of OSC. The magnitude of this error can be more than two orders of magnitude higher than expected [31].

The conventional four-probe method has been slightly modified to create the van der Pauw method [32] for better accuracy in measurement. In this method, four electrodes are positioned at the corners of the semiconductor substrate. Current is supplied through two of the contacts, while the voltage is measured across the other two. While the four-probe method is applied to evaluate R_C, the van der Pauw method is specifically used to determine the resistivity of semiconductors. It provides highly accurate values of resistivity with minimal impact from R_C [32].

c. Scanning Kelvin Probe Microscopy

The direct measurement of potential drop across the *contact region* can be achieved using scanning Kelvin probe microscopy (SKPM). This technique, initially introduced by Bürgi et al. for OFETs [33], has since become widely used for measurement of contact resistances in an OFET. SKPM is integrated with all state-of-the-art atomic force microscopes.

In SKPM, a conducting tip is attached to the cantilever. During the SKPM scan, the combined system functions like a capacitor. When the tip touches the surface, some charge carriers move between tip and sample surface to equalize the Fermi levels. This leads to the formation of potential differences (V_{cpd}). The transfer of charge carriers induces a charge double layer at the interface between tip and sample surface. To neutralize V_{cpd}, a potential difference is externally applied to the tip or electrode. Therefore, the externally applied voltage needed to nullify V_{cpd} represents the voltage drop across the *contact region*. R_C is then determined by dividing the V_{cpd} by the current flowing through the contact. However, this is just one measurement, and the capacitor must be discharged before subsequent measurements. This is accomplished by introducing a vibrating tip on the cantilever, which acts as a reference electrode. The vibrating tip forms a variable capacitor with the sample. The tip scans at a constant height in the lateral direction during this process, as demonstrated in Figure 8.6.

The SKPM measurement consists of two stages. First, the surface topography of the sample is recorded while operating in tapping mode, where the tip oscillates at its resonant frequency just above the surface. When the tip makes contact with the surface, it deflects. The deflection is monitored by reflecting a laser beam off the cantilever. The deflected beam is then

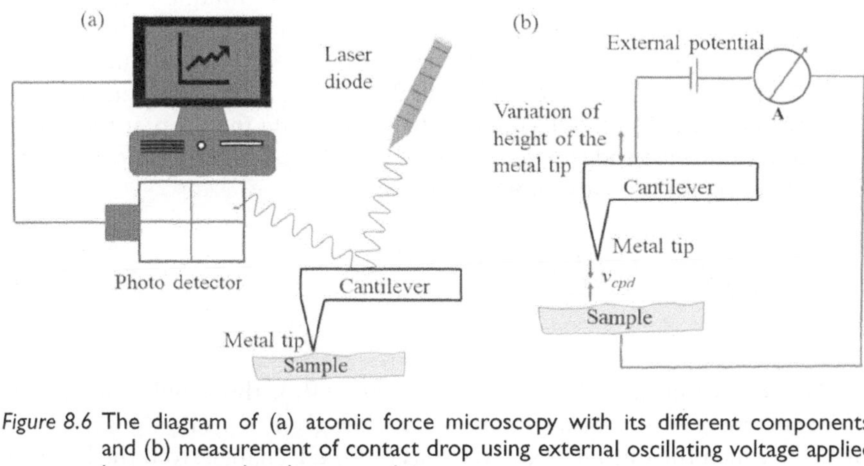

Figure 8.6 The diagram of (a) atomic force microscopy with its different components; and (b) measurement of contact drop using external oscillating voltage applied between sample substrate and tip.

detected by a four-quadrant photodetector. In the second step, the oscillation is stopped, and an alternating sinusoidal voltage (V_{ac} sinωt) is applied to the tip. This generates an oscillatory electrostatic force (F_{el}) between the tip and the surface of the sample. F_{el} is given by the following relation: $F_{el} = \frac{dC}{dz} V_{cpd} V_{ac} \sin\omega t$, where d$C$/d$z$ represents the change in tip/sample capacitance with respect to the tip height [34].

SKPM comes with its own limitations. The presence of traps on the sample surface can introduce errors in the topographic measurements. These traps can attract or repel the tip, altering the tip/sample distance and thereby affecting the accuracy of the topography determination [35,36]. In OFET measurements, this error is minimized by applying a gate voltage to clear the traps for accurate determination of topography [35].

Another intrinsic source of error arises from the complex interactions between the metallic tip and the sample [34]. This error can be reduced by generating a gradient of force, rather than maintaining a constant force, between the tip and the sample [35]. This approach enhances the accuracy of contact measurements. Further, incorporation of a gradient of force allows for the measurement of R_C at higher drain voltages, even exceeding 30 V [35].

d. Single Transistors Characteristics

The best method to determine the total R_C is by analyzing the characteristics of an OFET. Several methods have been proposed for extracting the R_C from the characteristics of an OFET. The details of this method are discussed below.

i. Transition Voltage Method (TVM) [37] is a technique used to determine the total R_C from the characteristics of an OFET. This method was proposed by S. D. Wang et al. and is used to determine the total R_C for pentacene-based OFETs in 2010 [37]. This technique modifies the classical MOSFET equation to incorporate the drop at the source and drain electrode contact ($I_D R_{C1}$ and $I_D R_{C2}$). The expression for total R_C using TVM is given by:

$$R_C = 2\frac{\sqrt{V_G - V_T}\left(\sqrt{2V_{Tr}} - \sqrt{V_G - V_T}\right)}{I_{D,sat}} \tag{8.7}$$

The expression involves transition voltage (V_{Tr}), threshold voltage, and drain current ($I_{D,sat}$) at saturation region for the given V_G. The V_{Tr} is the value of V_D at which the output characteristics of OFET change from the linear to

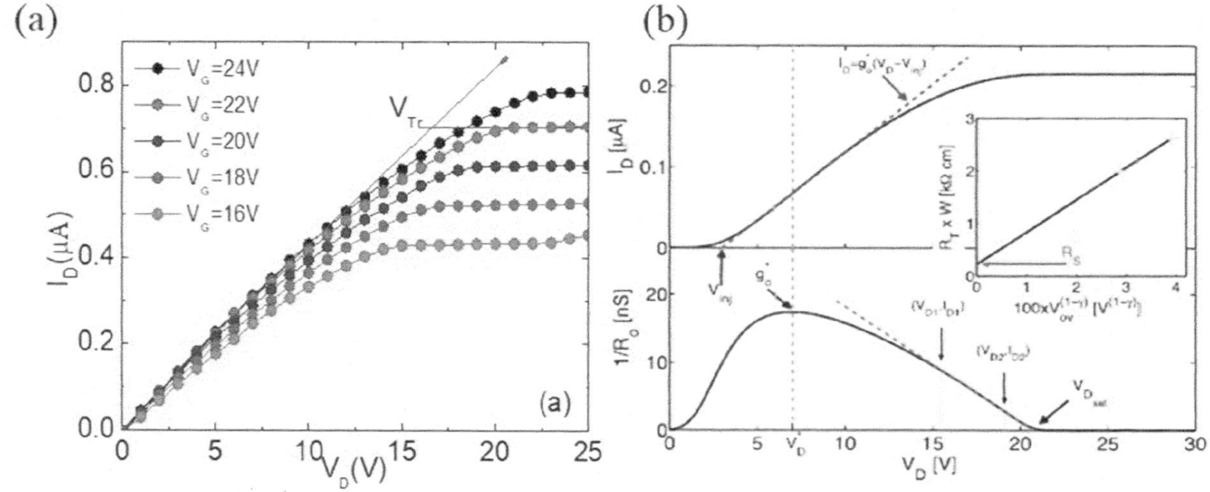

Figure 8.7 (a) Determination of transition voltage from the intersection of linear and saturation region. (b) Output characteristics and the variation of output conductance with drain voltage. Maximum conductance occurs at some value of drain voltage. ((a) Copyright 2024, Springer [39]; (b) Copyright 2014, American Institute of Physics [38].)

the saturation regime. This is shown in Figure 8.7(a). In this approach, it is presumed that the mobility stays constant throughout the operational range of an OFET. However, this assumption is not entirely accurate because in an OFET, mobility is dependent on the electric field. Additionally, some OFETs do not exhibit saturation at high gate and drain voltages, making this method inapplicable to such devices. Therefore, determination of V_{Tr} can be challenging for these OFETs. Despite these limitations and assumptions, TVM provides an estimate of R_C from a single OFET, offering a preliminary insight into the R_C.

ii. Fabrizio Torricelli et al. proposed a method called the Single Transistors Method (STM) [38], which requires only two output characteristics from a single transistor to evaluate R_C. In this method, the drain current (I_D) is given by:

$$I_D = \beta\left(\psi_s^\gamma - \psi_d^\gamma\right) \tag{8.8}$$

where ψ_s and ψ_d are two parameters which are proportional to the accumulated charge density near the source and drain contacts, respectively [38]. These are defined as: $\psi_s = V_G - V_T - V_S - V_C$ and $\psi_d = V_G - V_T - V_D$. Here, β is a pre-factor dependent on the physical and geometrical parameters of the transistor, and γ reflects the degree of disorder in the OSC [38]. The total contact drop V_C is expressed as:

$$V_C = I_D\, R_{C,bulk} + V_{inj} \tag{8.9}$$

where $R_{C,bulk}$ is the contribution of the OSC's bulk to the total R_C, and V_{inj} denotes the drop occurring at the metal – OSC interface. Before determining the R_C, the channel parameters including R_{ch}, γ, V_T, and β need to be extracted using the following relations:

$$R_{ch} = \left|\frac{\partial I_D}{\partial V_D}^{eq-1.4.1}\right|^{-1}_{V_S=V_S'} = \frac{1}{\beta\gamma\psi_d^{\gamma-1}} \tag{8.10}$$

$$R_C = R_0 - R_{ch} = R_0 - \frac{1}{\beta\gamma\left(V_G - V_T - V_D\right)^{\gamma-1}} \tag{8.11}$$

$$R_0 = \left[\frac{\partial I_D}{\partial V_D}\right]^{-1}_{meas} \tag{8.12}$$

$$\gamma = 1 + \ln\left[\frac{R_0\left(V_{D2}\right) - R_{C,bulk}}{R_0\left(V_{D1}\right) - R_{C,bulk}}\right] / \ln\left[\frac{\psi_{D1}}{\psi_{D2}}\right] \tag{8.13}$$

$$V_T = V_G - V_{D1} + \frac{g_0\left(V_{D1}\right)}{g_0\left(V_{D2}\right) - g_0\left(V_{D1}\right)}\left(V_{D2} - V_{D1}\right) \tag{8.14}$$

$$\beta = \left\{\frac{\gamma}{g_0^*}\left[\psi_{D^*}^{\gamma-1} - g_0^* R_{C,bulk}\,\psi_s^{\gamma-1}\right]\right\}^{-1} \tag{8.15}$$

At small values of V_D, the conductance $g_0 = 1/R_0$ increases with V_D . When V_D reaches V_{inj}, the drain current approaches zero, and the total resistance (R_{Tot}) at this point is attributed to the resistance resulting from the injection barrier (R_0 (V_{inj})). i.e.,

$$R_{Tot} = \left|\frac{\partial I_D}{\partial V_D}\right|^{-1}_{V_D = V_{inj}} = \frac{1}{\beta\gamma}\left(V_G - V_T - V_{inj}\right)^{1-\gamma} + R_{C,bulk} \tag{8.16}$$

In the equations (8.10) through (8.16), R_0 (V_D) is the measured values of total resistance at V_D, g_0 (V_D) is the conductance at V_D, g_0^* is the maximum output conductance at $V_D = V_D^*$ as presented in Figure 8.7(b).

The Equation (8.11) is used to evaluate the total R_C. However, this equation includes unknown parameters that must also be derived from the characteristics of the OFET. As a result, the method involves a series of calculations to extract the total R_C. First, the total resistance (R_0) is computed using equation (8.12). Next, the channel parameters, including, β, γ, and V_T are determined using equations (8.13) through (8.16).

The methods for finding the R_C from an OFET's output characteristics only need two measurements, which is convenient. However, there are drawbacks. First, equation (8.8) used for calculating drain current is empirical and not based on physical principles. Second, the method involves multiple calculations, which can enhance errors in the R_C values.

iii. S. Saikh et al. proposed a technique to simultaneously measure the total R_C and charge μ using the characteristics of an OFET [39]. The determination method relies on the conventional MOSFET equation. The presence of R_{C1} and R_{C2} causes partial drop of applied drain-source voltage across these regions. i.e.,

$$V_D = V_{sc} + V_{ch} + V_{dc} \tag{8.17}$$

In this case, V_{sc} and V_{dc} represent the voltage drops at the source and drain *contact regions*, respectively, while V_{ch} refers to the drop across the channel of an OFET. Thus, the accumulated charge per unit area induced by V_G is given by:

$$Q_G = C_i\left(V_G - V_T - V_C - V(x)\right) \tag{8.18}$$

where V_C = V_{sc} +V_{dc} represents the total voltage drop at *contact region*. Therefore, the drain current I_D is:

$$I_D = \mu_{ch} C_i W\left(V_G - V_T - V_C - V(x)\right)\frac{dV(x)}{dx} \tag{8.19}$$

The mobility in the channel region is denoted by μ_{ch}, W represents the width of the channel, and $dV(x)$ corresponds to the potential difference across a small segment of length dx. In the above expression, it is considered that the voltage drop at the *contact regions* does not vary with V_D, and $V_C = I_D R_C$. As the same current flows across the entire channel, the total drop of V_D along the channel can be expressed as:

$$I_D = \frac{\mu_{ch} W C_i}{L + \mu_{ch} W C_i R_C V_D}\left(V_G - V_T - \frac{V_D}{2}\right)V_D \tag{8.20}$$

At the linear region, $V_G - V_T >> V_D$, Therefore,

$$I_D = \frac{\mu_{ch} W C_i}{L + \mu_{ch} W C_i R_C V_D}\left(V_G - V_T\right)V_D \tag{8.21}$$

Hence the total resistance (R_{Tot}) is

$$R_{Tot} = \frac{V_D}{I_D} = \frac{L}{W\mu_{ch} C_i\left(V_G - V_T\right)} + \frac{R_C}{V_G - V_T}V_D \tag{8.22}$$

Equation (8.22) represents the overall resistance while OFET in the linear region and depends V_G, V_D, L and μ_{ch}. Therefore, the R_C can be calculated using the x-intercept of the equation (8.22). Additionally, the value of μ_{ch} can be determined from the point where the plot intersects the y-axis.

The derived expression is independent of the OFET configuration and can be used to extract R_C of a MOSFET. It is broad enough to be applicable to any active material or electrode configuration. While the expression provides the R_C in the linear operational regime of an OFET, it has some limitations. The derived equation is based on certain assumptions: first, that the total R_C is unaffected by the V_D, and second, that the channel mobility remains constant regardless of the drain voltage.

Thus, it could be observed that the determination of R_C cannot be done with the utmost accuracy. This is because all the above mentioned methods/procedure rely on certain assumptions and criteria. Moreover, device architecture and the morphological and electrical nature of contact also play vital roles in the decision-making process involved for the proper choice of the appropriate method for the determination of R_C. Needless to say, accurate extraction of R_C is a challenging problem and it is still open for further investigation and research.

8.5 REDUCTION OF CONTACT RESISTANCE

Since contact resistance significantly impacts the performance of an OFET, it is essential to address and minimize this parasitic effect. Several advanced techniques have been proposed to address this challenge, such as optimizing the electrode-OSC pair, doping the contacts, and incorporating self-assembled monolayers. This section provides a detailed discussion of these methods.

a. Electrode-Organic Semiconductor Combination

When two dissimilar materials come into atomic level contact, the nature of contact could be Schottky or Ohmic [40]. The result is influenced by the difference in the WF of the two materials. However, in the case of metal-OSC contact, the nature of the contact is almost always Schottky in nature. A Schottky contact poses a potential barrier to the charge injecting electrode. The height of this barrier determines the *injection* contact resistance by influencing the charge injection efficiency. For effective electron/hole injection, the electrode's WF should be proximity to the LUMO/HOMO of the OSC. Therefore, selecting an optimal metal-OSC contact is a crucial factor when designing an OFET. Thus, OFETs offer a broad range of options for electrode materials, spanning from inorganic metals to organic electrodes.

Metal electrodes like, such as calcium, zinc, lithium, and cerium, are well-suited for the effective injection of electrons into the LUMO level of an OSC as they have low WF. They are typically used for fabricating *n*-channel OFETs. Conversely, high WF metals like nickel, gold, palladium, iridium, and platinum are more compatible with hole injection into the HOMO level, making them ideal for *p*-channel OFETs. Additionally, metals like silver, which have WFs between the LUMO and HOMO levels, can serve as ambipolar electrodes and are suitable for fabricating both *p*- and *n*-channel OFETs.

The surface preparation method, cleanliness, and oxidation states can impact on the WF as it depends on the nature of the surface. Low WFs metal electrodes are sensitive to the ambient environment. They react with molecules or ions in the atmosphere easily. Fabricating *n*-channel OFETs presents significant challenges due to their inherent instability. This instability often leads to rapid degradation of the devices, making them less reliable and ever shortening their operational and shelf life.

High WF metal electrodes are well-suited for hole injection into organic semiconductors. However, during thermal or sputtering deposition, metal atoms can diffuse into the OSC due to its polycrystalline nature. This diffusion introduces additional disorders, which deteriorates the quality of contact and impairs effective charge injection. Furthermore, the metal electrode deposition process is often costly and it is associated with much less yield as most of the metal vapour is deposited on other exposed areas of the deposition chamber besides the sample.

Low-cost, large-area electronic devices can be fabricated utilizing OSC as an active material. Issues such as poor metal-OSC contact and the high cost of electrode deposition techniques drive researchers to seek alternative electrode materials. Organic electrodes are promising candidates for creating organic-organic contacts in these applications. Organic electrodes that closely match the composition of the organic semiconductor offer better compatibility, improved interface morphology, and minimized potential shifts at the organic/organic interface. A multitude of organic compounds have been explored and utilized as source/drain OFETs. Among these materials, several have shown particularly promising performance. Charge transfer complexes (CTs) [41,42] offer unique electronic properties due to their ability to facilitate charge transfer between different molecular components. Graphene and its derivatives are valued for their exceptional electrical conductivity, very high mobility, and high surface area, thus enhancing charge transport and stability [43–45]. Carbon nanotubes (CNTs) [46–48] also provide excellent conductivity and mechanical strength, contributing to improved electrode performance. Additionally, PEDOT (3,4-ethylenedioxythiophene) derivatives [49–51], known for their good electrical conductivity and processability, have emerged as strong candidates for organic electrodes. Each of these materials presents distinct advantages that can significantly enhance the functionality and efficiency of OFETs. All these organic materials show good OFET performance, lower R_C than their counterparts, as well as higher charge carrier mobility. Additionally, various organic devices such as organic solar cells and organic light-emitting diodes (OLEDs) are incorporating transparent organic electrodes, including graphene and graphene oxide.

Although organic electrodes can be easily fabricated by solution processes at low-cost, they offer some limitations too. Firstly, the performance of the device with organic electrodes can degrade due to their sensitivity to ambient conditions. And finally, the effective charge injection is limited by the relatively low conductivity of OSC at the contact interface.

b. Self-Assembled Monolayer

Inserting self-assembled monolayers (SAMs) between metals and OSC is a key technique for reducing contact resistance. SAMs enhance device performance by lowering the injection barrier height through modifications to the electrode's WF [52,53], leading to more efficient charge injection. A SAM has three main components: (1) a head group that forms a covalent bond with the atoms of substrate-surface, (2) a spacer or alkane chain that sets the thickness of the monolayer, and (3) at the end a terminal or functional group that interacts with the charge transport layer. A detailed schematic of the SAM is provided in Figure 8.8(a). The head group of SAMs varies depending on the reactivity of the substrate surface. For instance, thiol groups bond well covalently with gold (Au) surfaces, while silane groups are ideal for glass or silica substrates. The modification of substrate WF due to SAM occurs due to the interaction between local electric field with dipole of SAM [54]. Additionally, the dipole created by the newly formed covalent bond also contributes to the change in WF of the electrodes [54]. Typically, SAMs with electron-donor functional groups, such as -H or -NH_2, have dipoles oriented towards the end group. The orientation of dipoles helps to reduce the electrode's WF. In contrast, SAMs containing groups like -Cl, -F, or -NO_2 which are electron-acceptors have dipole moments directed towards the substrate surface. This results in increased WF of the electrodes. For example, the insertion of a thiophenol (TP) monolayer onto a silver electrode decreases the WF of the silver from 4.91 to 4.66 eV [55]. This reduced value is close to the LUMO level (–4.0 eV) of the Polyera ActivInk N2200, which is used in *n*-channel OFETs [55]. In contrast, the use of a pentafluorobenzene thiol (PFBT) SAM increases the WF to –5.24 eV, which is near the HOMO level of the active material, leading to *p*-channel transport [55,56]. Other fluorinated SAMs, such as 4-(trifluoromethyl) benzenethiol (TFBT) and 2,3,5,6-tetrafluoro-4-(trifluoromethyl) benzenethiol (TTFP) [55], have been demonstrated to increase the WF, reduce R_C, and reduce OFET performance [55, 57], as illustrated in Figure 8.8(b). Fluorinated SAMs are not the only means of modifying the WF of an electrode; dimers can also achieve similar effect. One such case is that the dimer bisjulolidyldisulfide reduces the WF of Au. The WF of Au reduces from 4.95 to 3.77 eV. This significant shift of 1.18 eV improved the performance of *n*-type OFETs using N2200 [55, 57]. Blending two SAMs has also been explored and has been found to enhance device performance. For instance, Xu et al. fine-tuned the WF by using

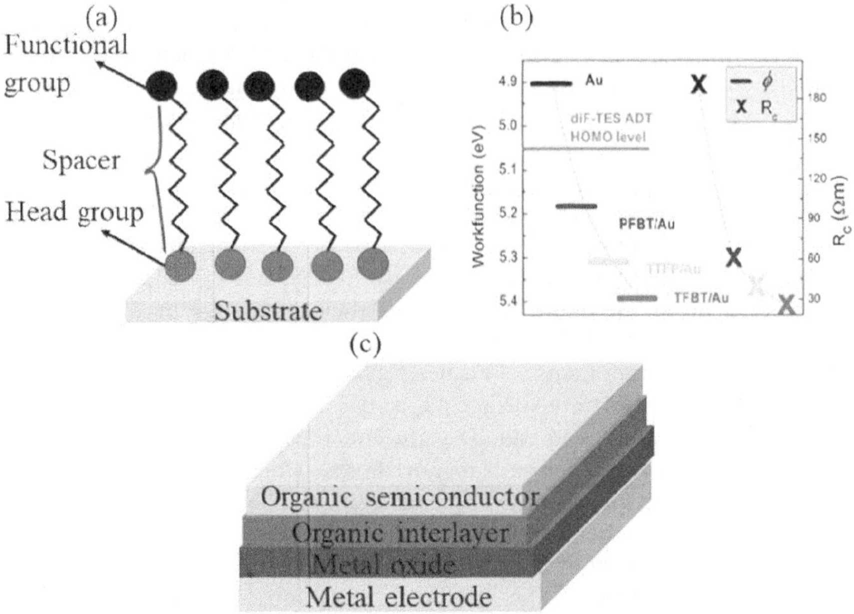

Figure 8.8 (a) Schematic and its various constituents of a self-assembled monolayer with different component. (b) Variation of work-function and contact resistance with different fluorinated self-assembled monolayer. (c) Schematic representation of bilayer contacts doping, where the first layer is a metal-oxide layer and the second layer is an organic layer. ((b) Copyright 2017, Elsevier [57].)

mixtures of two SAMs, namely, 1-decanethiol (1DT) and 1H,1H,2H,2H-perfluorodecanethiol (PFDT), in different proportion ratios [55]. This technique allowed them to progressively adjust the WF of Au from 4.5 to 5.7 eV. Consequently, these modified contacts facilitated both *n*-type and *p*-type transport in (Poly(9,9-di-noctylfluorene-alt-benzothiadiazole)) F8BT OFETs, which have HOMO/LUMO levels at −5.9 and −3.3 eV, respectively [55, 58]. The modifications of WF depend on degrees of order of the SAM and the strength and orientation of dipole moments. The degree of molecular arrangement in a SAM layer is influenced by several factors, such as the strength of the covalent bonds, the nature of the solvent, the intermolecular forces between SAM molecules, temperature, and others.

c. Doping at the Contact Region

Introducing dopants is an effective method to enhance the electrical conductivity of conjugated organic compounds. In organic semiconductor devices, trap states can arise at various stages during device fabrication. These trap states can be populated by the mobile carriers generated through bulk doping thereby decreasing both channel and contact resistance. By carefully selecting the sample-region for doping, one can achieve better control over device performance. Specifically, contact doping at the metal-OSC interface not only improves charge injection efficiency but also enhances several key parameters, including threshold voltage [13,59], mobility [60,61], and stability [62,63] in OFETs. Various techniques have been applied to achieve doping at the contact region, including photolithography, shadow masking, and the insertion of a SAM at the interface.

When a metal electrode contacts an organic semiconductor, a depletion region forms near the metal-OSC interface [64,65, 66]. This occurs due to relocation of charge carriers to equilibrate the Fermi levels between the metal and the OSC. Due to the relatively low carrier density in OSC, this depletion layer extends further into the OSC, resulting in a larger depletion width. Additionally, the inherent disorder in OSCs results in a significant number of traps within the material, which contribute to lower conductivity compared to the rest of the channel. Consequently, lower conductivity region contributes to the $R_{C,bulk}$ of the total high R_C.

To mitigate this R_C, various dopants have been tested in the *contact region*. The actual aim is to reduce the total R_C through either a decrease in $R_{C,inj}$ or $R_{C,bulk}$ or both. Dopants are categorized as either *p*-type or *n*-type. Notable *p*-type dopants include F_4TCNQ [67,68], F_6TCNQ [69,70], and $FeCl_3$ [71,72], while examples of *n*-type dopants are rhodocene dimer [73], cobaltocene [74], and various Lewis acids [75].

Careful selection of dopants can lower the injection barrier height by introducing intermediate states at the metal and contacting surface of OSC [55,76]. Additionally, the interfacial dipole formed between the dopants and the contact alters the WF [55,77] further reducing the barrier height. This decrease in barrier height contributes to a reduction in $R_{C,inj}$. Furthermore, dopants generate free carriers that fill trap states. The free carriers enhance the mobile carrier density at the *contact region* and, in turn, reduce $R_{C,bulk}$.

While doping is highly effective in inorganic electronic devices, it remains a challenging task when applied to OSC. Being weak van der Waals solids, OSCs show relatively weak intermolecular forces. As a result, when electrodes are deposited for contact, metal atoms can diffuse into the OSC. This diffusion of metal atoms can alter the properties of the OSC, potentially affecting its performance. Therefore, depositing electrodes in a precisely controlled manner is both difficult and challenging, especially at the *contact region*. Another challenge for effective doping in OSCs is their lower charge carrier density. As a result, a higher concentration of dopants is required to achieve effective doping. However, there is a flip side of using a high concentration of dopant. Firstly, higher dopant level poses the risk of migration of dopant molecules to regions away from the metal-OSC interface, thus doping the whole active layer. This further leads to a higher OFF current of the OFET which in turn hampers severely the ON/OFF ratio. Another risk of using high dopant concentration is potentially changing the composition of the active material to such an extent that the active material could no longer be regarded as a single moiety but a blend. The blend, thus formed, would show significantly shifted HOMO and LUMO levels along with the formation of a new energy level due to dopants. In conclusion, presence of a dopant can cause substantial changes in the properties of the OSC.

Another approach for contact-doping involves insertion of a metal oxide layer between the electrode material and the OSC. This method has shown significant reduction of R_C. Metal oxides exhibit a wide spectrum of WFs. It is ranges from 3.5 eV for zirconium oxide to 7 eV for vanadium oxide with a control over concentration of defect-state. This versatility enables the oxides to serve as an electron transport layer to the LUMO level or a hole transport layer to the HOMO level. Such flexibility makes metal oxides highly effective and advantageous as intermediate layers. For instance, the semiconductor C8-BTBT (2,7- didodecyl [1] benzothieno [3,2-b][1]benzothiophene) has a HOMO level around –5.7 eV, creating a significant injection barrier for Au contacts with a WF of approximately –5 eV. The addition of an MoO_3 interlayer significantly improved the R_C in C8-BTBT devices [55,78]. This resulted in a decrease of R_C by an order of magnitude, dropping from around 10^5 Ω cm for untreated Au contacts due to the large injection barrier to about 10^4 Ω cm for Au/MoO_3 contacts [55,78].

Other examples include vanadium oxide (V_2O_5, with a WF of approximately 7.0 eV), which has been employed to elevate the WF of molybdenum from 4.3 to 4.9 eV [79]. Transistors utilizing P3HT, dodecyl-substituted thienylenevinylene, and dodecylthiophene with Mo/V_2O_5 contacts demonstrated comparable performance to those with Au contacts, while also offering the benefit of reduced cost [55,79]. Darmawan et al. investigated various oxide films including those made from Al, Hf, Zr, Ta, Ti, and Si as 1-nm interlayers to enhance pentacene devices [80]. Among the materials tested, aluminium oxide (Al_2O_3) demonstrated the lowest R_C and μ [53]. Specifically, it achieved an R_C value of 1.9 kΩ·cm and a μ of 0.52 cm^2/V-s [53]. In contrast, devices with Au contacts exhibited a significantly higher R_C of 19.9 kΩ·cm and a lower μ of 0.32 cm^2/V-s [53]. By employing these optimized contacts, the R_C in pentacene-based OFETs is reduced to 8.1 kΩ·cm [53], a notable improvement over the 46 kΩ·cm observed with Au contacts and 25 kΩ·cm with MoO_3-treated Au contacts [53,55].

The physics behind the reduction of R_C due to the insertion of an oxide layer is not yet fully understood, and investigations are on track. However, several possible explanations could account for the observed decrease in R_C. One possibility is that the metal oxide layer improves alignment between its WF and the HOMO or LUMO levels of the OSC [56, 57, 81, 77,78]. Another potential reason is that the metal oxide might act as a dopant, which could enhance at the interfacial charge carrier density [82]. When the metal oxide is deposited onto a metal electrode, it can *pin* the Fermi level, potentially impacting R_C. However, this effect is more related to $R_{C,bulk}$ rather than $R_{C,inj}$ [82]. The observed reduction in R_C likely results from improved long-range charge transfer into the OSC, which decreases R_C [82].

Significant research has been carried out to tune the thickness of the oxide-layer in OFETs. The objective is to achieve the minimum R_C possible. The optimized value of oxide-thickness depends on the thickness-values of various constituents of OFET. This includes the contact pad, substrate, and semiconductor as well as the overall geometry of the device.

Most metal-OSC contact tuning strategies involve incorporating a single intermediate layer between the OSC and the electrode materials. However, Kotadiya et al. developed a versatile approach that uses two interlayers [55,82]. The first is a metal oxide layer, while the second is an OSC layer with an ionization energy greater than that of the active material [83], as shown in Figure 8.8(c). This second organic layer effectively decouples the contact from the oxide layer [83]. The organic/organic interface is more compatible than the organic/inorganic interface. This configuration enhances injection efficiency and results in improved device performance. The enhancement is due to the formation of injection pathways. Charge carriers traverse intermediate states within the bandgap through this pathway. Test devices featuring tris(4-carbazoyl-9-ylphenyl) amine (TCTA, ionization energy = 5.7 eV) sandwiched between a PEDOT electrode and a double-layer electrode exhibited improved injection with various organic interlayers on MoO_3 [55,83]. These interlayers included 4,4′-bis(N-carbazolyl) biphenyl (CBP), 1,2,3,5-tetrakis(carbazol-9-yl)-4,6-dicyanobenzene (4CzIPN), C_{60}, and 4,4″-bis(triphenylsilanyl)-(1,1′,4′,1″)-terphenyl (BST) [55]. The corresponding ionization energies are 6.0, 6.1, 6.4, and 7.0 eV, respectively [55]. A comparable trend is observed when CBP is used as the active material, despite its high ionization energy of 6.0 eV. In this case, C_{60}, 4CzIPN, and BST enhanced performance, while TCTA, with a lower ionization energy than CBP, led to reduced performance [55,83].

Although metal oxides show great promise for the reduction of R_C, challenges remain in the deposition process. Thin oxide layers are conductive, but their conductivity decreases as the layer thickens. Therefore, a thin layer allows for efficient tunnelling. However, if the layer becomes too thin, its WF may also decrease which may not be compatible for *smooth* tunnelling [84]. Another disadvantage is that when the metal oxide is deposited upon a metal, it can oxidize the metal underneath [84]. The degrees of oxidation are proportional to the of the oxide layer thickness. Consequently, an increase in the oxide layer's thickness can lead to inefficient charge injection.

Conclusively, there are various *tricks of the trade* as far as strategies for reduction of R_C is concerned. A researcher could think of any method, judge the pros and cons of the method and then put it into trial to test whether the method registers a yield as per expectation. Thus, there is no limit in devising and exercising any method of choice. However, irrespective of the method practiced, every method has its own advantages and disadvantages. Typically, the values of R_C in conventional silicon-based devices have been contained to very low values ~mΩ. However, OFETs still exhibit R_C of the order of a few kΩ or higher. Therefore, the very first step towards commercialization of the OFET is to reduce the contact resistance comparable to its inorganic counterpart. It indicates that future prospect of OFET research in this direction is enormous.

8.6 DIFFERENT MODEL OF CONTACT RESISTANCE

High contact resistance can profoundly impact the performance of an OFET. To evaluate and minimize these adverse effects, various measurement and reduction techniques have been discussed in the earlier section *albeit* a strong theoretical understanding is crucial. It helps to identify the root causes of high R_C and its dependence on various factors. This could assist guided efforts for the reduction of R_C. R_C is influenced by various factors such as the injection barrier, OFET architecture, and charge carrier mobility. Therefore, understanding how these factors interrelate is crucial for designing improved contacts, novel architectures, advanced methods, and exploration of new materials. To address these challenges, several models of contact resistance have been developed. This section is dedicated to a detailed discussion of various models of contact resistance, including their dependencies and limitations.

The total R_C is a combination of the $R_{C,inj}$ and $R_{C,bulk}$. Therefore, it can be expressed as: $R_C = R_{C,inj} + R_{C,bulk}$ [1–3]. The contributions of barrier height and bulk affect R_C in a varied manner for staggered and coplanar OFETs. In a staggered OFET, charge injection occurs perpendicular to the channel, as can be confirmed from Figure 8.9(a). Perpendicular injection requires the injected carriers to traverse through a few layers of OSC in order to reach the accumulation channel. Figure 8.9(b) illustrates charge injection and carrier movement across various OFET configurations. In contrast, in a planar OFET, charge injection happens parallel to the accumulated channel, resulting in a substantially shorter distance for carriers to reach the accumulation-channel. Consequently, various models for $R_{C,inj}$ and $R_{C,bulk}$ have been proposed. Since, both the above mentioned components are present in an OFET, therefore it is natural to expect an interplay of these models. The interplay amounts to interdependency of various device parameters. This section first discusses the contact resistance model for the staggered OFET geometry and then covers the model for the planar OFET geometry.

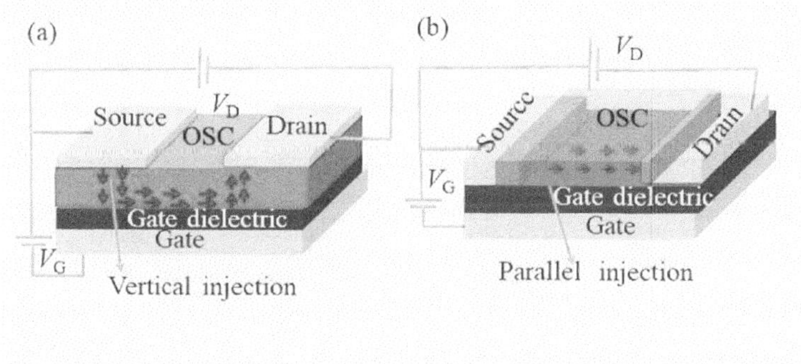

Figure 8.9 (a) Schematics of the charge injection for staggered OFET geometry; and (b) Schematics of the charge injection for planar OFET geometry. Red arrows indicate direction and movement of the carriers.

a. Contact Resistance Model for Staggered Configuration

Under *staggered* configuration, three models are discussed. It might be possible that there could exist other models that are not covered in this chapter. These models do not carry any specific name.

i. The very first model for R_C is based on the *current crowding model*, which was proposed by Chiang et al. [85]. The expression for the R_C in this model is written as:

$$R_C = L_T R_{ch} \coth\left(\frac{d}{L_T}\right) \tag{8.23}$$

where L_T is the transfer length as shown in Figure 8.10. L_T represents the distance over which majority of charge carriers are transferred and is expressed as follows:

$$L_T = \sqrt{\frac{R_{ch}}{W\rho_c}} \tag{8.24}$$

In this model, d, ρ_{ch}, ρ_c, and W represent the OSC layer thickness, the channel resistivity, *contact region* resistivity, and the contact width, respectively. In this model, it is assumed that ρ_c is independent of V_G. The model explains the V_G dependence of R_C by considering the V_G-dependent R_{ch}. Accumulation of carriers increases with an increase of V_G which reduces

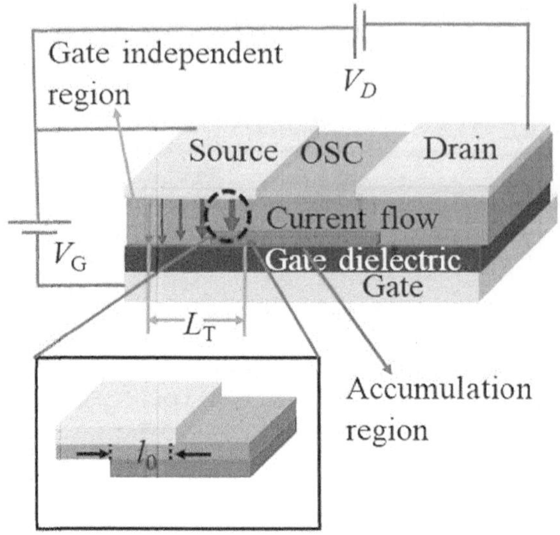

Figure 8.10 A detailed schematic illustrating the effective charge injection in an organic field-effect transistor is presented, where L_T denotes the effective characteristic length of the injected electrode. l_0 is the channel length under source contact. (Figure adapted from T.J. Richards and H. Sirringhaus, 2007, Journal of Applied Physics [85].)

both the R_{ch} and the R_C. However, it is observed by researchers that ρ_c is V_G dependent and this is represented by the following relation [85]:

$$\rho_c = \frac{A}{V_G - V_T} + \rho_{c0} \tag{8.25}$$

In the above expression, A and ρ_{co} are the two fitting parameters. A depends on the dielectric constant of the gate dielectric, thickness of the OSC and mobility [85]. ρ_{c0} is the resistivity of the *contact region* depicted in Figure 8.1(a) and Figure 8.1(b), where the carrier concentration is independent of the V_G [85].

Although the reduction in R_C with V_G is explained by the modified contact resistivity, there is still a lack of explanation on how $R_{C,inj}$ depends on V_G .

ii. The second model is an empirical model for V_G dependent R_C. It was introduced by Luan and Neudeck for amorphous silicon thin-film transistors [86]. This model has been extended for *staggered* OFETs. The empirical relation for total R_C is given by the following relation [86]:

$$R_C = \frac{l_0}{W\mu C_i (V_G - V_T)} + R_{C0} \tag{8.26}$$

where l_0 represents the characteristic length of the accumulation channel under the source/drain contact, and R_{C0} denotes the residual R_C in the gated negligible region shown in Figure 8.10. Since l_0 is much smaller than the channel length, Kanicki adjusted for this by incorporating l_0 into L, defining it as the effective length of the channel $l_{0\text{eff}}$. Therefore, the $l_{0\text{eff}}$ is expressed as $l_{0\text{eff}} = l_0 + L$.

iii. Although the two proposed models account for the dependency of R_C on V_G, they do not consider the effect of traps during transport through the OSC. The influence of traps is incorporated into the empirical formula for R_C, which is based on the Meyer-Neldel Rule (MNR) for trap distribution [8]. The empirical model of R_C is expressed as [8]:

$$R_C \propto \exp\left(\frac{E_B}{kT} - \frac{E_B}{E_{MN}}\right) \tag{8.27}$$

In this model, E_B represents the average trap energy, E_{MN} denotes the width of the trap distribution known as Meyer-Neldel energy, k is the Boltzmann constant, and T is the absolute temperature.

This empirical model doesn't account for how the gate field influences R_C. Additionally, Jin et al. use the Meyer-Neldel Rule to study R_C, but the model fails to clarify the source of the trap energy E_B or its connection to the trapping processes [8]. This is because specifying the nature of the trapping process is crucial for determining the trap energy E_B. While it is generally challenging to achieve this specification, it becomes more feasible when examining the trapping process in the context of a polycrystalline OSC thin film within a staggered OFET configuration. In a polycrystalline OFET, charge trapping processes predominantly occur at the grain boundaries (GBs) [8]. These processes can be effectively described by the grain boundary trapping model (GBTM) [8]. In a polycrystalline OSC thin film, grain boundaries also come into contact with metal electrodes [8]. Consequently, charge trapping occurs at these grain boundaries, which can act as a barrier to carriers, ultimately affecting the $R_{C,\text{bulk}}$.

Vinciguerra et al. addressed the V_G dependency of bulk contact resistance by applying the Grain Boundary Trapping Model [8]. In this model, the hopping between the adjacent grain is controlled by an energy barrier E_B.

A relation exists between density of trap states N_t and E_B in GBTM and it is given by [8]:

$$E_B = \frac{q^2 N_t}{8\epsilon_s N_D} \tag{8.28}$$

In this equation, ϵ_s represents the dielectric constant of the bulk OSC, and N_D is the concentration of shallow donors. When a V_G is applied, it induces a surface charge density N_G, calculated as $N_G = \frac{C_i V_G}{q}$. This induced surface charge modifies the barrier height caused by the traps. Consequently, the donor concentration is adjusted from N_D to a value of $N_D - N_G/t$, with t denoting the OSC thickness. The reduction in barrier height is represented by ΔE_B and it is given by the following expression:

$$\Delta E_B = \frac{E_B N_G}{t N_D - N_G} \tag{8.29}$$

Thus, the effective barrier height of the trap states is given by:

$$E_B + \Delta E_B = \frac{q^2 N_t}{8\epsilon_s} \frac{t}{t N_D - V_G} = \frac{q^2 N_t^2}{8\epsilon_s} \frac{t}{|N_G| - t|N_D|}$$

or,

$$E_B + \Delta E_B = \frac{q^3 N_t^2 t}{8\epsilon_s C_i} \frac{1}{|V_G| - \frac{qt}{C_i}|N_D|} \tag{8.30}$$

Therefore, the expression for contact resistance, considering the contribution of traps, is:

$$R_C = R_{C0} \exp\left\{\left(\frac{1}{kT} - \frac{1}{E_{MN}}\right)(E_B + \Delta E_B)\right\}$$

or

$$R_C = R_{C0} \exp\left\{\left(\frac{1}{kT} - \frac{1}{E_{MN}}\right)\left(\frac{q^3 N_t^2 t}{8\epsilon_s C_i} \frac{1}{|V_G| - |V_T|}\right)\right\} \tag{8.31}$$

Where $V_T = \frac{qt}{C_i}|N_D|$. Equation (8.31) is valid only for $R_{C,bulk}$ and does not account for interfacial effects. When the electrode atoms interact with the OSC, an injection barrier forms at the contact. The applied field influences the height of this barrier by generating image charge within the electrode material. Moreover, application of V_G creates an accumulation channel of mobile charges. These charges create an image charge which reduces the interfacial barrier and makes the injection smooth. This barrier lowering effect can be either a Schottky barrier lowering effect [3] or a Poole-Frenkel lowering effect [87]. This model accounts for the reduction in Schottky barrier height, which affects the effective interfacial barrier height as expressed in the following relation:

$$\phi_{eff} = \phi_{bo} - \Delta\phi_s, \text{ where } \Delta\phi_s = \sqrt{\frac{q^3 V_G}{4\pi\epsilon_\varepsilon t_{ox}}} \tag{8.32}$$

where ϕ_{b0} is the barrier height at zero electric field depicted in Figure 8.2(c), t_{ox} is the thickness of gate insulator. Therefore, the expression for the total R_C in a *staggered* OFET is:

$$R_C = R'_{C0} \exp\left\{\left(\frac{1}{kT} - \frac{1}{E_{MN}}\right)\left(\frac{q^3 N_t t}{8\epsilon_S C_i}\frac{1}{|V_G| - |V_T|}\right)\right\} \exp\left(\frac{q\phi_{b0}}{kT} - \frac{1}{kT}\sqrt{\frac{q^3 V_G}{4\pi\epsilon_S t_{ox}}}\right) \tag{8.33}$$

In the above expression, R'_{C0} is a constant with the dimension of resistance and it is based on the thermionic injection mechanism:

$$R'_{C0} = \frac{h^3}{4\pi q^2 m^* kT}\frac{1}{Wt} \tag{8.34}$$

Here, m^* represents the charge carrier effective mass and h is Planck's constant. In the above expression, it is assumed that charge accumulation occurs throughout the entire bulk of the organic semiconductor. However, device physics analysis indicates that charge actually accumulates only within a few monolayers of the organic semiconductor [88,89].

b. Contact Resistance Model for Planar Configuration

So far, we have explored various models of R_C for *staggered* OFET geometries. However, the *planar* OFET geometry, source, drain, and gate electrodes are located on the same side of the OSC. This preference is due to its simpler fabrication process as compared to *staggered* and the reduced risk of damaging the semiconductor during electrode deposition, which is a common issue in *staggered* configurations. Additionally, lithography techniques can achieve increasingly smaller electrode dimensions. Therefore, there is a pressing need to have a R_C model specifically for OFETs with bottom contact configuration. In this regard, three R_C models have been discussed. The models don't bear any specific name as before.

i. The very first model is proposed by Chang Hyun Kim et al. and it focuses on the charge distribution at the interface between the source electrode and the OSC [9]. This model examines how charge carriers accumulate and interact near the interface. By analyzing the charge distribution, a model for R_C is developed. The source contact injects charge carriers into the OSC upon application of drain bias. When a V_G is applied, these carriers distribute and accumulate at the OSC-gate dielectric interface. At the metal-OSC interface, there is a sharp transition between the electrode material and OSC that leads to significant disorder in the region near the interface. The abrupt transition causes non-uniformity in the distribution of charge carriers. Specifically, the accumulation density near the interface is much lower with respect to the rest of the *effective* region of the accumulation channel. Therefore, the carrier density changes between the contact region and the channel. The region where the carrier density changes gradually is called the *transition region*. The distribution of charge carrier is demonstrated in Figure 8.11(a). For a given V_G, the carrier density at the end of this region aligns with the density in the accumulation channel, which remains constant. The derived model is based on the average carrier density within this transition region.

   The carrier density in the transition zone can be expressed as [9]:

$$p(x) = p_s \exp\left(\frac{x}{x_s}\right) \tag{8.35}$$

The transition region extends from $x = 0$ to $x = b$, with $x = 0$ being the position of the source contact. At the transition point $x = b$, the carrier density is:

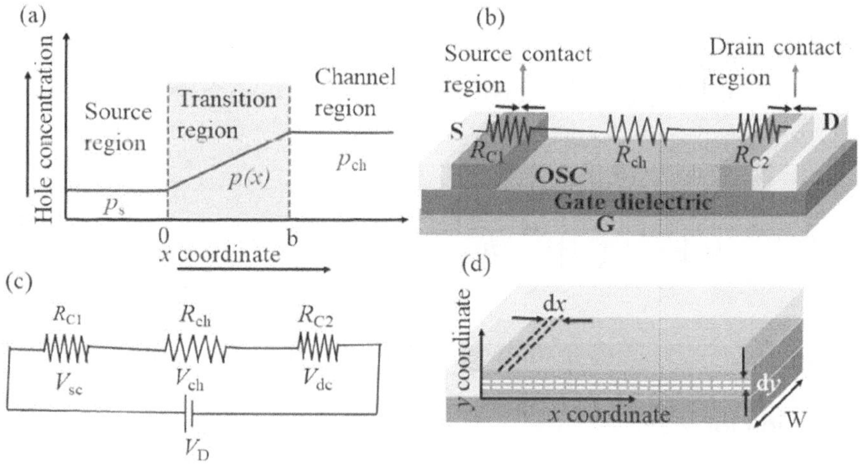

Figure 8.11 A comprehensive diagram illustrates the carrier density in both the contact and channel regions. The transition zone extends from the source contact into the OSC, marking the boundary between the source and the OFET channel. Here, $p(x)$ is the charge carrier density in the transition region, while p_{ch} denotes the in the channel carrier density. (Figure 8.11a adapted from C.H. Kim, Y. Bonnassieux, and G. Horowitz. 2012. IEEE [9].)

$$p(b) = p_s \exp\left(\frac{b}{x_s}\right) \tag{8.36}$$

At this transition point, the carrier density aligns to the in the accumulated channel density, p_{ch}. Solving equation (8.36), one obtains:

$$b = x_s \ln\left(\frac{p_{ch}}{p_s}\right) \tag{8.37}$$

The mean carrier concentration in the transition region is expressed as:

$$p_{avg} = \frac{1}{b}\int_0^b p(x)\,dx \tag{8.38}$$

Substituting $p(x)$ into the equation (8.38) yields:

$$p_{avg} = \frac{p_s}{b}\int_0^b \exp\left(\frac{x}{x_s}\right)dx$$

$$p_{avg} = \frac{p_s x_s}{b}\left[\exp\left(\frac{x}{x_s}\right)\right]_0^b = \frac{p_s x_s}{b}\left[\exp\left(\frac{b}{x_s}\right) - 1\right] \tag{8.39}$$

Substituting $b = x_s \ln\left(\frac{p_{ch}}{p_s}\right)$ into equation (8.39), one obtains the final expression for p_{avg}:

$$p_{avg} = \frac{p_s x_s}{x_s\, ln\left(\frac{p_{ch}}{p_s}\right)}\left[\frac{p_{ch}}{p_s} - 1\right] = \frac{p_{ch} - p_s}{ln\left(\frac{p_{ch}}{p_s}\right)} \tag{8.40}$$

Since the carrier density of channel is much higher than that in transition region, i.e., $p_{ch} >> p_s$, therefore, the expression for p_{avg} boils down to:

$$p_{avg} = \frac{p_{ch}}{\ln\left(\frac{p_{ch}}{p_s}\right)} \tag{8.41}$$

Since p_s and p_{ch} are functions of the thickness of the OSC, the average carrier density p_{avg} also depends on the thickness of the OSC. To evaluate the R_C, consider an elemental slice of the OSC along its thickness. Let the length and width of this elemental slice be W and dy, respectively. Then, the conductance of this elemental slice is given by:

$$dG = \frac{q\mu W p_{avg}}{b}dy \tag{8.42}$$

Hence, the overall resistance of the transition region, which represents total R_C, is expressed as follows:

$$R_C = \frac{1}{G} = \frac{\sqrt{\epsilon_s kT}}{q^2 \mu W}\left[\int_0^d \frac{p_{ch}(y)\sqrt{p_s(y)}}{\left[\ln\left(\frac{p_{ch}(y)}{p_s(y)}\right)\right]^2} dy\right]^{-1} \tag{8.43}$$

The equation (8.43) describes the total R_C, which is based on the charge distribution near the source contact region. Since the expression cannot be solved analytically, only numerical methods can be used to find solutions. In this expression, there are three unknown parameters: the $p(x)$ in the transition region, p_{ch}, and the μ in the transition region. Numerical solutions can only be obtained when these distribution functions are known. These distribution functions are determined through approximations based on the finite thickness of the semiconductor [9].

ii. Another model for R_C proposed by Jun Li et al. is based on the thermionic injection mechanism [2]. The injected current density at the metal-OSC interface is [2],

$$J = \frac{4\pi q^2 m^*}{h^3} T^2 \exp\left(-\frac{q\phi_b}{kT}\right) \tag{8.44}$$

Therefore, J depends strongly on the $q\phi_b$ present at the metal-OSC interface. The injected charge induces an image charge at the source. The image charge and injected charge forms a dipole. The dipole creates a strong local field at the metal-OSC interface which causes a reduction of barrier height. The reduction of barrier height improves the efficiency of charge injection. The reduction barrier height can be a Schottky lowering effect or a Poole-Frenkel barrier lowering effect. This model accounts for the reduction in Schottky barrier height, which in turn affects the effective interfacial barrier height as expressed in the following relation:

$\phi_{eff} = \phi_{bo} - \Delta\phi_s$, where

$$\Delta\phi_s = \sqrt{\frac{q^3 V_G}{4\pi\epsilon_s t_{ox}}} \tag{8.45}$$

So, the equation for injected current density,

$$J = \frac{4\pi qm^* k^2}{h^3} T^2 \exp\left(-\frac{\phi_{b0} + \Delta\phi_s}{kT} \right) \tag{8.46}$$

$$J = \frac{4\pi qm^* k^2}{h^3} T^2 \exp\left(-\frac{\phi_{b0}}{kT} + \frac{q}{kT}\sqrt{\frac{qE_d}{4\pi\epsilon_s}} \right) \tag{8.47}$$

Here E_d is the external field applied between the source-drain electrodes. The resistivity of OSC at the immediate vicinity of interface is given by:

$$\rho = \frac{E}{J} = \frac{h^3}{4\pi qm^* k^2 T^2} \left(\frac{kT}{\beta} \right)^2 \frac{x^2}{\exp(x)} \exp\left(\frac{\Delta\phi_s}{kT} \right) \tag{8.48}$$

where x is associated with the local field in the following manner: $x = \frac{\beta\sqrt{E}}{kT}$ and $\beta = \frac{q}{kT}\sqrt{\frac{q}{4\pi\epsilon_s}}$. The local electric field is the resultant electric field E between the externally applied field and the field generated within the OSC. Therefore, the net electric field,

$$E = \frac{\epsilon_i}{\epsilon_d t_s + \epsilon_s t_d} (V_G - V_T) \tag{8.49}$$

where ε_s and ε_d, are the relative dielectric constant OSC and gate insulator, respectively, and t_s and t_d are thicknesses of OSC and gate insulator, respectively. It is to be noted that the OSC and gate insulator are considered as two dielectric layers arranged in a series configuration. Hence, the normalized R_C is given by the following expression [2]:

$$R_C W = A \frac{\left(B\sqrt{V_G - V_T} \right)^2}{\exp\left(B\sqrt{V_G - V_T} \right)} + C \tag{8.50}$$

Here A, B, and C are fitting parameters. These fitting parameters bear physical significances. The A parameter has the unit of resistance and it represent the resistance due to injection barrier. B has the unit of $V^{-0.5}$ and it represents the deviation of the external electric field from a strong local electric field at the metal-OSC interface. C represents $R_{C,bulk}$.

iii. S. Saikh et al. proposed an R_C model for *planar* OFET configuration [3]. This model is also based on the thermionic injection mechanism. In this model, a Poole-Frenkel barrier lowering effect is considered. It assumes a *contact region* near the interface between the metal and OSC, as shown in Figure 8.11(b). In this region, the morphology and electrical properties differ from those in the channel region away from the interface.

The current density equation due to carrier injection can be expressed as:

$$J = n_{if} q \mu_c E \tag{8.51}$$

$$\text{where } n_{if} = \frac{N_h}{1 + \exp\left(\frac{q\phi_{bp}}{kT}\right)} \tag{8.52}$$

where N_h, k, T, and $q\varphi_{bp}$ are the hopping sites density at the *contact region*, Boltzmann constant, absolute temperature, and the injection barrier, respectively.

In this model, the injection into the OSC is influenced by the carrier mobility of the *contact region* [3]. When carrier mobility is low, carriers spend more time near the metal-OSC interface. The extended duration raises the probability of repulsive interactions between carriers in the channel and those near the metal-OSC interface, just before injection, ultimately leading to an increase in R_C [3]. This model considers the V_G-dependent mobility (μ_C) in the *contact region*, which is based on the multiple trapping and release hopping mechanism, and is expressed as follows [3]:

$$\mu_C = f_0 \left(V_G - V_T\right)^{\gamma} \tag{8.53}$$

Here, f_0 is the pre-factor with units of $cm^2V^{-(\gamma+1)}\ s^{-1}$ which is associated with the *contact region* density states, while γ is related to the hopping process between localized states and the characteristic temperature of the traps.

Thus, the total injected current can be determined from the above current density equation.

$$I_{inj} = J_{inj} Wt \tag{8.54}$$

where W and t is the width and thickness of the electrode.

By incorporating equations (8.51), (8.52), and (8.53) into equation (8.54), the expression for the injected current is obtained as follows.

$$I_{inj} = \frac{qN_h f_0 Wt\left(V_G - V_T\right)^{\gamma} V_D}{L\left[1 + \exp\left(\frac{q\phi_{bp}}{kT}\right)\right]} \tag{8.55}$$

The source-contact resistance R_{C1} is given by:

$$R_{C1} = \frac{V_D}{I_{inj}} = \frac{L\left(1 + \exp\left(\frac{q\phi_{bp}}{kT}\right)\right)}{qN_h f_0 Wt\left(V_G - V_T\right)^{\gamma}} \tag{8.56}$$

Assuming R_{C2} to be identical to R_{C1}, the total R_C is:

$$R_C = R_{C1} + R_{C2} = 2R_{C1} = \frac{2L\left(1 + \exp\left(\frac{q\phi_{bp}}{kT}\right)\right)}{qN_h f_0 Wt\left(V_G - V_T\right)^{\gamma}} \tag{8.57}$$

The expression for total R_C illustrates that R_C is a highly complex phenomenon. Therefore, R_C is a function of many factors, such as the energy and width of the trap states, temperature, applied electric field at the metal-OSC interface, grain boundary size, and the effective mass of the charge carrier. However, the mismatch of WF between the metal and OSC is even more intricate. Besides trap states and interfacial effects, factors such as interfacial dipole, surface charge density, charge carrier mobility, and charge carrier density also significantly impact the total R_C. Consequently, these models are still incomplete. A universal model that addresses all these factors and is applicable to all OFET device architectures is urgently needed.

8.7 CONCLUSIONS

This chapter presents an in-depth exploration of the origin and impact of contact resistance between metal and organic semiconductors. The impact of contact resistance on organic semiconductor devices is enormous. Proper numerical estimation of contact resistance is the very first step to start with. In this regard, the chapter has discussed various measurement techniques, such as transfer line method, four probe method, scanning Kelvin probe method, etc. Since the presence of contact resistance in any semiconductor-based device is considered to be a detrimental feature of the device, therefore, it is very important to get rid of this parasitic element. In this purview, a number of methods for reduction of contact resistance have been discussed in the context of organic semiconductor devices. Key reduction strategies, such as the use of self-assembled monolayers, contact doping, and incorporation of various electrode-organic semiconductor combinations, are discussed in the relevant sections. Additionally, the limitations of these approaches are also considered. Even with these advancements, the contact resistance is still approximately four orders of magnitude larger compared to its inorganic counterpart.

A proper understanding of the behaviour of contact resistance is as important as the origin of this parasitic element. This chapter has also discussed various models for contact resistance in organic field effect transistors (OFETs). Since operation of OFETs encompasses various current-voltage parameters, such as gate voltage, drain voltage, and drain current, it is necessary to know the dependence of contact resistance on the current-voltage parameters. From this perspective, six models are discussed in the chapter. Nevertheless, the models do show dependence on the architecture of the OFET device, namely, the staggered and planar geometry. In summary, this chapter presents a concise and up-to-date description of contact resistance in organic semiconductor devices.

8.8 FUTURE PERSPECTIVE OF AN ORGANIC FIELD EFFECT TRANSISTOR

OSC are becoming increasingly recognized as promising materials for wearable plastic electronics, as they offer low-cost device fabrication techniques. They have garnered considerable attention because of their capability for large-area coverage, and potential for integration into a broad spectrum of applications. Among the various organic electronic devices, OFETs stand out as fundamental components. OFETs serve as a crucial building block for the development of electronics devices. OFETs can enable the creation of flexible and affordable devices. These devices can be used in diverse fields such as displays, sensors, and electronic circuits. Their mechanical flexibility and capacity for adaptation serve as a driving force towards flexible electronics. Since the inception of the first OFET, significant efforts have been made to enhance their performance. Despite their potential, several challenges persist, a major issue being the high contact resistance between the metal and OSC. Reducing high contact resistance remains a crucial challenge for organic electronics researchers, necessitating ongoing efforts in both theoretical and experimental domains. Current methods, such as the TLM and the four-probe method, are being refined for the accurate determination of contact resistance. The refined methods incorporate the bias-dependent carrier mobility and contact resistance throughout the operational range of OFETs. Additionally, improvements are being made to address issues in current-voltage relationships, such as non-linearity and lack of saturation in output characteristics, to accurately measure important parameters. These parameters are vital for the development of new techniques.

Recent research has highlighted distinct charge transport mechanisms in various organic electronic devices, such as organic light emitting diodes (OLEDs), organic photovoltaics (OPVs), and organic field-effect transistors (OFETs) [90]. The physical principles underlying OLEDs and OPVs do not fully explain the behaviour observed in OFETs, indicating the need for device-specific insights.

While much research is focused on minimizing contact resistance, high contact resistance can also be advantageous in specific applications. For example, organic Schottky transistors utilize high contact resistance for unipolar charge carrier transport, operating efficiently in the subthreshold region. This allows them to function at lower voltages and offers potential for amplifying weak signals. Moreover, the high R_C can be deliberately leveraged to restrict specific carriers and regulate recombination processes, which is particularly useful in designing *n*-type OFETs.

Despite extensive research into measurement and reduction techniques, some aspects of OFETs remain underexplored. Charge injection into OSC and the transport in OFETs require further investigation to achieve significant breakthroughs. We hope this chapter provides valuable insights into the physics of contact resistance of OFETs.

REFERENCES

1. M. Weis, M. Nakao, J. Lin, T. Manaka, and M. Iwamoto, “Thermionic emission model for contact resistance in organic field-effect transistor,” *Thin Solid Films*, vol. 518, no. 2, pp. 795–798, 2009.
2. J. Li, W. Ou-Yang, and M. Weis, “Electric-field enhanced thermionic emission model for carrier injection mechanism of organic field-effect transistors: Understanding of contact resistance,” *Journal of Physics D: Applied Physics*, vol. 50, no. 3, p. 35101, 2016.
3. S. Saikh, N. Rajan, and A. K. Mukherjee, “Thermionic injection and contact resistance model for bottom contact organic field-effect transistors,” *Journal of Electronic Materials*, vol. 53, no. 6, pp. 1–11, 2024.
4. B. K. Sarker and S. I. Khondaker, “Thermionic emission and tunneling at carbon nanotube–organic semiconductor interface,” *ACS Nano*, vol. 6, no. 6, pp. 4993–4999, 2012.
5. P. V Pesavento, R. J. Chesterfield, C. R. Newman, and C. D. Frisbie, “Gated four-probe measurements on pentacene thin-film transistors: Contact resistance as a function of gate voltage and temperature,” *Journal of Applied Physics*, vol. 96, no. 12, pp. 7312–7324, 2004

6. Y. Zheng, A. T. S. Wee, C. Troadec, and N. Chandrasekhar, "Temperature-dependent transition from injection-limited to space-charge-limited current in metal-organic diodes," *Applied Physics Letters*, vol. 95, no. 14, p. 143303, 2009.
7. E. H. Rhoderick and R. H. Williams, *Metal-semiconductor contacts*, vol. 129. Clarendon press Oxford, 1988.
8. V. Vinciguerra, M. La Rosa, D. Nicolosi, G. Sicurella, and L. Occhipinti, "Modeling the gate bias dependence of contact resistance in staggered polycrystalline organic thin film transistors," *Organic Electronics*, vol. 10, no. 6, pp. 1074–1081, 2009.
9. C. H. Kim, Y. Bonnassieux, and G. Horowitz, "Charge distribution and contact resistance model for coplanar organic field-effect transistors," *IEEE Transactions on Electron Devices*, vol. 60, no. 1, pp. 280–287, 2012.
10. C. Liu *et al.*, "Device physics of contact issues for the overestimation and underestimation of carrier mobility in field-effect transistors," *Physical Review Applied*, vol. 8, no. 3, p. 34020, 2017
11. M. Nikolka *et al.*, "Performance improvements in conjugated polymer devices by removal of water-induced traps," *Advanced Materials*, vol. 30, no. 36, p. 1801874, 2018.
12. J. N. Haddock, X. Zhang, S. Zheng, Q. Zhang, S. R. Marder, and B. Kippelen, "A comprehensive study of short channel effects in organic field-effect transistors," *Organic Electronics*, vol. 7, no. 1, pp. 45–54, 2006.
13. J.-L. Hou, D. Kasemann, J. Widmer, A. A. Günther, B. Lüssem, and K. Leo, "Reduced contact resistance in top-contact organic field-effect transistors by interface contact doping," *Applied Physics Letters*, vol. 108, no. 10, p. 103303, 2016.
14. Y. Shi, J. Liu, Y. Hu, W. Hu, and L. Jiang, "Effect of contact resistance in organic field-effect transistors," *Nano Select*, vol. 2, no. 9, pp. 1661–1681, 2021.
15. P.-C. Shen *et al.*, "Ultralow contact resistance between semimetal and monolayer semiconductors," *Nature*, vol. 593, no. 7858, pp. 211–217, 2021.
16. K. Hong, S. Y. Yang, C. Yang, S. H. Kim, D. Choi, and C. E. Park, "Reducing the contact resistance in organic thin-film transistors by introducing a PEDOT: PSS hole-injection layer," *Organic electronics*, vol. 9, no. 5, pp. 864–868, 2008.
17. Y. Kim *et al.*, "Highly stable contact doping in organic field effect transistors by Dopant-Blockade method," *Advanced Functional Materials*, vol. 30, no. 28, p. 2000058, 2020.
18. Y. Xu *et al.*, "Extraction of low-frequency noise in contact resistance of organic field-effect transistors," *Applied Physics Letters*, vol. 97, no. 3, p. 033503, 2010.
19. E. G. Bittle, J. I. Basham, T. N. Jackson, O. D. Jurchescu, and D. J. Gundlach, "Mobility overestimation due to gated contacts in organic field-effect transistors," *Nature Communications*, vol. 7, no. 1, p. 10908, 2016.
20. V. Podzorov, "Organic single crystals: Addressing the fundamentals of organic electronics," *MRS Bulletin*, vol. 38, no. 1, pp. 15–24, 2013.

21. S. Jung, J. W. Jin, V. Mosser, Y. Bonnassieux, and G. Horowitz, "A compact model and parameter extraction method for a staggered OFET with power-law contact resistance and mobility," *IEEE Transactions on Electron Devices*, vol. 66, no. 11, pp. 4894–4900, 2019.
22. G. Horowitz, P. Lang, M. Mottaghi, and H. Aubin, "Extracting parameters from the current–voltage characteristics of organic field--effect transistors," *Advanced Functional Materials*, vol. 14, no. 11, pp. 1069–1074, 2004.
23. N. Padma, "Channel length influenced contribution of hole and electron trapping effect to threshold voltage stability in organic field effect transistors," *Materials Science in Semiconductor Processing*, vol. 39, pp. 384–389, 2015.
24. Y. Xu, R. Gwoziecki, I. Chartier, R. Coppard, F. Balestra, and G. Ghibaudo, "Modified transmission-line method for contact resistance extraction in organic field-effect transistors," *Applied Physics Letters*, vol. 97, no. 6, p. 063302, 2010.
25. Weis, K. Lee, D. Taguchi, T. Manaka, and M. Iwamoto, "Modified transmission-line method for evaluation of the contact resistance: Effect of channel-length-dependent threshold voltage," *Japanese Journal of Applied Physics*, vol. 53, no. 1, p. 11601, 2013.
26. E. von Hauff, N. Spethmann, and J. Parisi, "A gated four probe technique for field effect measurements on disordered organic semiconductors," *Zeitschrift für Naturforschung A*, vol. 63, no. 9, pp. 591–595, 2008.
27. R. J. Chesterfield *et al.*, "Variable temperature film and contact resistance measurements on operating n-channel organic thin film transistors," *Journal of applied physics*, vol. 95, no. 11, pp. 6396–6405, 2004.
28. J. Cho and T. Mori, "Low-temperature band transport and impact of contact resistance in organic field-effect transistors based on single-crystal films of Ph-BTBT-C10," *Physical Review Applied*, vol. 5, no. 6, p. 64017, 2016.
29. W. Wu *et al.*, "High-Performance, low-operating-voltage organic field-effect transistors with low pinch-off voltages," *Advanced Functional Materials*, vol. 18, no. 5, pp. 810–815, 2008.
30. S. Tanwar, R. Millan-Solsona, S. Ruiz-Molina, M. Mas-Torrent, A. Kyndiah, and G. Gomila, "Nanoscale Operando Characterization of electrolyte-gated organic field-effect transistors reveals charge transport bottlenecks," *Advanced Materials*, vol. 36, no. 13, p. 2309767, 2024
31. H. H. Choi, K. Cho, C. D. Frisbie, H. Sirringhaus, and V. Podzorov, "Critical assessment of charge mobility extraction in FETs," *Nature Materials*, vol. 17, no. 1, pp. 2–7, 2018.
32. C. Rolin, E. Kang, J.-H. Lee, G. Borghs, P. Heremans, and J. Genoe, "Charge carrier mobility in thin films of organic semiconductors by the gated van der Pauw method," *Nature Communications*, vol. 8, no. 1, p. 14975, 2017.
33. L. Bürgi, H. Sirringhaus, and R. H. Friend, "Noncontact potentiometry of polymer field-effect transistors," *Applied Physics Letters*, vol. 80, no. 16, pp. 2913–2915, 2002.
34. C. Maragliano, S. Lilliu, M. S. Dahlem, M. Chiesa, T. Souier, and M. Stefancich, "Quantifying charge carrier concentration in ZnO thin films by Scanning Kelvin Probe Microscopy," *Scientific reports*, vol. 4, no. 1, p. 4203, 2014.

35. G. de Tournadre, F. Reisdorffer, R. Rödel, O. Simonetti, H. Klauk, and L. Giraudet, "High voltage surface potential measurements in ambient conditions: Application to organic thin-film transistor injection and transport characterization," *Journal of Applied Physics,* vol. 119, no. 12, p. 125501, 2016.
36. N. B. Bercu, L. Giraudet, O. Simonetti, and M. Molinari, "Development of an improved Kelvin probe force microscope for accurate local potential measurements on biased electronic devices," *Journal of Microscopy*, vol. 267, no. 3, pp. 272–279, 2017.
37. S. D. Wang, Y. Yan, and K. Tsukagoshi, "Transition-voltage method for estimating contact resistance in organic thin-film transistors," *IEEE Electron Device Letters*, vol. 31, no. 5, pp. 509–511, 2010.
38. F. Torricelli, M. Ghittorelli, L. Colalongo, and Z.-M. Kovacs-Vajna, "Single-transistor method for the extraction of the contact and channel resistances in organic field-effect transistors," *Applied Physics Letters*, vol. 104, no. 9, p. 093303, 2014.
39. S. Saikh, N. Rajan, and A. K. Mukherjee, "A method for direct determination of voltage dependent contact resistance and mobility of an organic field effect transistor," *Journal of Materials Science: Materials in Electronics*, vol. 35, no. 5, pp. 1–11, 2024.
40. M. Hanic, T. Vincze, V. Rezo, and M. Weis, "Nature of Ohmic and Schottky contacts on pentacene-based organic field-effect transistor," *Synthetic Metals*, vol. 305, p. 117590, 2024
41. R. Pfattner, C. Rovira, and M. Mas-Torrent, "Organic metal engineering for enhanced field-effect transistor performance," *Physical Chemistry Chemical Physics*, vol. 17, no. 40, pp. 26545–26552, 2015.
42. Y. Takahashi, T. Hasegawa, Y. Abe, Y. Tokura, K. Nishimura, and G. Saito, "Tuning of electron injections for n-type organic transistor based on charge-transfer compounds," *Applied Physics Letters*, vol. 86, no. 6, p. 063504, 2005.
43. S. Pang, Y. Hernandez, X. Feng, and K. Müllen, "Graphene as transparent electrode material for organic electronics," *Advanced Materials*, vol. 23, no. 25, pp. 2779–2795, 2011.
44. S. Basu, M. C. Lee, and Y.-H. Wang, "Graphene-based electrodes for enhanced organic thin film transistors based on pentacene," *Physical Chemistry Chemical Physics*, vol. 16, no. 31, pp. 16701–16710, 2014.
45. P. H. Wöbkenberg *et al.*, "Reduced graphene oxide electrodes for large area organic electronics," *Advanced Materials*, vol. 23, no. 13, pp. 1558–1562, 2011.
46. C. M. Aguirre, C. Ternon, M. Paillet, P. Desjardins, and R. Martel, "Carbon nanotubes as injection electrodes for organic thin film transistors," *Nano Letters*, vol. 9, no. 4, pp. 1457–1461, 2009.
47. F. Cicoira, C. M. Aguirre, and R. Martel, "Making contacts to n-type organic transistors using carbon nanotube arrays," *ACS Nano*, vol. 5, no. 1, pp. 283–290, 2011.
48. Y. Y. Zhang *et al.*, "Poly (3, 3‴-didodecylquarterthiophene) field effect transistors with single-walled carbon nanotube based source and drain electrodes," *Applied Physics Letters*, vol. 91, no. 22, p. 223512, 2007.

49. N. Stutzmann, R. H. Friend, and H. Sirringhaus, "Self-aligned, vertical-channel, polymer field-effect transistors," *Science*, vol. 299, no. 5614, pp. 1881–1884, 2003.
50. M. Hasan *et al.*, "Uniform resistive switching with a thin reactive metal interface layer in metal-La0. 7Ca0. 3MnO3-metal heterostructures," *Applied Physics Letters*, vol. 92, no. 20, p. 202102, 2008.
51. M.-W. Lee, M.-Y. Lee, J.-C. Choi, J.-S. Park, and C.-K. Song, "Fine patterning of glycerol-doped PEDOT: PSS on hydrophobic PVP dielectric with ink jet for source and drain electrode of OTFTs," *Organic Electronics*, vol. 11, no. 5, pp. 854–859, 2010.
52. D. J. Gundlach *et al.*, "Contact-induced crystallinity for high-performance soluble acene-based transistors and circuits," *Nature Materials*, vol. 7, no. 3, pp. 216–221, 2008.
53. S. Choi *et al.*, "A study on reducing contact resistance in solution-processed organic field-effect transistors," *ACS Applied Materials & Interfaces*, vol. 8, no. 37, pp. 24744–24752, 2016.
54. O. Fenwick *et al.*, "Modulating the charge injection in organic field-effect transistors: Fluorinated oligophenyl self-assembled monolayers for high work function electrodes," *Journal of Materials Chemistry C*, vol. 3, no. 13, pp. 3007–3015, 2015.
55. M. Waldrip, O. D. Jurchescu, D. J. Gundlach, and E. G. Bittle, "Contact resistance in organic field-effect transistors: Conquering the barrier," *Advanced Functional Materials*, vol. 30, no. 20, p. 1904576, 2020.
56. J. Roh *et al.*, "Injection-modulated polarity conversion by charge carrier density control via a self-assembled monolayer for all-solution-processed organic field-effect transistors," *Scientific Reports*, vol. 7, no. 1, p. 46365, 2017.
57. Y. Mei *et al.*, "Interface engineering to enhance charge injection and transport in solution-deposited organic transistors," *Organic Electronics*, vol. 50, pp. 100–105, 2017.
58. D. M. Alloway *et al.*, "Tuning the effective work function of gold and silver using ω-functionalized alkanethiols: Varying surface composition through dilution and choice of terminal groups," *The Journal of Physical Chemistry C*, vol. 113, no. 47, pp. 20328–20334, 2009.
59. X. Liu, D. Kasemann, and K. Leo, "Top-gate organic depletion and inversion transistors with doped channel and injection contact," *Applied Physics Letters*, vol. 106, no. 10, p. 103301, 2015.
60. A. F. Paterson *et al.*, "Addition of the Lewis acid Zn (C6F5) 2 enables organic transistors with a maximum hole mobility in excess of 20 cm2 V− 1 s− 1," *Advanced Materials*, vol. 31, no. 27, p. 1900871, 2019.
61. J. Panidi *et al.*, "Remarkable enhancement of the hole mobility in several organic small-molecules, polymers, and small-molecule: Polymer blend transistors by simple admixing of the Lewis acid P-dopant B (C6F5) 3," *Advanced Science*, vol. 5, no. 1, p. 1700290, 2018.
62. M. Nikolka *et al.*, "High operational and environmental stability of high-mobility conjugated polymer field-effect transistors through the use of molecular additives," *Nature Materials*, vol. 16, no. 3, pp. 356–362, 2017.

63. M. P. Hein *et al.*, "Molecular doping for control of gate bias stress in organic thin film transistors," *Applied Physics Letters*, vol. 104, no. 1,p. 013507, 2014.
64. Y. Kim *et al.*, "Enhanced charge injection properties of organic field-effect transistor by molecular implantation doping," *Advanced Materials*, vol. 31, no. 10, p. 1806697, 2019.
65. R. R. Schliewe, F. A. Yildirim, W. Von Emden, R. Witte, W. Bauhofer, and W. Krautschneider, "Static model for organic field-effect transistors including both gate-voltage-dependent mobility and depletion effect," *Applied Physics Letters,* vol. 88, no. 23, p. 233514, 2006.
66. A. Takshi, A. Dimopoulos, and J. D. Madden, "Depletion width measurement in an organic Schottky contact using a metal-semiconductor field-effect transistor," *Applied Physics Letters*, vol. 91, no. 8, p. 083513, 2007.
67. A. Yamamura *et al.*, "Wafer-scale, layer-controlled organic single crystals for high-speed circuit operation," *Science advances*, vol. 4, no. 2, p. eaao5758, 2018.
68. S. Braun and W. R. Salaneck, "Fermi level pinning at interfaces with tetr afluorotetracyanoquinodimethane (F4-TCNQ): The role of integer charge transfer states," *Chemical Physics Letters*, vol. 438, no. 4–6, pp. 259–262, 2007.
69. R. Janneck *et al.*, "Integration of highly crystalline C8-BTBT thin-films into simple logic gates and circuits," *Organic Electronics*, vol. 67, pp. 64–71, 2019.
70. H. Kleemann, C. Schuenemann, A. A. Zakhidov, M. Riede, B. Lüssem, and K. Leo, "Structural phase transition in pentacene caused by molecular doping and its effect on charge carrier mobility," *Organic Electronics*, vol. 13, no. 1, pp. 58–65, 2012.
71. P. Darmawan *et al.*, "Optimal structure for high-performance and low-contact-resistance organic field-effect transistors using contact-doped coplanar and pseudo-staggered device architectures," *Advanced Functional Materials*, vol. 22, no. 21, pp. 4577–4583, 2012.
72. T. Minari, P. Darmawan, C. Liu, Y. Li, Y. Xu, and K. Tsukagoshi, "Highly enhanced charge injection in thienoacene-based organic field-effect transistors with chemically doped contact," *Applied Physics Letters*, vol. 100, no. 9, p. 093303, 2012.
73. S. Singh *et al.*, "Reduction of contact resistance by selective contact doping in fullerene n-channel organic field-effect transistors," *Applied Physics Letters*, vol. 102, no. 15,p. 153303, 2013.
74. C. Liu *et al.*, "Effect of doping concentration on microstructure of conjugated polymers and characteristics in N-Type polymer field-effect transistors," *Advanced Functional Materials*, vol. 25, no. 5, pp. 758–767, 2015.
75. J. Kim *et al.*, "Systematic study of widely applicable N-Doping strategy for high-performance solution-processed field-effect transistors," *Advanced Functional Materials*, vol. 26, no. 43, pp. 7886–7894, 2016.
76. Y. Xu *et al.*, "Doping: A key enabler for organic transistors," *Advanced Materials*, vol. 30, no. 46, p. 1801830, 2018.
77. S. Olthof, W. Tress, R. Meerheim, B. Lüssem, and K. Leo, "Photoelectron spectroscopy study of systematically varied doping concentrations in

an organic semiconductor layer using a molecular p-dopant," *Journal of Applied Physics*, vol. 106, no. 10, p. 103711, 2009.
78. M. Kano, T. Minari, and K. Tsukagoshi, "Improvement of subthreshold current transport by contact interface modification in p-type organic field-effect transistors," *Applied Physics Letters*, vol. 94, no. 14, p. 143304, 2009.
79. K.-J. Baeg, G.-T. Bae, and Y.-Y. Noh, "Efficient charge injection in p-type polymer field-effect transistors with low-cost molybdenum electrodes through V2O5 interlayer," *ACS Applied Materials & Interfaces*, vol. 5, no. 12, pp. 5804–5810, 2013.
80. P. Darmawan, T. Minari, A. Kumatani, Y. Li, C. Liu, and K. Tsukagoshi, "Reduction of charge injection barrier by 1-nm contact oxide interlayer in organic field effect transistors," *Applied Physics Letters*, vol. 100, no. 1, p. 013303, 2012.
81. M. Alt *et al.*, "The Swiss-Army-Knife Self-Assembled monolayer: Improving electron injection, stability, and wettability of metal electrodes with a one-minute process," *Advanced Functional Materials*, vol. 26, no. 18, pp. 3172–3178, 2016.
82. N. B. Kotadiya *et al.*, "Universal strategy for Ohmic hole injection into organic semiconductors with high ionization energies," *Nature Materials*, vol. 17, no. 4, pp. 329–334, 2018
83. M. T. Greiner, L. Chai, M. G. Helander, W. Tang, and Z. Lu, "Metal/metal-oxide interfaces: How metal contacts affect the work function and band structure of MoO3," *Advanced Functional Materials*, vol. 23, no. 2, pp. 215–226, 2013.
84. M. T. Greiner and Z.-H. Lu, "Thin-film metal oxides in organic semiconductor devices: Their electronic structures, work functions and interfaces," *NPG Asia Materials*, vol. 5, no. 7, pp. e55–e55, 2013.
85. T. J. Richards and H. Sirringhaus, "Analysis of the contact resistance in staggered, top-gate organic field-effect transistors," *Journal of Applied Physics*, vol. 102, no. 9, p. 094510, 2007.
86. S. Luan and G. W. Neudeck, "An experimental study of the source/drain parasitic resistance effects in amorphous silicon thin film transistors," *Journal of Applied Physics*, vol. 72, no. 2, pp. 766–772, 1992.
87. J. G. Simmons, "Poole-Frenkel effect and Schottky effect in metal-insulator-metal systems," *Physical Review*, vol. 155, no. 3, p. 657, 1967.
88. M. Kiguchi, M. Nakayama, K. Fujiwara, K. Ueno, T. Shimada, and K. Saiki, "Accumulation and depletion layer thicknesses in organic field effect transistors," *Japanese Journal of Applied Physics*, vol. 42, no. 12A, p. L1408, 2003.
89. M. Uno, Y. Tominari, and J. Takeya, "Three-dimensional organic field-effect transistors: Charge accumulation in the vertical semiconductor channels," *Applied Physics Letters*, vol. 93, no. 17, p. 173301, 2008.
90. P. Tantitarntong *et al.*, "High sensitivity tuning of work function of self-assembled monolayers modified electrodes using vacuum ultraviolet treatment," *ACS Applied Materials & Interfaces*, vol. 9, no. 34, pp. 28151–28156, 2017.

Chapter 9

Synergistic Integration of High Electron Mobility Transistors (HEMT) and Quantum Dots for Spintronics-Based Quantum Computing

Swastik Kumar Sahu and Kaushik Mazumdar

9.1 INTRODUCTION

Quantum computing offers a paradigm shift in processing capacity, utilizing quantum mechanics to process data in fundamentally novel ways. Unlike traditional computers, which utilize bits as the smallest unit of data, quantum computers use qubits, which can exist in superpositions of states. This enables quantum computers to tackle certain problems exponentially faster than classical computers, with numerous practical applications [1].

Spintronics, or spin-based electronics, is an emerging field that exploits the intrinsic spin of the electron, in addition to its charge, for information processing. Spintronics offers advantages such as non-volatility, lower power consumption, and faster data processing compared to conventional electronics. In quantum computing, spintronics has the potential to provide robust and scalable qubits, enabling more efficient quantum information processing [2].

The integration of spintronic principles into quantum computing is essential for the development of scalable and energy-efficient quantum devices. Spin-based qubits can be manipulated using well-established semiconductor technology, making spintronics a promising approach for building practical quantum computers [3]. The synergy of quantum mechanics and spintronics is seen as a key enabler for advancing quantum computing.

This chapter aims to analysis the current state of research on the integration of High Electron Mobility Transistors (HEMTs) and quantum dots within the framework of spintronics-based quantum computing. The primary focus is on understanding the roles of HEMTs and quantum dots individually and how their integration can enhance the performance and scalability of quantum computing systems. The analysis will summarize key findings, identify existing challenges, and highlight potential solutions from the literature.

HEMTs are known for their high-speed performance and low noise, making them valuable components in various electronic devices [4]. In spintronics-based quantum computing, HEMTs can be used for efficient

DOI: 10.1201/9781003533535-9

spin manipulation and detection. Quantum dots, on the other hand, serve as potential qubits due to their discrete energy levels and ability to confine electrons [5]. The integration of HEMTs with quantum dots is a promising approach to creating high-performance, scalable quantum computing systems. Recent developments related to HEMT and quantum dots for spintronics-based quantum computing are mentioned in Table 9.1.

Table 9.1 Recent developments related to HEMT, quantum dots and spintronics for quantum computing

Ref.	*First Author (Year) Journal*	*Remark*
[6]	Cha (2020) IEEE Electron Device Letters	This research shows that these InP HEMTs can function at 6 GHz in cryogenic LNAs with close to 100 W power dissipation. The InP HEMT LNA should be placed closer to the quantum processor, ideally below 1 K, to reduce connection losses and improve qubit readout sensitivity.
[7]	Faramehr (2017) IEEE Transactions on Electron Devices	GaN-based devices have attracted commercial interest as a result of improvements such as power switching GaN-on-silicon. During operation, GaN devices can experience carrier trapping, which reduces power efficiency and causes current collapse.
[8]	Zhang (2018) IEEE Microwave and Wireless Components Letters	GaN is a promising material for electronic devices due to its large band gap (3.4 eV) and a number of advantages. GaN HEMTs in the L-band have seen improvements in power, operating voltage, and power-added efficiency.
[9]	Kargarrazi (2019) IEEE Journal of the Electron Devices Society	High-temperature SiC devices such as NMOS, CMOS, and BJT have been examined, however they face issues such as gate oxide degradation and constant base current requirements. HEMTs based on AlGaN/GaN have been proposed for low-power, high-temperature electronics.
[10]	Jeong (2021) IEEE Transactions on Electron Devices	Si CMOS scalability has almost reached its limit, prompting the exploration of new options. InGaAs-based HEMTs are well-known for their fast electron mobility and large sheet carrier density, which makes them a good choice for high-speed and millimeter-wave applications.
[11]	Okazaki (2016) Nature Communications	This research uses a gate-controlled Quantum Dot based on the GaAs/AlGaAs combination, which provides better control over confined electron states, and the material's intrinsic piezoelectric characteristics efficiently couple electrons to mechanical motion.

(*continued*)

Table 9.1 (Cont.)

Ref.	*First Author (Year) Journal*	*Remark*
[12]	Blokhina (2021) IEEE Open Journal of Circuits and Systems	Noisy Intermediate Scale Quantum (NISQ) is the name given to the current generation of quantum computers. Quantum dots are used in semiconductor qubit technology to build qubits that use electron spin, charge, or both.
[13]	Stuyck (2020) IEEE Electron Device Letters	Qubits formed from isolated electron spin states hold great potential for Quantum Information Processing technology. The SET functions as a charge sensor in a qubit device, reading the spin state of an electron.
[14]	Tadokoro (2021) Scientific Reports (Nature)	Semiconductor spin qubits based on electron spins are thought to be viable candidates for quantum processors.
[15]	Hendrickx (2020) Nature communications	Hole spins in group-IV materials, particularly germanium, have several advantages, including strong spin-orbit coupling and a high excited state energy.
		Early studies showed the feasibility of employing spin-orbit coupling to drive hole spins all-electrically, but challenges persisted due to nuclear spins.
[16]	Graziano (2023) IEEE Transactions on Electron Devices	Power dissipation is an important issue when it comes to integrating electronics close to the quantum processor, particularly at the 4 K temperature level, where active electronics can be inserted due to available cooling power.

9.2 HEMTS IN SPINTRONICS AND QUANTUM COMPUTING

9.2.1 Overview of HEMTs

HEMTs have been a significant advancement in semiconductor technology since their inception in the late 20th century [4]. Developed to improve performance in high-frequency and high-speed applications, HEMTs are based on heterojunctions, interfaces between different semiconductor materials. The key breakthrough was the introduction of the AlGaAs/GaAs heterojunction, which enabled the creation of two-dimensional electron gases (2DEGs) with high mobility [17]. Figure 9.1 shows the schematic diagram of an emerging GaN based HEMT with sapphire substrate and AlGaN/GaN heterojunction.

HEMTs have since evolved, with improvements in material quality, device design, and fabrication techniques. Advances include the development of GaN-based HEMTs, which offer high-power and high-frequency capabilities [18].

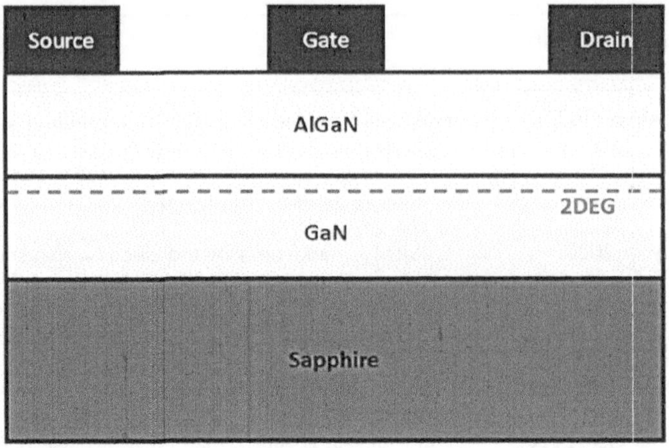

Figure 9.1 Schematic of a GaN HEMT with Sapphire substrate and AlGaN/GaN heterojunction.

HEMTs are characterized by their use of a heterojunction to form a 2DEG, which is a high-mobility channel for electron transport. The basic structure of a HEMT consists of:

Source and Drain Contacts: These electrodes are where the external voltage is applied to control the current flow.

Gate Electrode: This is used to modulate the conductivity of the 2DEG channel by applying a voltage.

Barrier Layers: These layers, typically composed of materials like AlGaAs or AlN, create the heterojunction that forms the 2DEG.

The gate controls the current flow across the source and drain by modulating the density of the 2DEG whenever a voltage is applied to it. The high mobility of the 2DEG results in low resistance and high-speed operation.

HEMTs have several advantages that make them suitable for spintronic applications:

High-Speed Performance: The high electron mobility of HEMTs enables rapid switching and high-frequency operation, essential for spintronic devices that require fast spin manipulation.

Low Noise: HEMTs exhibit low noise characteristics, which is crucial for detecting weak spin signals and minimizing errors in spintronic systems.

Versatility: HEMTs can be integrated with various spintronic materials and structures, making them versatile components in spintronic circuits.

Several spintronic devices utilize HEMTs for spin injection, detection, and manipulation. For instance, HEMTs can be used in spintronic sensors to measure magnetic fields or spin polarization with high precision.

Despite their advantages, HEMTs face several challenges in spintronic applications:

Material Compatibility: Integrating HEMTs with other spintronic materials can be challenging due to differences in lattice constants and thermal expansion coefficients.

Spin Injection Efficiency: Achieving efficient spin injection into the 2DEG remains a challenge, as spin polarization needs to be maintained over the length of the channel.

Thermal Management: High-speed operation of HEMTs can generate significant heat, which can affect the performance and reliability of spintronic devices [19].

9.2.2 Applications of HEMTs in Spintronic Devices

i. Spintronic Sensors and Detectors

HEMTs can be used in spintronic sensors and detectors for applications such as magnetic field sensing and spin current detection. Their high-speed and low-noise characteristics enable the precise measurement of weak magnetic fields and spin signals.

Magnetic Field Sensors: HEMTs can be employed in magneto-resistive sensors to detect magnetic fields with high sensitivity. These sensors are used in various applications, including industrial and medical fields.

Spin Current Detectors: HEMTs can detect spin currents generated in spintronic devices, providing insights into spin transport and manipulation.

ii. Spintronic Logic Devices

HEMTs are also being explored for use in spintronic logic devices, where they can act as components in logic gates and circuits. The high-speed operation of HEMTs can enhance the performance of spintronic logic gates, enabling faster and more efficient computation.

Spintronic Gates: HEMTs can be integrated into spintronic logic gates to control and manipulate spin states, contributing to the development of spin-based computing architectures.

Spintronic Circuits: HEMTs can be used in spintronic circuits for various applications, including signal processing and data storage.

9.3 QUANTUM DOTS IN QUANTUM COMPUTING

9.3.1 Introduction to Quantum Dots

Quantum dots (QDs) are basically semiconductor nanoparticles with quantum confinement effects that arise when their dimensions are comparable to or smaller than the exciton Bohr radius [20]. The concept of quantum dots emerged from the study of nanoscale semiconductor structures in the 1980s and 1990s. Figure 9.2 shows the simple illustration of a quantum dot.

Advancements in synthesis techniques, such as colloidal chemistry and molecular beam epitaxy, have significantly enhanced the control over the size, shape, and composition of quantum dots [21]. These advancements have made it possible to engineer quantum dots with specific optical and electronic properties, making them suitable for a wide range of applications,

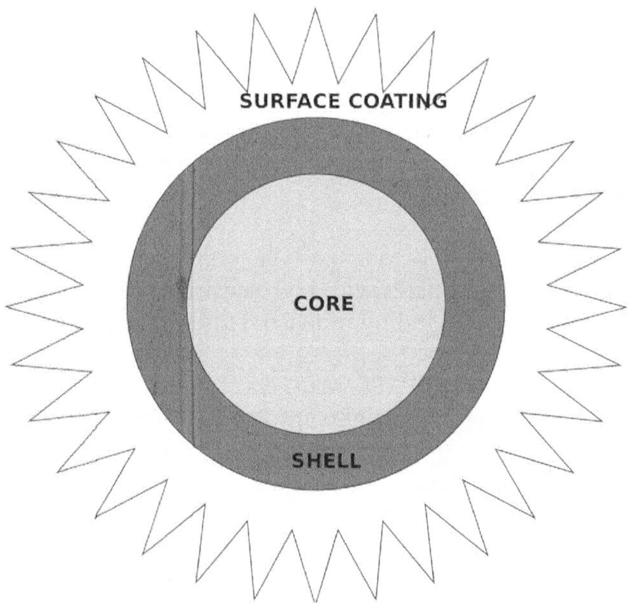

Figure 9.2 Schematic representation of a quantum dot.

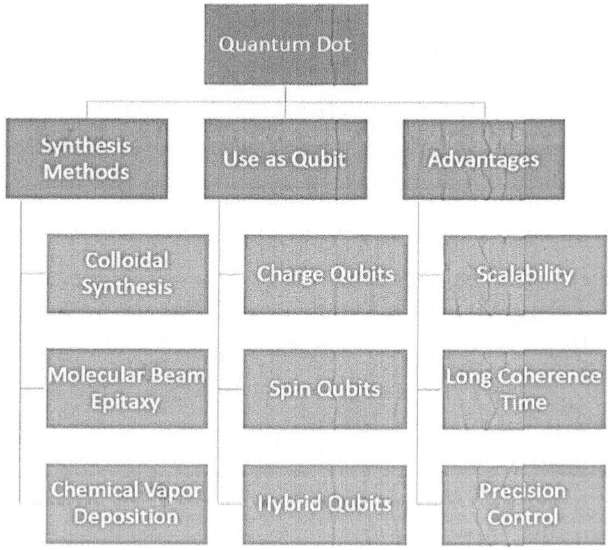

Figure 9.3 An overview of quantum dots.

including quantum computing. A general overview on the synthesis and applications of the quantum dot has been shown in Figure 9.3.

Quantum dots are characterized by their discrete energy levels, which result from quantum confinement effects. These energy levels are similar to atomic orbitals, which gives quantum dots their "artificial atom" properties. Key properties of quantum dots include:

Size-Dependent Optical Properties: The size of quantum dots determines their energy levels, which affects their absorption and emission spectra. Smaller quantum dots emit light at shorter wavelengths, while larger dots emit at longer wavelengths.

Discrete Energy Levels: Quantum dots have well-defined energy levels due to quantum confinement. This property makes them ideal candidates for use as qubits in quantum computing.

High Photostability and Brightness: Quantum dots exhibit high photostability and brightness, making them useful for various imaging and sensing applications.

Quantum dots can be synthesized using various methods, each offering different advantages in terms of size control, material composition, and scalability:

Colloidal Synthesis: This method involves the chemical synthesis of quantum dots in a solution. It allows for precise control over size and composition and is widely used for producing quantum dots with tailored properties [22].

Molecular Beam Epitaxy (MBE): MBE is a high-precision technique used to grow quantum dots on a substrate. It allows for the creation of high-quality quantum dots with well-defined sizes and interfaces [23].

Chemical Vapor Deposition (CVD): CVD is used to deposit thin films of semiconductor materials, which can then be used to form quantum dots. It is suitable for producing quantum dots with specific compositions and structures [24].

9.3.2 Quantum Dots as Qubits

i. Charge Qubits

Charge qubits use the charge state of a quantum dot to represent qubit states [25]. In this approach, the quantum dot can hold either one or no electron, which corresponds to the two states of the qubit, typically denoted as $|0\rangle$ and $|1\rangle$.

Basic Operation: The fundamental principle behind charge qubits is the manipulation of the charge state of the quantum dot. When a single electron is added or removed, it changes the quantum dot's charge state. This charge state can be detected using capacitively coupled nearby sensors, such as single-electron transistors.

Control and Readout: To control a charge qubit, gate voltages are applied to tune the energy levels of the quantum dot and to induce transitions between the charge states. The readout of the charge state is achieved through quantum dot sensors or other measurement techniques that detect the presence or absence of the electron.

ii. Spin Qubits

Quantum information can be encoded using the superposition of up ($|\uparrow\rangle$) and down ($|\downarrow\rangle$) states for the electron spin [26].

Basic Operation: Spin-orbit coupling, or magnetic fields are used to change the electron's spin. Controllable energy splitting between spin states is produced by the quantum dot and can be achieved through the application of gate voltages or external forces.

Methods of Control: Spin qubits are controlled by methods like spin-orbit interactions to produce coherent spin rotations or electron spin resonance

(ESR), which involves applying a resonant microwave field to flip the spin state.

Readout: Measurement of spin qubits is often performed using spin-to-charge conversion. This technique involves transferring the spin state into a charge state that can be detected with high precision. For example, the spin state can be converted to a different charge configuration that affects the conductance of a nearby sensor.

iii. Hybrid Qubits

Hybrid qubits combine both charge and spin degrees of freedom to leverage the advantages of both approaches. This integration can offer improved qubit performance and functionality.

Charge-Spin Qubits: One hybrid approach is to use the spin state of an electron in a quantum dot while also considering its charge state. This method allows for more robust qubit operations and error correction.

Coupled Qubits: Hybrid systems may also involve coupling charge and spin qubits to enable interactions and gate operations that leverage both types of qubits. This approach can enhance the performance of quantum gates and improve the overall fidelity of quantum operations.

Advantages of Quantum Dots as Qubits

Scalability: Quantum dots can be fabricated using well-established semiconductor technology, making it feasible to scale up to large arrays of qubits. This scalability is essential for building practical quantum computers.

Long Coherence Time: Spin qubits in quantum dots often exhibit long coherence times compared to other qubit types. This is due to the isolation of the spin from external noise sources and the potential for effective error correction.

Precision Control: Advances in gate technologies allow precise control over quantum dots, enabling accurate manipulation of qubit states. This precision is crucial for implementing complex quantum algorithms and error correction schemes.

9.3.3 Recent Developments and Advances

Recent advancements in synthesis techniques have enhanced the quality and uniformity of quantum dots:

Colloidal Synthesis Improvements: Innovations in colloidal synthesis methods have led to the production of quantum dots with better size control and optical properties. This has improved the performance of quantum dots in various applications, including quantum computing.

Molecular Beam Epitaxy (MBE): Advances in MBE techniques have enabled the growth of high-quality quantum dots with well-defined sizes and interfaces. This method is crucial for achieving the precision required for quantum computing applications.

Efforts to address decoherence and improve qubit reliability include:

Error-Correction Codes: Researchers are developing error-correction codes specifically designed for quantum dots. These codes aim to correct errors caused by decoherence and other noise sources, enhancing the reliability of quantum computations.

Decoherence Reduction Techniques: New methods for reducing decoherence, such as shielding techniques and material improvements, are being explored to improve qubit performance and stability.

Future research in quantum dots for quantum computing is focused on several key areas:

Exploring New Materials: Researchers are investigating alternative semiconductor materials that may offer better performance or new functionalities for quantum dots.

Scaling Up: Efforts are underway to develop methods for scaling up quantum dot technology to build large-scale quantum computing systems.

Interdisciplinary Approaches: Collaboration between materials scientists, quantum physicists, and engineers is crucial for advancing quantum dot technology and overcoming existing challenges.

9.4 INTEGRATION OF HEMTS AND QUANTUM DOTS

Integrating HEMTs with quantum dots for spintronics-based quantum computing benefits the unique properties of both devices to enable precise control of quantum states, particularly the spin of electrons. HEMTs are known for their high electron mobility, which is achieved by confining electrons in a 2DEG at the interface of two semiconductor materials with different band gaps, such as AlGaN/GaN. This confinement allows for the rapid and efficient movement of electrons with minimal scattering, making HEMTs ideal for high-speed and low-power applications in quantum devices [27].

Quantum dots, on the other hand, are nanoscale semiconductor particles that confine electrons in all three dimensions, effectively trapping single or few electrons in discrete energy levels [21]. These confined electrons can have their spin states manipulated, which is crucial for quantum computing since the spin of an electron can serve as a quantum bit (qubit). By integrating quantum dots with HEMTs, it becomes possible to combine the benefits of high electron mobility with the quantum confinement needed for spin manipulation.

In spintronics-based quantum computing, the spin of an electron is used to represent quantum information, as the electron's spin can be in a superposition of "up" and "down" states, akin to the 0s and 1s in classical computing [28]. HEMTs provide the necessary platform for fast and efficient control of electrons, while quantum dots ensure the precise trapping and manipulation of these electrons. The high electron mobility in HEMTs aids in reducing decoherence, a major challenge in quantum computing, by allowing electrons to move and be manipulated with minimal energy loss. The quantum dots, embedded in or near the 2DEG of a HEMT, act as the storage and processing units for qubits, where electron spins can be initialized, manipulated, and read out using electrical and magnetic fields.

This integration of HEMTs and quantum dots offers a promising approach to scalable quantum computing, as it combines the well-established semiconductor fabrication techniques used for HEMTs with the emerging capabilities of quantum dots for spin-based qubit control. By utilizing both electron mobility and spin, this hybrid system can potentially overcome some of the key disadvantages of the quantum computing technologies, like coherence times and qubit scalability. The flowchart for the integration of HEMT and quantum dots for spintronics-based quantum computing is shown in Figure 9.4.

Now we discuss the combination of HEMTs and quantum dots for spintronics-based quantum computing, which involves several intricate interactions between these components, as mentioned in Figure 9.5.

9.4.1 Spin Injection and Manipulation [29]

Spin Injection:

HEMTs create a 2DEG at the interface between two different semiconductor materials. This 2DEG is highly conductive and has low scattering, making it ideal for transporting electrons. Figure 9.6 shows the flowchart for Spin Injection and Detection in a HEMT-QD System.

- In spintronics, the goal is to inject electrons with a specific spin (spin-polarized electrons) into a quantum dot, where the electron's spin state will represent a quantum bit (qubit).

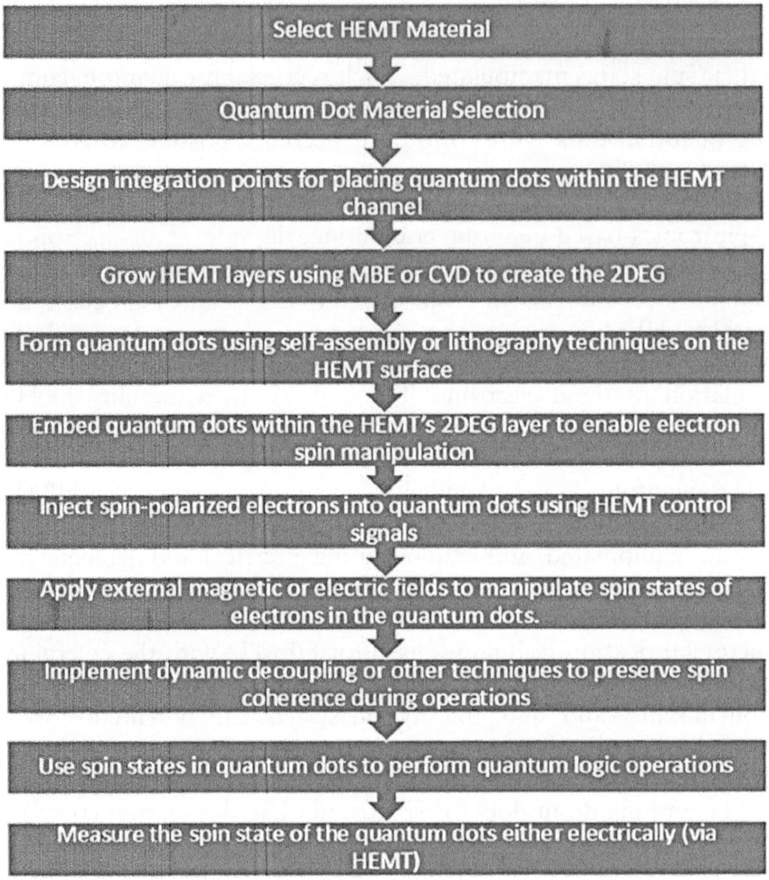

Figure 9.4 Flowchart for Integration of HEMT and quantum dots for spintronics-based quantum computing.

- The HEMT's 2DEG can be engineered using materials with strong spin-orbit coupling, such as InGaAs or InAlAs. This strong spin-orbit coupling allows for the control of the electron's spin orientation by applying an electric field, rather than a magnetic field.

Spin injection into the quantum dot is achieved by creating a difference in spin populations at the interface, often using ferromagnetic contacts or by exploiting spin-orbit interaction in the HEMT.

Spin Manipulation: Once the electron is confined in the quantum dot, its spin can be manipulated using external electric or magnetic fields.

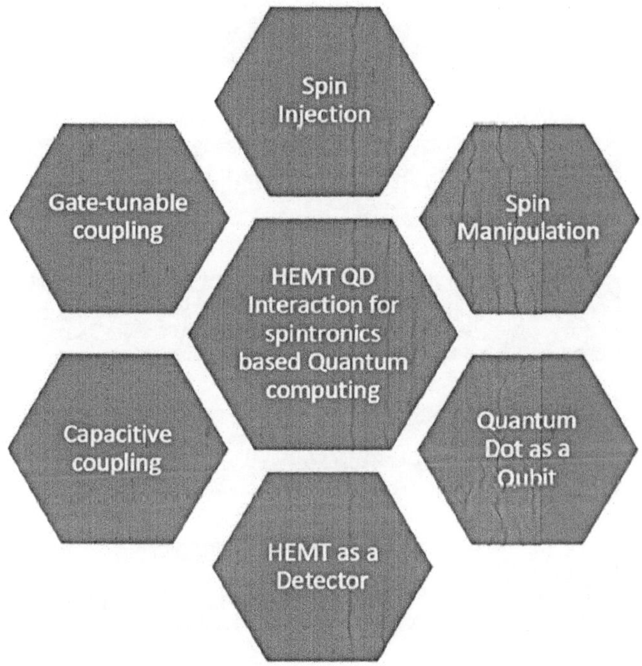

Figure 9.5 An overview on HEMT and quantum dots integration methods for spintronics based quantum computing.

Electric Field Manipulation (via the Rashba effect): In materials with strong spin-orbit coupling, applying an electric field changes the effective magnetic field experienced by the electron, causing its spin to rotate. This is particularly useful for fast spin manipulation without needing large magnetic fields.

Magnetic Field Manipulation (via the Zeeman effect): A magnetic field splits the spin states into different energy levels (spin-up and spin-down). By applying microwave radiation or using fast pulses, it's possible to flip the spin state between these two levels, effectively performing a quantum gate operation.

9.4.2 Spin-Qubit Control

Quantum Dot as a Qubit:

- A quantum dot can trap a single electron, and the electron's spin state (up or down) represents a qubit. In quantum computing, this binary state serves as the fundamental informational unit.

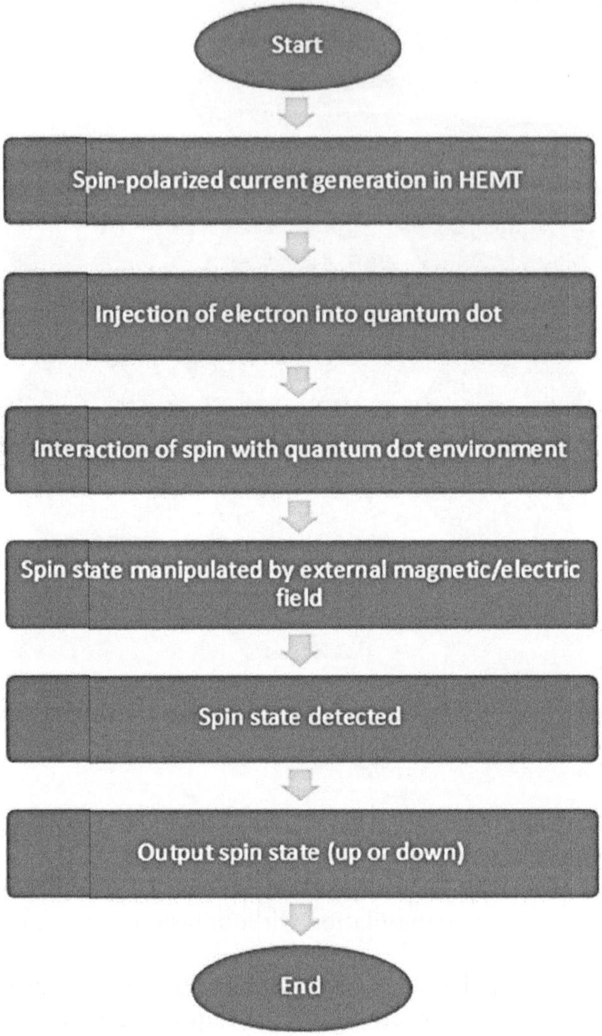

Figure 9.6 Flowchart for spin injection and detection in HEMT-QD system.

Spin control in a quantum dot involves precise manipulation of the electron's energy levels and spin states. The energy levels are quantize.

External Controls:

Electric gates placed near the quantum dot allow for fine control over the dot's potential landscape. By regulating the voltages on these gates, you can control the number of electrons in the dot and their spin states.

Microwave pulses can be applied to induce transitions between spin states (e.g., from spin-up to spin-down), which are used to implement quantum logic gates (e.g., the Pauli-X gate).

Spin-orbit coupling in the HEMT allows for electric-field control of spin states in the quantum dot, reducing the reliance on magnetic fields for spin manipulation.

9.4.3 Spin Readout

Readout Mechanisms:

Spin-to-charge conversion: One of the most common methods for reading out the spin state of a quantum dot qubit is to convert the spin information into a charge signal, which can be more easily detected.

Single-electron transistors (SETs) can be used in conjunction with the HEMT structure to detect whether an electron with a specific spin state has tunneled out of the quantum dot.

HEMT as a Detector:

- The HEMT can serve as a highly sensitive charge detector due to the high mobility of the 2DEG. When the spin state of an electron in the quantum dot influences its tunneling probability into a nearby reservoir, the resulting charge fluctuation can be detected as a change in the current through the HEMT [30].

Spin-dependent tunneling: By tuning the quantum dot's energy levels, you can make the tunneling of an electron (into or out of the dot) spin-dependent. This allows the HEMT to indirectly measure the spin state by detecting the change in charge.

9.4.4 Coupling of Qubits

Quantum Dot Coupling:

Inter-dot coupling is essential for creating two-qubit gates, which are necessary for universal quantum computing. In HEMT-based systems, quantum dots can be coupled via the 2DEG.

Capacitive coupling [31]: Two quantum dots can be coupled capacitively, where the charge state of one dot influences the potential of the other. This allows the spin states in the two dots to interact, enabling two-qubit operations like the CNOT gate.

Exchange interaction: If two quantum dots are brought close enough, the spins of electrons in neighboring dots can interact via the exchange interaction, which is a quantum mechanical effect where the spins become entangled.

Tunable Coupling [32]:
Gate-tunable coupling: The coupling strength between two quantum dots can be controlled by adjusting the voltages on nearby gates. This tunability is critical for switching between independent qubit operations (single-qubit gates) and coupled operations (two-qubit gates).

On-demand entanglement: By tuning the coupling strength appropriately, you can entangle the spins of two electrons in different quantum dots, which is a crucial resource for quantum computation.

Example of Qubit Coupling:

- Suppose you have two quantum dots, each containing one electron. By adjusting the gate voltages, you can bring the quantum dots into a regime where the electron spins interact strongly via the exchange interaction.
- In this regime, a quantum operation (like the CNOT gate) can be implemented by controlling the interaction time and the applied fields, entangling the spins of the two electrons and performing a quantum computation.

9.4.5 Synergistic Benefits of Integration

The integration of High Electron Mobility Transistors (HEMTs) with quantum dots (QDs) represents a significant advancement in spintronics and quantum computing. Each technology offers distinct advantages that, when combined, can enhance the performance and functionality of quantum computing systems.

Enhanced Spin Manipulation: HEMTs are renowned for their ability to operate at high speeds with minimal noise, making them ideal for applications requiring rapid and precise control of electron spins. Quantum dots, as potential qubits, rely heavily on the accurate manipulation of spin states. Integrating HEMTs can significantly improve the efficiency and precision of spin manipulation in quantum dots, enabling more reliable and faster quantum operations.

Efficient Spin Detection: HEMTs' low noise and high sensitivity make them excellent candidates for detecting weak signals, such as the subtle changes in spin states of quantum dots. This integration facilitates high-fidelity spin readout, which is crucial for maintaining the integrity of quantum information and ensuring accurate measurement of qubit states.

Scalability: Both HEMTs and QDs are based on semiconductor technologies, which simplifies the integration process and supports the scalability of quantum computing systems. The use of established semiconductor fabrication techniques allows for the development of large-scale arrays of quantum dots and associated HEMT-based electronics, paving the way for practical and scalable quantum computing architectures.

The integration of HEMTs with quantum dots is an area of active research, with ongoing efforts to combine these technologies into functional devices capable of performing quantum computing operations.

Experimental Demonstrations: Various experimental studies have demonstrated the feasibility of integrating HEMTs with quantum dots. These demonstrations often involve the use of HEMTs to control and measure the spin states of quantum dots, showcasing the potential benefits of such integration. For instance, researchers have successfully used HEMTs to manipulate spin states with high precision and to detect spin signals with improved sensitivity.

Theoretical Models: Theoretical models and simulations play a crucial role in understanding the interactions between HEMTs and quantum dots. These models help predict the performance of integrated systems and guide the design of experimental setups. They also provide insights into the potential benefits and limitations of integration, helping to refine the approach and optimize device performance.

9.4.6 Comparative Analysis of Integration Methods

Each integration method has distinct trade-offs:

Direct Integration: Provides tight coupling between HEMTs and quantum dots, which can enhance device performance but requires precise fabrication techniques and careful management of interface quality.

Coupled Structures: Offers flexibility in design and allows for optimization of interaction without the need for direct integration. However, it may involve more complex coupling mechanisms and potential challenges related to alignment and distance.

Hybrid Devices: Combine the strengths of both methods and can offer improved performance and functionality. However, they may require advanced design and fabrication techniques to achieve optimal results.

9.4.7 Challenges in Integration

Figure 9.7 shows the various challenges and solutions in the integration process.

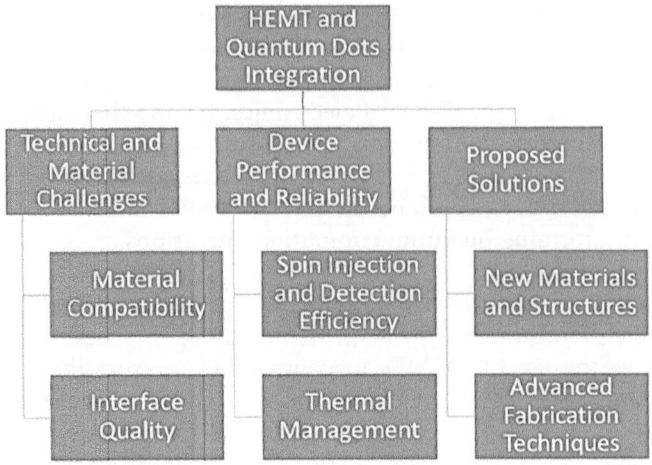

Figure 9.7 Various challenges faced during the integration process and their solutions.

Technical and Material Challenges

Integrating HEMTs and quantum dots presents several technical and material challenges that need to be addressed to achieve successful integration:

Material Compatibility: HEMTs and quantum dots are often made from different semiconductor materials, which can create challenges related to lattice matching and thermal expansion. Ensuring compatibility between these materials is crucial for maintaining high device performance and reliability. Techniques such as the use of intermediate layers or buffer materials can help address these issues.

Interface Quality: The quality of the interface between HEMTs and quantum dots is critical for device performance. Defects or imperfections at the interface can affect spin manipulation and detection, leading to reduced efficiency and reliability. Research is focused on developing methods to improve interface quality and minimize defects.

Device Performance and Reliability: Maintaining high performance and reliability in integrated HEMT-QD systems involves addressing several key issues:

Spin Injection and Detection Efficiency: Efficient spin injection into quantum dots and accurate detection of spin states are essential for the performance of spintronic devices. Improving these aspects involves optimizing the interaction between HEMTs and quantum dots, reducing sources of noise and interference, and developing advanced techniques for spin control and measurement.

Thermal Management: High-speed operation of HEMTs can generate significant heat, which can affect the performance of quantum dots and the overall device. Effective thermal management is necessary to ensure stable operation and prevent thermal-related issues. Strategies such as improved cooling systems and thermal insulation materials are being explored to address these challenges.

9.4.8 Proposed Solutions and Ongoing Research

Research is actively addressing the challenges associated with integrating HEMTs and quantum dots:

New Materials and Structures: The new semiconductor materials and structures that are compatible with both HEMTs and quantum dots can help overcome material compatibility issues. For example, the use of intermediate layers or buffer materials can improve lattice matching and reduce interface defects [33].

Advanced Fabrication Techniques: Innovations in fabrication techniques, such as atomic layer deposition and precision etching, are enhancing the quality of the interface between HEMTs and quantum dots. These techniques enable more precise control over device properties and improve overall performance.

Error Correction and Optimization: Implementing error correction techniques and optimizing device parameters are crucial for improving spin injection and detection efficiency. Research is focused on developing methods for reducing noise, enhancing qubit stability, and increasing the reliability of integrated systems.

Machine learning-based device optimization: Various machine learning algorithms like ResNet can be used for device structure optimization based on training and testing of the dataset for optimal; performance [34, 35].

9.5 APPLICATIONS AND FUTURE PERSPECTIVES

9.5.1 Current Applications in Quantum Computing

The integration of HEMTs with quantum dots is a burgeoning field with promising implications for quantum computing. This integration enables the development of advanced spintronic devices that offer enhanced

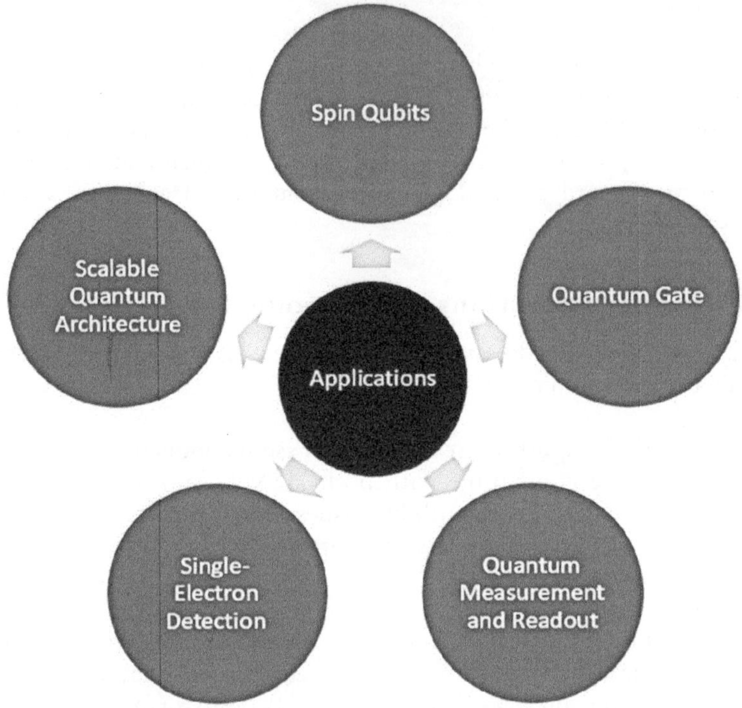

Figure 9.8 Applications of the integrated system.

performance and functionality for quantum computation. Currently, the applications of this integration focus on several key areas, as shown in Figure 9.8.

Spin Qubits and Quantum Gates: Quantum dots are employed as spin qubits due to their ability to confine individual electrons and manipulate their spin states. The integration of HEMTs allows for precise control and readout of these spin qubits. Quantum gates rely on the ability to manipulate qubits with high accuracy. HEMTs enhance this capability by providing low-noise, high-speed control over spin states, enabling faster and more reliable quantum gate operations.

Quantum Gate Implementation: In practice, HEMTs can be used to apply electric fields that influence the spin states of quantum dots, effectively implementing quantum gates. This capability is crucial for executing quantum algorithms, as gates are the fundamental building blocks of quantum circuits.

Quantum Measurement and Readout: One of the significant challenges in quantum computing is the accurate measurement of qubit states. HEMTs' sensitivity and low noise make them ideal for readout applications, where they detect the spin states of quantum dots with high precision. This capability is essential for extracting information from quantum systems and verifying quantum computations [36].

Single-Electron Detection: HEMTs enable the detection of single-electron states in quantum dots, which is crucial for applications involving spin qubits. The ability to measure individual electron spins accurately supports the development of quantum computers that rely on spin-based qubits.

Scalable Quantum Architectures: The integration of HEMTs and quantum dots contributes to the scalability of quantum computing architectures. By using semiconductor-based technologies, researchers can build larger arrays of quantum dots and associated electronics, facilitating the development of scalable quantum systems.

9.5.2 Emerging Applications in Spintronics

Beyond quantum computing, the integration of HEMTs and quantum dots has significant implications for spintronics, a field that focuses on the manipulation of electron spins for information processing. Key emerging applications are shown in Figure 9.9.

Spintronic Memory Devices: Spintronics offers the potential for developing memory devices that utilize the spin state of electrons to store information [37]. Integrated HEMTs and quantum dots can enhance the performance of spintronic memory by providing precise control and readout of spin states.

MRAM: MRAM is a type of spintronic memory that uses magnetic properties to store data [38]. Integration with quantum dots can improve the efficiency and speed of MRAM devices, enabling faster and more reliable data storage.

Spin-Based Sensors: Spintronic sensors leverage the spin properties of electrons to detect magnetic fields and other environmental changes. HEMTs' sensitivity and low noise can enhance the performance of spin-based sensors, making them suitable for applications in medical imaging, environmental monitoring, and other fields.

Magnetic Field Sensors: Integration of quantum dots with HEMTs can improve the sensitivity and accuracy of magnetic field sensors. These

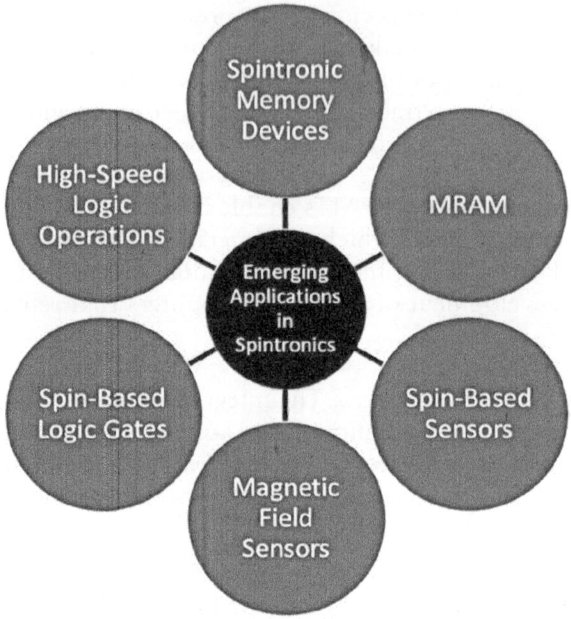

Figure 9.9 Emerging applications in spintronics.

sensors are used in applications such as detecting weak magnetic fields and monitoring magnetic materials.

Spin-Based Logic Gates: Spintronic logic gates utilize electron spins to perform logical operations, offering potential advantages in terms of speed and power efficiency [39]. HEMTs can contribute to the development of spin-based logic gates by providing efficient spin manipulation and readout.

High-Speed Logic Operations: The combination of HEMTs and quantum dots enables high-speed spin-based logic operations, which could lead to faster and more efficient computing systems.

9.6 FUTURE PROSPECTS AND DIRECTIONS

The integration of HEMTs and quantum dots for spintronics-based quantum computing represents a dynamic and rapidly evolving field. As research progresses, several promising future prospects and directions are emerging. Figure 9.10 shows the anticipated advancements and key areas of focus that are likely to shape the future of this integration.

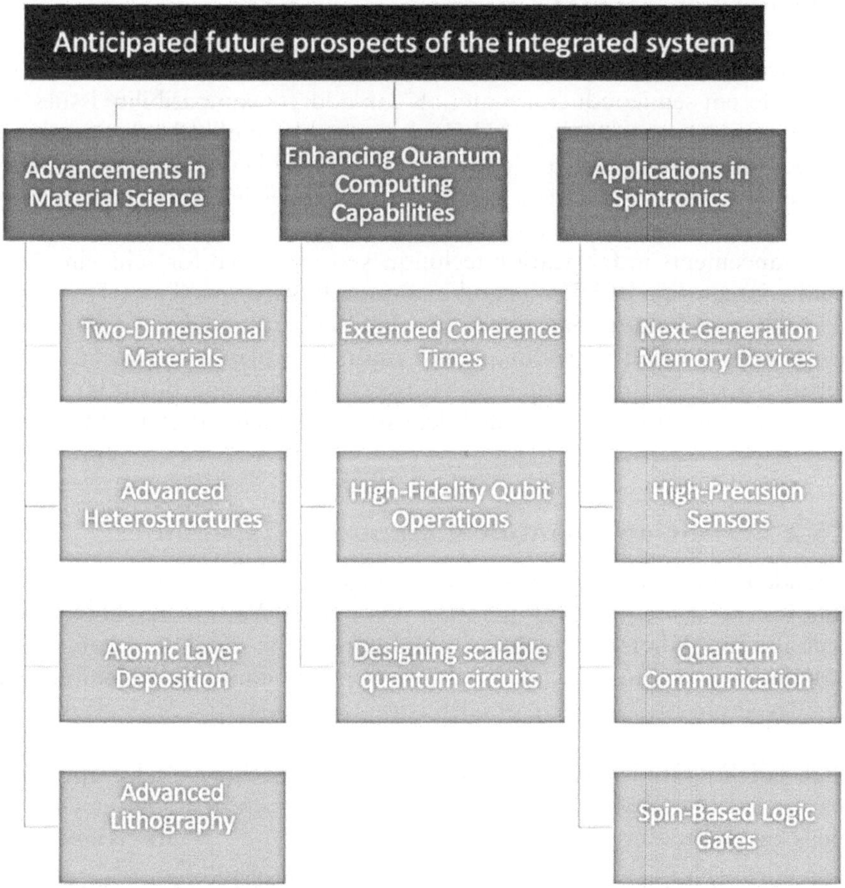

Figure 9.10 An overview on the anticipated future prospects and directions.

9.6.1 Advancements in Material Science and Technology

The success of integrating HEMTs and QDs heavily depends on the materials used. Future advancements in material science will play a crucial role in enhancing the performance and scalability of these integrated systems.

Two-Dimensional Materials: Research into two-dimensional (2D) materials, such as graphene and transition metal dichalcogenides (TMDs), offers exciting possibilities for improving integration [40]. These materials exhibit unique electronic properties and can potentially enhance the functionality of quantum dots and HEMTs. For instance, 2D materials could provide better lattice matching and reduced interface defects, improving the overall performance of integrated devices.

Advanced Heterostructures: Developing novel heterostructures that combine different semiconductor materials can address compatibility issues and improve integration quality [33]. Techniques like molecular beam epitaxy (MBE) and chemical vapor deposition (CVD) are being explored to create high-quality heterostructures with tailored properties, enhancing the performance of both HEMTs and quantum dots.

Advancements in fabrication techniques are essential for achieving high-quality integration of HEMTs and quantum dots. Atomic Layer Deposition (ALD) is a technique that allows for the precise deposition of thin layers of materials [41]. This technique is expected to play a significant role in improving the quality of interfaces between HEMTs and quantum dots by providing uniform and controlled deposition, reducing defects, and enhancing device performance.

9.6.2 Enhancing Quantum Computing Capabilities

Extended Coherence Times: Achieving longer coherence times for quantum dots is crucial for practical quantum computing. Various methods can be explored to minimize decoherence and improve the stability of qubits. This includes investigating new materials, optimizing quantum dot confinement, and developing advanced error correction techniques.

High-Fidelity Qubit Operations: Enhancing the fidelity of qubit operations is essential for reliable quantum computations. Future advancements will focus on improving the precision of quantum gate operations and reducing error rates. Techniques such as dynamical decoupling and quantum error correction will play a significant role in achieving high-fidelity qubit operations.

Scaling up the integration of HEMTs and quantum dots to create large-scale quantum computing systems is a key future direction. Efforts in this area include:

Quantum Circuit Design: Designing scalable quantum circuits that utilize integrated HEMTs and quantum dots is essential for building practical quantum computers. Research will focus on developing efficient circuit architectures, optimizing qubit connectivity, and integrating multiple qubit types to enhance computational power.

Mass Production Techniques: Developing mass production techniques for integrated HEMT-QD devices will be crucial for scaling up quantum computing systems. This involves creating reliable and reproducible fabrication processes that can produce large quantities of high-quality devices.

9.6.3 Expanding Applications in Spintronics

The integration of HEMTs and quantum dots opens up new possibilities for spintronic devices beyond quantum computing. Future research will explore:

Next-Generation Memory Devices: Spintronic memory devices, such as magnetic random-access memory (MRAM), will benefit from the integration of quantum dots and HEMTs. Future developments will focus on improving the speed, reliability, and storage capacity of these devices, making them more competitive with traditional memory technologies.

High-Precision Sensors: Spintronic sensors with integrated HEMTs and quantum dots will offer enhanced sensitivity and accuracy for various applications. Research will focus on developing sensors for applications such as medical imaging, environmental monitoring, and industrial process control.

The integration of HEMTs and quantum dots is likely to lead to the development of new technologies and applications:

Quantum Communication: Integrated HEMT-QD systems could play a role in quantum communication networks, where secure and high-speed data transmission is required. Research will explore how these systems can be used to develop quantum repeaters and secure communication protocols.

Spin-Based Logic Gates: The development of spin-based logic gates that leverage the integration of HEMTs and quantum dots will offer potential advantages in terms of speed and power efficiency. Future research will focus on designing and optimizing these logic gates for practical applications.

9.6.4 Interdisciplinary Collaboration and Research

The integration of HEMTs and quantum dots requires collaboration across multiple disciplines. Future research directions include:

Materials Science and Engineering: Collaboration between materials scientists and engineers will drive innovations in materials and fabrication techniques. This interdisciplinary approach will address challenges related to material compatibility, interface quality, and device performance.

Quantum Physics and Computing: Quantum physicists and computer scientists will work together to explore the theoretical and practical aspects of integrating HEMTs and quantum dots. This collaboration will focus on optimizing qubit performance, developing error correction techniques, and designing scalable quantum circuits.

9.6.5 Ethical and Societal Considerations

As quantum computing and spintronics technologies advance, it is important to address ethical considerations:

Data Privacy and Security: The potential impact of quantum computing on data privacy and security requires careful consideration. Research into quantum-resistant cryptography and secure communication protocols will be essential to address these concerns and protect sensitive information.

Responsible Innovation: Ensuring that advancements in HEMT-QD integration align with ethical standards and societal values is crucial. Responsible innovation involves considering the potential societal impact of new technologies and developing them in ways that benefit society as a whole.

Evaluating the environmental and economic impact of integrated HEMT-QD technologies will guide sustainable development:

Sustainability Practices: Implementing sustainability practices in the development and production of quantum and spintronic devices will contribute to minimizing environmental impact. This includes using eco-friendly materials, optimizing energy consumption, and reducing waste.

Economic Opportunities: The commercialization of integrated HEMT-QD systems has the potential to create new economic opportunities and drive growth in various industries. Research will explore the economic impact of these technologies and identify opportunities for commercialization and market adoption.

9.7 CONCLUSION

The integration of HEMTs with quantum dots QDs represents a significant advancement in the fields of quantum computing and spintronics. This integration has opened new avenues for improving qubit performance, enhancing spintronic devices, and scaling up quantum computing architectures. By leveraging the high-speed, low-noise characteristics of HEMTs and the precise control of spin states offered by quantum dots, researchers have made notable strides in developing more reliable and efficient quantum systems.

Throughout this analysis, we have highlighted the key achievements and ongoing challenges associated with HEMT-QD integration. Successful case studies have demonstrated the potential for precise spin control, enhanced sensor sensitivity, and effective quantum gate operations. These advancements underscore the promise of integrating HEMTs and quantum dots to create high-performance quantum computing systems and advanced spintronic devices.

Looking to the future, the continued development of new semiconductor materials, innovations in fabrication techniques, and advancements in quantum computing capabilities will be critical. The exploration of two-dimensional materials and advanced heterostructures, along with improvements in atomic layer deposition and lithography, will likely lead to further enhancements in integration quality and device performance. Additionally, the scalability of quantum computing architectures and the expansion of spintronic applications hold great promise for transforming various industries and technologies.

As we move forward, interdisciplinary collaboration will play a vital role in addressing the complex challenges of integration and advancing the field. Materials scientists, quantum physicists, and engineers must work together to overcome issues related to material compatibility, interface quality, and fabrication complexity.

The progress made thus far provides a solid foundation for future research and development, and the ongoing efforts to overcome existing challenges will pave the way for transformative advancements in these fields. As the technology evolves, it is essential to maintain a focus on both technical excellence and ethical responsibility, ensuring that the benefits of these advancements are realized in a manner that aligns with societal values and contributes to the greater good.

REFERENCES

[1] Preskill, John. "Quantum computing 40 years later." In Anthony JG. Hey, Anthony JG. Hey and Robin W. Allen (Eds.) *Feynman Lectures on Computation*, pp. 193–244. CRC Press, 2023.

[2] Dieny, Bernard, Ioan Lucian Prejbeanu, Kevin Garello, Pietro Gambardella, Paulo Freitas, Ronald Lehndorff, Wolfgang Raberg et al. "Opportunities and challenges for spintronics in the microelectronics industry." *Nature Electronics* 3, no. 8 (2020): 446–459.

[3] Burkard, Guido, Thaddeus D. Ladd, Andrew Pan, John M. Nichol, and Jason R. Petta. "Semiconductor spin qubits." *Reviews of Modern Physics* 95, no. 2 (2023): 025003.

[4] He, Jiaqi, Wei-Chih Cheng, Qing Wang, Kai Cheng, Hongyu Yu, and Yang Chai. "Recent advances in GaN-based power HEMT devices." *Advanced Electronic Materials* 7, no. 4 (2021): 2001045.

[5] García de Arquer, F. Pelayo, Dmitri V. Talapin, Victor I. Klimov, Yasuhiko Arakawa, Manfred Bayer, and Edward H. Sargent. "Semiconductor quantum dots: Technological progress and future challenges." *Science* 373, no. 6555 (2021): eaaz8541.

[6] Cha, Eunjung, Niklas Wadefalk, Giuseppe Moschetti, Arsalan Pourkabirian, Jörgen Stenarson, and Jan Grahn. "InP HEMTs for sub-mW cryogenic low-noise amplifiers." *IEEE Electron Device Letters* 41, no. 7 (2020): 1005–1008.

[7] Faramehr, Soroush, and Petar Igić. "Analysis of gan hemts switching transients using compact model." *IEEE Transactions on Electron Devices* 64, no. 7 (2017): 2900–2905.

[8] Zhang, Li-Jiang, Jiang-Hui Mo, Yu-Xing Cui, Xing-Chang Fu, Guang Qian, Xian-Jie Li, and Tong Zhang. "A 750-W AlGaN/GaN HEMT operating at 80 V for $ L $-Band applications." *IEEE Microwave and Wireless Components Letters* 28, no. 5 (2018): 440–442.

[9] Kargarrazi, Saleh, Ananth Saran Yalamarthy, Peter F. Satterthwaite, Scott William Blankenberg, Caitlin Chapin, and Debbie G. Senesky. "Stable operation of AlGaN/GaN HEMTs for 25 h at 400° C in air." *IEEE Journal of the Electron Devices Society* 7 (2019): 931–935.

[10] Jeong, Jaeyong, Seong Kwang Kim, Jongmin Kim, Dae-Myeong Geum, Juyeong Park, Jae-Hyung Jang, and Sanghyeon Kim. "Stackable InGaAs-on-insulator HEMTs for monolithic 3-D integration." *IEEE Transactions on Electron Devices* 68, no. 5 (2021): 2205–2211.

[11] Okazaki, Yuma, Imran Mahboob, Koji Onomitsu, Satoshi Sasaki, and Hiroshi Yamaguchi. "Gate-controlled electromechanical backaction induced by a quantum dot." *Nature Communications* 7, no. 1 (2016): 11132.

[12] Blokhina, Elena, Andrii Sokolov, Panagiotis Giounanlis, Xutong Wu, Imran Bashir, Dirk Leipold, Robert Bogdan Staszewski, Angelo Brambilla, and Federico Bizzarri. "Towards the co-simulation of charge qubits: A methodology grounding on an equivalent circuit representation." *IEEE Open Journal of Circuits and Systems* 2 (2021): 548–563.

[13] Stuyck, Nard Dumoulin, Roy Li, Stefan Kubicek, Fahd A. Mohiyaddin, Julien Jussot, B. T. Chan, George Simion, Bogdan Govoreanu, Marc Heyns, and Iuliana Radu. "An integrated silicon MOS single-electron transistor charge sensor for spin-based quantum information processing." *IEEE Electron Device Letters* 41, no. 8 (2020): 1253–1256.

[14] Tadokoro, Masahiro, Takashi Nakajima, Takashi Kobayashi, Kenta Takeda, Akito Noiri, Kaito Tomari, Jun Yoneda, Seigo Tarucha, and Tetsuo Kodera. "Designs for a two-dimensional Si quantum dot array with spin qubit addressability." *Scientific Reports* 11, no. 1 (2021): 19406.

[15] Hendrickx, N. W., W. I. L. Lawrie, L. Petit, A. Sammak, G. Scappucci, and M. Veldhorst. "A single-hole spin qubit." *Nature Communications* 11, no. 1 (2020): 3478.

[16] Graziano, Giacomo, Alberto Ferraris, Eunjung Cha, and Cezar B. Zota. "The effect of cryogenic temperatures on the lateral heat spreading in InGaAs/InP HEMTs." *IEEE Transactions on Electron Devices* (2023).

[17] Külah, Elcin, Christian Reichl, Jan Scharnetzky, Luca Alt, Werner Dietsche, and Werner Wegscheider. "The improved inverted AlGaAs/GaAs interface: Its relevance for high-mobility quantum wells and hybrid systems." *Semiconductor Science and Technology* 36, no. 8 (2021): 085013.

[18] Dubey, Shashank Kumar, Meena Mishra, and Aminul Islam. "Characterization of AlGaN/GaN based HEMT for low noise and high frequency application." *International Journal of Numerical Modelling: Electronic Networks, Devices and Fields* 35, no. 1 (2022): e2932.

[19] Sahebghalam, Nika, Majid Shalchian, Amirali Chalechale, and Farzan Jazaeri. "High-temperature HEMT model." *IEEE Transactions on Electron Devices* 69, no. 9 (2022): 4821–4827.

[20] Sahu, Swastik Kumar, and Kaushik Mazumdar. "State-of-the-art Analysis of Single Electron Transistor." In *2024 IEEE 3rd International Conference on Electrical Power and Energy Systems (ICEPES)*, pp. 1–6. IEEE, 2024. https://doi.org/10.1109/ICEPES60647.2024.10653610

[21] Liu, Mengxia, Nuri Yazdani, Maksym Yarema, Maximilian Jansen, Vanessa Wood, and Edward H. Sargent. "Colloidal quantum dot electronics." *Nature Electronics* 4, no. 8 (2021): 548–558.

[22] Yang, Junjie, Zizhuo Liu, Pamela Jurczak, Mingchu Tang, Keshuang Li, Shujie Pan, Ana Sanchez et al. "All-MBE grown InAs/GaAs quantum dot lasers with thin Ge buffer layer on Si substrates." *Journal of Physics D: Applied Physics* 54, no. 3 (2020): 035103.

[23] Ha, Jun Mok, Na Eun Lee, Young Jun Yoon, Seoung Ho Lee, Yong Seok Hwang, Jae Kwon Suk, Chan Young Lee, Cho Rong Kim, and Sunmog Yeo. "Universal dry synthesis and patterning of high-quality and-purity graphene quantum dots by ion-beam assisted chemical vapor deposition." *Carbon* 186 (2022): 28–35.

[24] Kratochwil, Benedikt, Jonne V. Koski, Andreas J. Landig, Pasquale Scarlino, Jose C. Abadillo-Uriel, Christian Reichl, Susan N. Coppersmith et al. "Charge qubit in a triple quantum dot with tunable coherence." *Physical Review Research* 3, no. 1 (2021): 013171.

[25] Cogan, Dan, Giora Peniakov, Zu-En Su, and David Gershoni. "Complete state tomography of a quantum dot spin qubit." *Physical Review B* 101, no. 3 (2020): 035424.

[26] Khan, Md Abdul Kaium, Mohammad Abdul Alim, and Christophe Gaquiere. "2DEG transport properties over temperature for AlGaN/GaN HEMT and AlGaN/InGaN/GaN pHEMT." *Microelectronic Engineering* 238 (2021): 111508.

[27] Bhat, Hilal Ahmad, Farooq Ahmad Khanday, Brajesh Kumar Kaushik, Faisal Bashir, and Khurshed Ahmad Shah. "Quantum computing: Fundamentals, implementations and applications." *IEEE Open Journal of Nanotechnology* 3 (2022): 61–77.

[28] Rajput, Priti J., Sheetal U. Bhandari, and Girish Wadhwa. "A review on—Spintronics an emerging technology." *Silicon* 14, no. 15 (2022): 9195–9210.

[29] Chaturvedi, Nidhi, Kuldip Singh, Pharyanshu Kachhawa, Richard Lossy, Shivanshu Mishra, Ashok Chauhan, Dheeraj K. Kharbanda et al. "AlGaN/GaN HEMT based sensor and system for polar liquid detection." *Sensors and Actuators A: Physical* 302 (2020): 111799.

[30] Sun, Fu-Li, Yuan-Dong Wang, Jian-Hua Wei, and Yi-Jing Yan. "Capacitive coupling induced Kondo–Fano interference in side-coupled double quantum dots." *Chinese Physics B* 29, no. 6 (2020): 067204.

[31] Tong, Chuyao, Rebekka Garreis, Angelika Knothe, Marius Eich, Agnese Sacchi, Kenji Watanabe, Takashi Taniguchi et al. "Tunable valley splitting and bipolar operation in graphene quantum dots." *Nano Letters* 21, no. 2 (2021): 1068–1073.

[32] Sahu, Swastik Kumar, Kaushik Mazumdar, Bahn Kitmo, Yosef Berhan Jember, and Sima Das. "Design and investigation of InGaAs/InP/InAlAs MOSFET with optimized switching efficiency." *IEEE Access* (2024). https://doi.org/10.1109/ACCESS.2024.3401851

[33] Sahu, Swastik Kumar, and Ram Narayan Yadav. "Performance Analysis of ResNet in Facial Emotion Recognition." In *International Conference on Machine Intelligence and Signal Processing*, pp. 639–648. Springer Nature Singapore, 2022. https://doi.org/10.1007/978-981-19-2165-0_16

[34] Sahu, Swastik Kumar, and Ram Narayan Yadav. "Key facial points recognition using ResNet." *Materials Today: Proceedings* 66 (2022): 3651–3656. https://doi.org/10.1016/j.matpr.2022.07.342

[35] Zeng, Yin, Jörgen Stenarson, Peter Sobis, Niklas Wadefalk, and Jan Grahn. "Sub-mW cryogenic InP HEMT LNA for qubit readout." *IEEE Transactions on Microwave Theory and Techniques* (2023).

[36] Yang, Hyunsoo, Sergio O. Valenzuela, Mairbek Chshiev, Sébastien Couet, Bernard Dieny, Bruno Dlubak, Albert Fert et al. "Two-dimensional materials prospects for non-volatile spintronic memories." *Nature* 606, no. 7915 (2022): 663–673.

[37] Shi, Yuhan, Sangheon Oh, Zhisheng Huang, Xiao Lu, Seung H. Kang, and Duygu Kuzum. "Performance prospects of deeply scaled spin-transfer torque magnetic random-access memory for in-memory computing." *IEEE Electron Device Letters* 41, no. 7 (2020): 1126–1129.

[38] Ghising, Pramod, Chandan Biswas, and Young Hee Lee. "Graphene spin valves for spin logic devices." *Advanced Materials* 35, no. 23 (2023): 2209137.

[39] Mbayachi, Vestince B., Euphrem Ndayiragije, Thirasara Sammani, Sunaina Taj, and Elice R. Mbuta. "Graphene synthesis, characterization and its applications: A review." *Results in Chemistry* 3 (2021): 100163.

[40] Richey, Nathaniel E., Camila De Paula, and Stacey F. Bent. "Understanding chemical and physical mechanisms in atomic layer deposition." *The Journal of Chemical Physics* 152, no. 4 (2020).

Chapter 10

Evaluation

CMOS Technology for the Estimation of Digital PLL Building Blocks

Rachana Arya and B.K. Singh

10.1 INTRODUCTION

A feedback device called a lock loop setup is utilized to equalize the input and output signals. Applications can categorize a locked loop as a software PLL, DPLL, ADPLL, or analogue PLL. PLL is APLL if it is built using analog blocks. It creates a feedback voltage directed toward the input after analyzing the circuit's throughput. The evaluation of both signals produces a difference signal [1, 2]. Furthermore, this control signal will adjust the output signal by the input signals. All PLLs do, however, have several disadvantages, such as jitter supervision, noise consequence, coupling effect, and clock secure time. ASIC technologies are significantly complicated by the usage of low supply voltage. A further way to get around the analog PLL is to use a digital one. The main building blocks of digital anatomy are sequential, utilizing a clock for each circuit or block at the same time [3]. All of the components in this circuit are digital, except the voltage-controlled oscillator. Both active and passive parts can be used to create this oscillator. The most commonly used oscillator in PLL applications is the LC oscillator. In comparison to other oscillators, it offers good noise rejection characteristics [4, 11]. However, it has a big chip size, a short tuning bandwidth, and a high power consumption. An alternative is a ring oscillator made with CMOS transistors.

Compressed design, exceptional packing stability, and low power consumption are the main advantages of ring oscillators. Its noise cancellation is not great, though. The bit slip ratio in the communication link has the potential to raise it. Phase noise plus the oscillators' simulated orientation can be used to calculate the sensitivity limit [5, 6]. Sensitivity is the input-restrained minimum detecting R.F. power. Scalability, integration possibilities, and reduced power consumption compared to their analog equivalents. Performance and efficiency must sometimes be sacrificed in favor of conventional designs, which frequently call for substantial resource requirements. A potential solution to this problem is the development of approximation

DOI: 10.1201/9781003533535-10

techniques, which allow for the efficient design of digital PLLs without sacrificing crucial performance measures.

Principal Approximation Methods:

Phase Recognition:
Diminished accuracy: Using a lower resolution when quantifying the phase difference lowers complexity but increases jitter (e.g., [1, 2]).

Estimated circuits: Lower power and area are traded for precision when using simpler designs, such as XOR-based techniques (e.g., [3, 4]).

Low Pass Filter:
Lower-order filters: Although they use fewer resources, lower-order filter implementations might affect stability and settling times (e.g., [5, 6]).

Distributed arithmetic (DA): DA-based architectures provide compact design; however, there may be accuracy constraints (see, for example, [7, 8]).

Reference Dividers:
Coarse-to-fine dividers: Using multi-stage structures that split data first and then fine-tune it saves resources, but also adds delay (e.g., [9, 10]).

Algorithmic dividers: Lower area can be obtained at the expense of higher calculation complexity by employing algorithms rather than conventional counters (e.g., [11, 12]).

Quantitative Evaluation: Numerous studies have applied these approximation techniques in CMOS and assessed their influence on important performance metrics:

- Lock time: Research like [13] and [14] examines the effect of approximate phase detectors and reduced precision on lock time.
- Jitter: [15] and [16] study the jitter performance of various approximation techniques in loop filters and reference dividers.
- Settling time: [17] and [18] examine the impact of reduced order filters on settling time.
- Resource utilization: [19] and [20] compare the area, power, and gate count of conventional and approximated designs.
- Resource incorporation: [19] and [20] compare the area, power, and gate count of estimated and conventional architectures.

Alternatives and Difficulties: Although approximation has advantages, it is important to carefully weigh the trade-offs.

- Performance vs. accuracy: worse resource consumption frequently results in worse accuracy, which may have an effect on jitter, lock time, or stability.
- Application-specific optimization: Depending on the needs of the application and the required levels of performance, different approximation approaches are used.
- Stability and robustness: Careful planning and analysis are required to ensure system stability and resilience with approximations.

Current Developments: With new developments, the field of digital PLL is growing.

- Estimates derived from machine learning: Phase detection and loop filtering using machine learning have the potential to increase accuracy and efficiency.
- Flexible approximation techniques: Further optimizing of performance and resource utilization can be achieved with methods that dynamically modify approximation levels according to operational conditions.

Design Resources and Frameworks: Innovation in PLL design can be accelerated by creating frameworks and tools that enable systematic investigation of approximation. In the design of high-performance, resource-efficient digital PLLs in CMOS technology, approximation approaches have become a potent tool. Not abandoning critical performance measures, we may accomplish considerable savings in area, power consumption, and gate count by selectively introducing controlled mistakes in certain components.

Important Conclusions:

There are several ways to approximate phase detection. Loop filters and reference dividers are some of the methods available. Each has trade-offs related to accuracy, efficiency, and resource conservation.

- Quantitative analysis is essential. Thoroughly characterizing lock time, jitter, settling time, and resource use facilitates the comparison of several approximations and the selection of the best method for particular applications.
- Trade-offs must be made: Application needs and performance criteria must be carefully considered to balance accuracy with resource efficiency.
- Prospective developments include ongoing efforts to push efficiency and performance limits in the areas of adaptive schemes, machine learning-based approximations, and design automation tools.

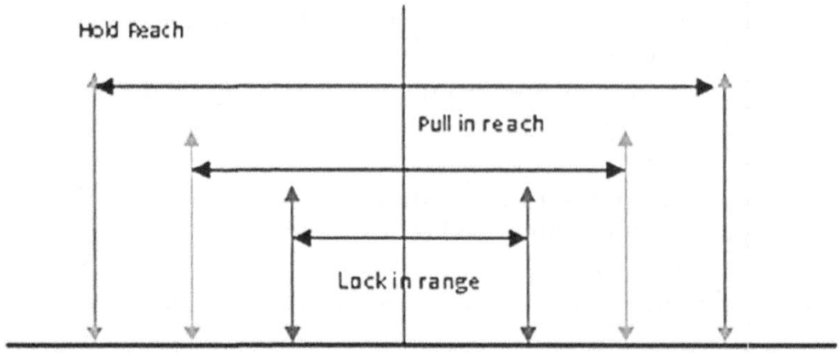

Figure 10.1 Simple DPLL block diagram.

Furthermore, the sensitivity of a trans-receiver is determined by both active and passive components. A loop filter can remove the oscillator noise, but the PLL is unable to remove the unusual noise. Furthermore, this modifies the circuit's inclusive performance. The DPLL building block diagram is covered in Figure 10.1. It is composed of these primary parts: Phase detector, voltage-controlled oscillator, loop filter, and N counter division. The phase detector is an essential component of the PLL. It is connected to the frequencies of input and feedback. By evaluating both signals (IN_{ref}, $IN_{feedback}$), the difference signal may be produced. Up-and-down signals are generated by the error deviation. The loop filter now uses this differential voltage as its input to create a DC voltage.

The voltage-controlled oscillator is altered in phase and frequency by this voltage [7, 13]. A divide-by-n counter is used in the output signal to provide feedback to the input, changing the input signal. A clock pulse with a set duty cycle is produced by an ideal PLL. In contrast, real PLL clocks have minute deviations in the timing pulse from their ideal time. We call this jitter. The responsiveness of the system can be made smoother by filtering high-frequency jitters with a loop filter.

10.2 ACTING OF PLL OPERATION

The Phase Lock Loop (PLL) can function in three main zones of action, as seen in Figure 10.2. The PLL is in a dynamic state if it is not locked, meaning that the input and output are on separate phases. It is sometimes referred to as a static state if both phases are the same.

Hold Range

This is the range of frequencies in which the PLL can maintain phase tracking between the input and output. If the reference frequency is gradually

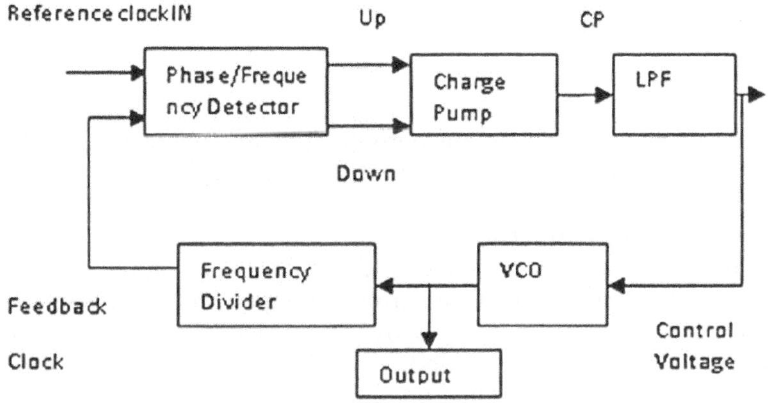

Figure 10.2 PLL region of operation.

condensed or enlarged, the PLL will be unable to find a lock. This is called the hold range. It will be temporarily stable just in this range [9].

Pull-In Range

In a dynamic state, it will be constantly locked. Initially, PLL is unlocked. It will obtain a locked state if a reference frequency is applied in the pull-in range. If the input frequency is outside, then the PLL will never acquire a lock condition.

Lock Range

The frequency range that the loop can maintain the locked state. Figure 10.2 displays every operational area.

Pull-Out Range

The frequency makes it easier to disengage the PLL. When the frequency of the input falls below the withdrawal range, the PLL stays locked. Conversely, should it surpass, PLL will be unable to track the output signal. It might eventually get a lock. It is known as the PLL's static state.

10.3 DPLL Schematic

A locked loop in DPLL is produced by comparing the input and output signals. The output signal's phase and frequency are kept constant by combining the charge pump and VCO. Assume that the phases of the input and output signals are ref and out. PLL is mostly used to determine the phase difference between two signals. However, because of the negative feedback loop, it is necessary to make additional adjustments to the output phase to synchronize with the input phase [5]. The following analysis leads to the following conclusion:

$$\varnothing_i = \varnothing_o \tag{10.1}$$

$$f_i = Nf_o \tag{10.2}$$

where Φ_i, Φ_o, and N represent the output, input, and reference signal phases, respectively, and the numerical value of the Φ_i divide by N circuit.

Phase-Frequency Detector

Phase and frequency detectors may be derived from a wide variety of logical circuits in the digital PLL. Comparing the input and feedback signals is this block's primary task. Phase frequency detectors, flip-flop counter-phase detectors, J-K flip-flop kinds, Ex-OR phase detectors, and double-edge-triggered PFDs are the most significant varieties of PFDs. Below is an explanation of a few of the phase detectors:

a. Ex-OR type

A linear multiplier is comparable to this detector. It is assumed that the signals v_1 and v_2 are symmetrical square waves, as the signals for DPLLs are always digital waves. The block diagram of the EX-OR phase detector is shown in Figure 10.3. Because both of the signal values are high-end, this phase detector resets its internal states and is insensitive to the reference duty cycle [10, 14]. It will provide zero output if both inputs are high or low; nevertheless, it will produce high output if there is a difference in the inputs.

$$CK_{pd} = CK_{input} - CK_{feedback} \tag{10.3}$$

$$= \frac{\Delta_t}{T_{input}} 2\pi \text{radians} \tag{10.4}$$

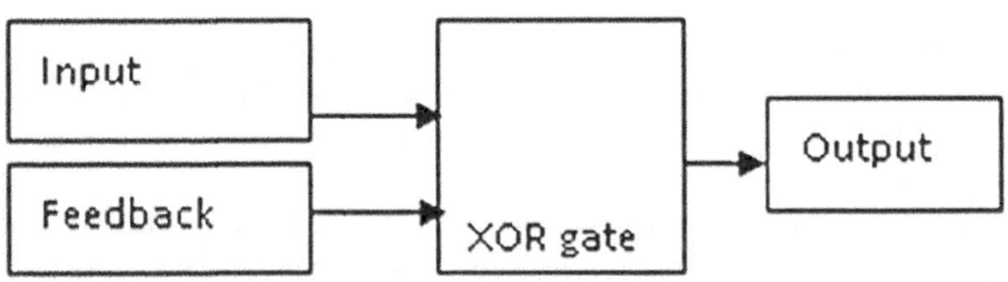

Figure 10.3 Basic EX-OR type phase detector.

The phase difference between the locked data input and the feedback clock may be expressed as

$$CK_{pd} = \frac{\pi}{2} \tag{10.5}$$

For this phase detector, the average voltage is represented as

$$CK_{out} = V_{dd} \frac{CK_{pd}}{\pi} = k_{out}\ CK_{pd} \tag{10.6}$$

where the gain k_{out} may be expressed as

$$k_{Out} = \frac{V_{dd}}{\pi} \text{ V/radians} \tag{10.7}$$

As seen in Figure 10.4, up and down signals are produced by comparing the oscillator and reference frequencies, which are equivalent to input signals. The DN and UP signals, which are sent to the PLL's subsequent state, are the frequencies that differ between the two. This kind of phase detector is very basic, straightforward, and efficient.

b. The edge-triggered J-K flip-flop

Another type of phase detector is the J-K Flip-Flop. This is not responsive to waveform irregularity because the J-K flip-flop is edge-triggered, as shown in Figure 10.5.

Two signals, V_1 and V_2, were applied to the J-K flip-flop and produced output from Q. The Q' is the inverted output signal. So the next state of the flip-flop is Q*.

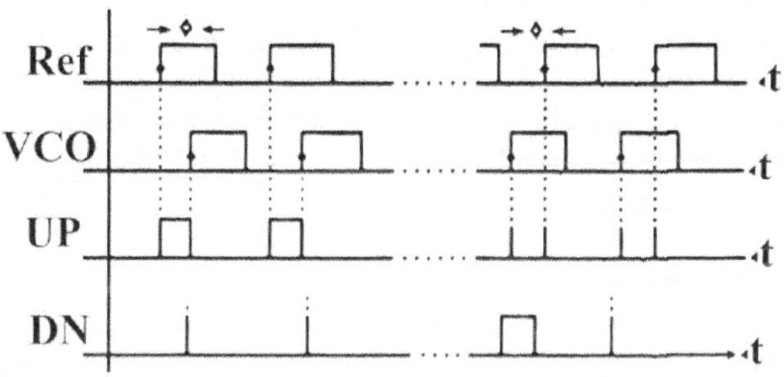

Figure 10.4 Phase detector waveform of the EX-OR type.

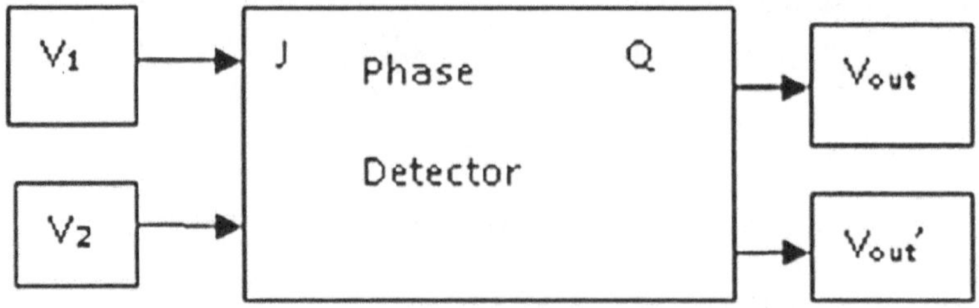

Figure 10.5 J-K type phase detector.

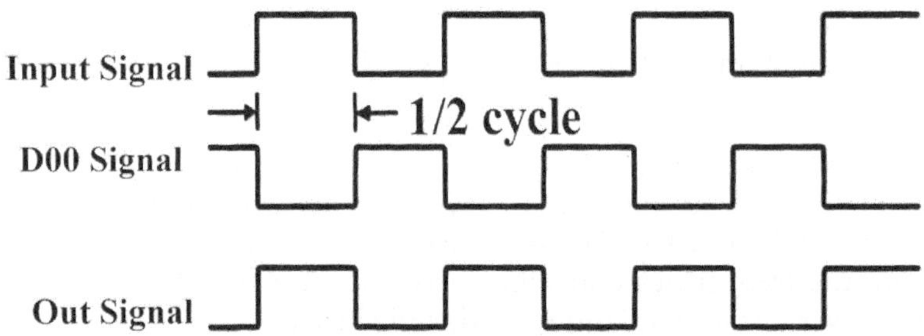

Figure 10.6 Waveform of J-K type phase detector.

$$Q^{*} = JQ' + K'Q \tag{10.8}$$

When a positive edge-triggered input is applied at the J terminal, the output is high, and at Q', the output is low. The phase error of the phase detector "θ_{pd}" is initially zero, and input and output are on opposite phases, as shown in Figure 10.6. The output signal V_{out} is the same as the input phase [16, 21]. When the input duty cycle is less than half, θ_{pd} approaches 180°, and Q' becomes positive.

And when θ_{pd} approaches –180°, the Q' becomes negative. When the phase detector error is between $-180° < \theta_{pd} < 180°$, the JK phase detector sustains phase tracking with the input.

c. Phase frequency detector

This circuit in Figure 10.7 uses two flip-flops, D_1 and D_2, with a reset clock terminal (the output of the AND gate) [7]. The input terminals of both flip-flops are connected to VDD. The outputs Q_A and Q_B are called UP and

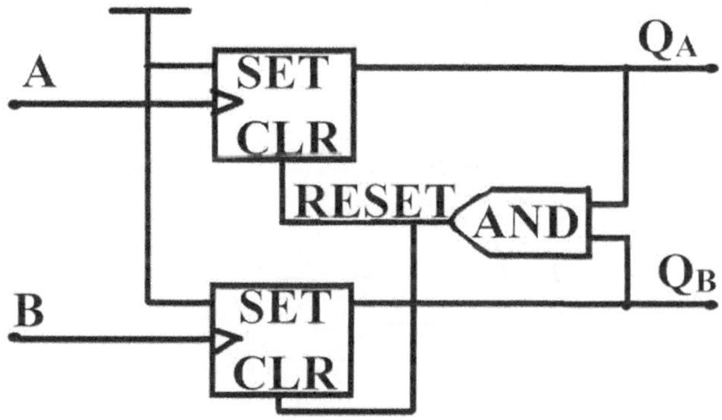

Figure 10.7 Circuit diagram of a phase frequency detector.

DOWN signals, respectively, as shown in Figure 10.8. If PFD is in "0" state and the circuit changes state "A" to state "1", then the values of the outputs are $Q_A = 1$ and $Q_B = 0$. Until a positive change comes, the circuit remains in this condition [8].

The changeover from state "0" to state "2" is similar to the changes from state "0" to state "1". The mere dissimilarity is a positive change at "B" occurring in its place of "A".

Low Pass Filter

Another block of the DPLL is the loop filter. This is a low-pass filter that converts the PFD's output to the DC control signal. This control signal will alter the output of the VCO. This filter may be passive or active. Loop filters manage the characteristics of the PLL, like locking range, confinement range, and bandwidth. The first-order filter is generally common in DPLL because of its fast sensitivity. However, it enhances frequency jitter because of the intrinsic voltage jump at resistor R_2 [12].

By including a supplementary capacitor C_3 comparable through the RC branch, as shown in Figure 10.9, this jitter can be suppressed. The total range of frequencies over which the loop can track or respond to changes in the input frequency is known as the tracking range or lock-in range. When the circuit attains the phase-locked condition, it is in capture range. The transfer function of this filter system is shown in Figure 10.10.

$$\frac{V_o}{V_i} = \frac{Z_2}{Z_2 + Z_1} \tag{10.9}$$

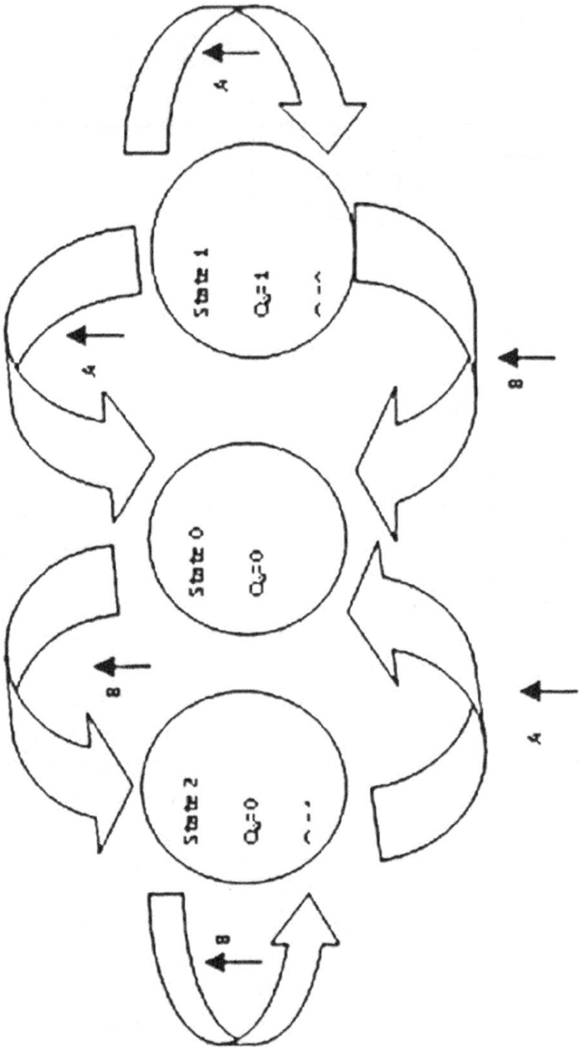

Figure 10.8 State diagram of PFD.

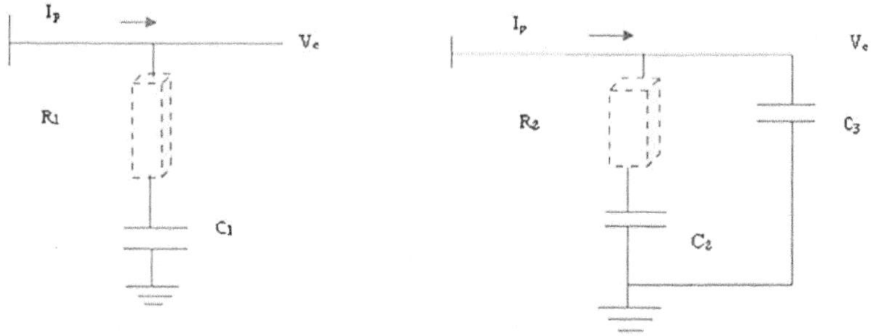

Figure 10.9 1st order and 2nd order loop filter.

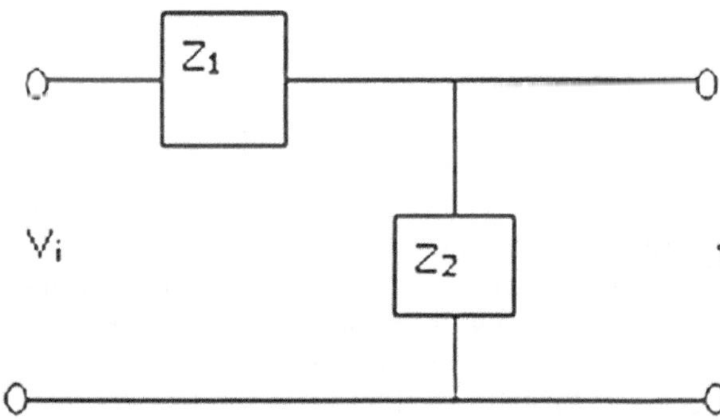

Figure 10.10 Simple low pass filter structure.

where Z_1 and Z_2 are the resistors and capacitors of the circuit. So gain could be calculated using the following equation:

$$\frac{V_o}{V_i} = \frac{1/sC_2}{\frac{1}{sC_2} + R_2} = \frac{1/j\omega C_2}{\frac{1}{j\omega C_2} + R_2} \tag{10.10}$$

$$\frac{V_o}{V_i} = \frac{1}{j\omega C_2 R_2 + 1} \tag{10.11}$$

The magnitude response is given by

$$\left|\frac{V_o}{V_{in}}\right| = \left|\frac{1}{sC_2R_2 + 1}\right| = \left|\frac{1}{j\omega_c C_2 R_2 + 1}\right|$$

$$= \left| \frac{1}{\sqrt{[1^2 + (\omega_c C_2 R_{2)}{}^2]}} \right| \tag{10.12}$$

The expression of the cut-off angular frequency is given as

$$\omega_c = \frac{1}{C_2 R_2} \tag{10.13}$$

Loop filtering introduces poles to the PLL transfer function. That will enhance the PLL bandwidth. As the order of the filter increases, the noise cancellation would be better, so a second order or more is required for ASIC technology [9].

Voltage Controlled Oscillator

VCO is conceivably the most significant block of the PLL. It is used to manage the mid-range frequency so that bandwidth can be controlled. The tuning range will follow the center frequency changes. As the gain of the VCO changes with the error voltage, non-linearity in the tuning range will occur. The main characteristic of VCO is to achieve a constant oscillation so that the generated voltage will maintain the lock condition. Yet when a constant error voltage is given to the input of the VCO, the VCO output is not ideally periodic. Mainly, all the electronic devices suffer from supply noise that adds up to the output and causes jitter [8, 12]. Mainly, oscillators are of two types:

Resonant oscillator

The resonant oscillator can be divided into two categories, i.e., the LC oscillator and the crystal oscillator. The waveform oscillator can also be divided into two categories, the ring oscillator and the relaxation oscillator.

a. Ring oscillator

The most frequent oscillator used in PLL is the ring oscillator, as shown in Figure 10.11. In this, an odd number of inverters is connected in series, and the output is fed back to the input of the first inverter.

By using odd numbers of inverters, the circuit cannot have a fixed operating point. As a result, the output should be oscillating. The propagation delay is shown in Figure 10.11. The total oscillation period is given by the following equation:

$$T = 2\tau_p N \tag{10.14}$$

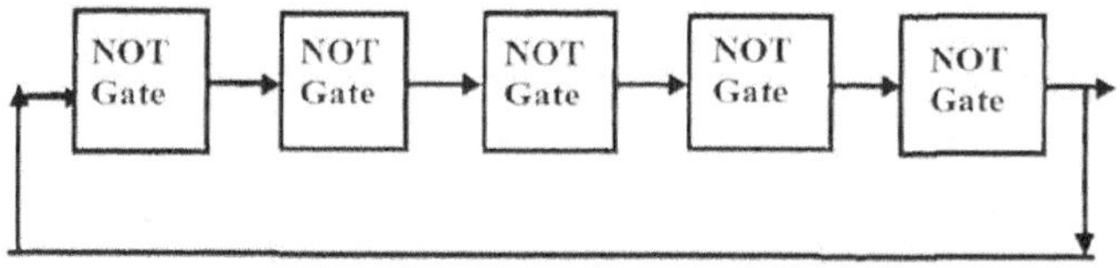

Figure 10.11 Five-stage ring oscillator

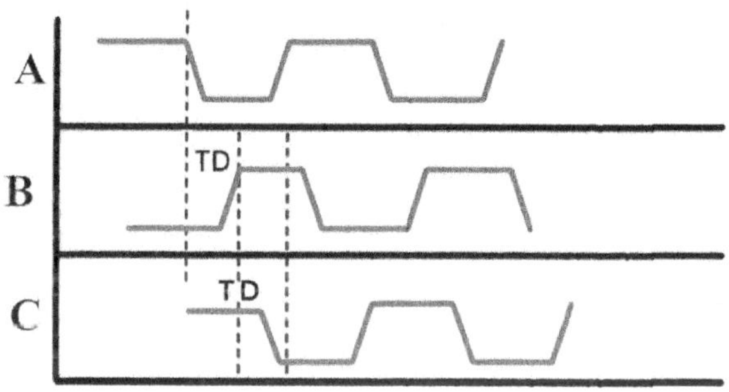

Figure 10.12 Output response of ring oscillator.

where T is total period of the circuit, τ_p is total delay of the circuit, and N is total number of stages.

As described in Figure 10.12, A, B, and C are the clock, and input and output waves are shown, respectively. The delay of input and output throughput is shown. For all the inverter circuits, the fan-out is one [10]. The main benefit of a ring oscillator is the design of the circuit. With transistor design, it uses very little area, so the power density is very low. This oscillator is easy to tune, so it provides a wide bandwidth. However, the major problem with this oscillator is poor phase noise. As a result, this is mainly used for clock recovery, generation, and synchronization.

b. LC Oscillators

This type of oscillator is also known as a tank oscillator (in Figure 10.13), where capacitors and inductors are running in parallel. If the tank circuit is providing an impulse current, the energy is transferred backwards and forwards between the inductor and the capacitor. Nevertheless, this happens simply when L and C are lossless.

The inductor and capacitor quality factors can be computed by the following equation:

$$Q_L = \frac{\omega L}{R_s} \tag{10.15}$$

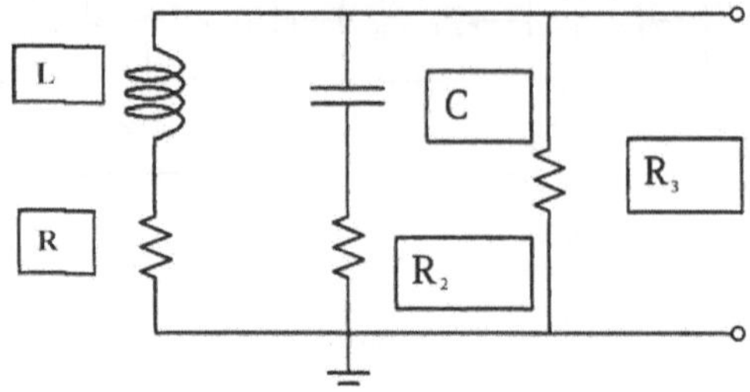

Figure 10.13 LC circuit.

$$Q_C = \frac{1}{\omega C R_{sc}} \tag{10.16}$$

The characteristic impedance Z_0 of the tank circuit is as follows:

$$Z_0 = \frac{1}{\omega C} = \omega L = \sqrt{\frac{L}{C}} \tag{10.17}$$

The total quality factor for the circuit is given by the following equation:

$$\frac{1}{Q_T} = \frac{Z_0}{R_p} + \frac{1}{Q_L} + \frac{1}{Q_C} \tag{10.18}$$

The value of Q_T is the lowest quality factor in the circuit. The value of the center frequency can be dependent on the values of L and C. The Z_0 is directly proportional to the value of the inductor. To balance the losses at R_{sc}, a parallel resistance R_p is connected to sustain oscillations [8, 11].

10.4 COMPARISON OF DPLL PARAMETERS

Tables 10.1–10.3 show the different comparison parameters of all the components.

10.5 SIMULATED DIAGRAM OF ALL BUILDING BLOCKS

The simulated design of PFD, LPF, and VCO is shown in Figures 10.14 and 10.15. By using 120 nm CMOS technology, the duration of the glitch in

Table 10.1 Comparison of different types of phase detectors

Phase Detector			
S.N.	*Type*	*Advantages*	*Disadvantages*
1	EX-OR	It will generate the output voltage for both cycles of the input pulse.	The data losses occurred because that is not sensitive to both edges.
2	JK flip flop	The data losses will not occur because that is sensitive for both clock edges.	It produced larger phase tracking, and the angle is −180° to 180° phase error.
3	PFD	The phase error shifted from −360° to 360°. It will lock the input for different conditions.	Circuit complexity is high as the number of components is higher than that of other phase detectors.

Table 10.2 Comparison of different types of loop filters

Loop Filter			
S.no.	*Type*	*Advantages*	*Disadvantages*
1	Active filter	In this filter load, resistance does not affect the operation of the circuit.	At the high frequency circuit output is not fixed.
		The bandwidth of the circuit can be easily adjusted by varying the parameters of the circuit.	The throughput is unreliable at high frequencies.
		It provides good isolation between the load and the circuit components.	As the change in the input voltage occurs, it will openly change the output of the circuit.
		The Q factor is high.	It becomes costly as active components are used.
2	Passive filter	Easy to design and compact, light in weight.	Q factor is low.
		Works on very high frequency, so it will suffer from noise.	It cannot reduce the ripples.
		Enhances the circuit complexity.	It will easily alter the control voltage.
		No isolation required between the load and the circuit.	Modulate the clock frequency so the clock pulse will vary.

the PFD is 1.016–1.426 ns. For the first-order filter, the value of R1 is 300 kΩ and C_1 is 10 pF. Therefore, by the calculation using these parameters, the center frequency is 98.76 MHz. The ring oscillator's total delay time for five stages is t_d = 20.25 ns, and the oscillation frequency is 333.4 MHz (Figure 10.16).

Table 10.3 Advantages and disadvantages of VCO

		Voltage Control Oscillator	
S.no.	*Type*	*Advantages*	*Disadvantages*
1	LC	Squat phase noise	Outsized layout area->larger area for inductor
		Excellent stability	Tank networks may have resultant resonant frequencies, which cause unwanted performance loss to the circuit.
		Low jitter performance of the circuit	Constricted tuning range, requires a lot of classification to tune, and it has a supplementary complex design.
2	Ring	Simple to integrate	Reduced phase noise
		Superior phase noise	Hardly ever worn for RF systems
		Phase generation is 360° at the time of oscillation, and it is used in elevated data links.	It will suffer from jitter problems

10.6 RESULT AND DISCUSSION

From the above figures, it is clear that the purpose of limiting the center frequency is to scale the transistors at different levels. As a result, the locking time of the PLL is reduced as compared to the conventional design of PLL [19]. As the same supply voltage is used in both papers, the transistor's size would change all parameters. The oscillation frequency is far better than in the conventional design of PLL. The power consumption is almost 1.08 mW with five stages of the oscillator. The comparison parameters are shown in Table 10.4. By varying the low-pass filter parameters, the oscillation frequency can be changed.

The phase frequency detector is designed by the D flip flop, therefore one glitch is observed at 1.06 ns. After that, the PFD generates the up and down pulses shown in Figure 10.14b. These pulses can change the control voltage for VCO. The maximum peak occurs at the 98.76 MHz frequency.

10.7 CONCLUSION AND FUTURE WORK

DPLLs are extremely open to future CMOS scaling techniques. They have outstanding performance over a wide range of frequencies. The relative study of DPLL building blocks has been done in this chapter. The DPLLs are good in terms of phase noise and low power consumption, and they enhance the tuning range for mid-frequency. Locking time mostly depends on the parameters used for the PFD, VCO, and loop filter. So by changing the parameters, an improved lock range can be achieved. By scaling

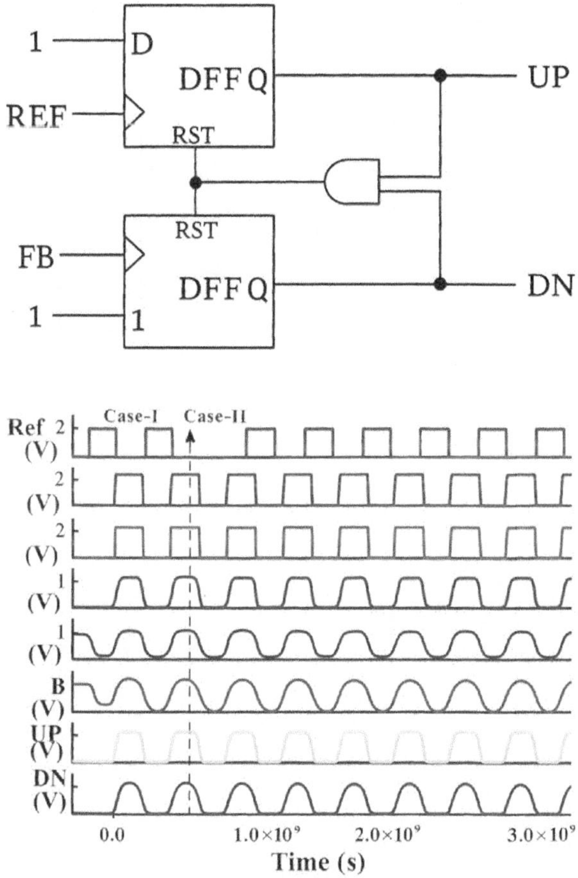

Figure 10.14 a) Simulated diagram of PFD. b) Simulated waveform of PFD.

the transistors, a better center frequency would be attained. The order of the filter will control the circuit bandwidth. The locking time is inversely proportional to the bandwidth of the circuit. As soon as the PLL achieves the locking range, the BW and noise jitter will also reduce. Ample-range applications are available for DPLLs. Due to their small size with respect to analog PLLs, they are widely used in clock generation and recovery [14]. By using this technique, analog to digital and D to A techniques would be reduced. Revival of low-frequency signals is easy; otherwise, they vanish in noise. As they are used to match the frequency and phase of the input signal, input sampling can be done digitally. They can also be used in clock generation for different microprocessor units. They can also be used to control speed of the motor.

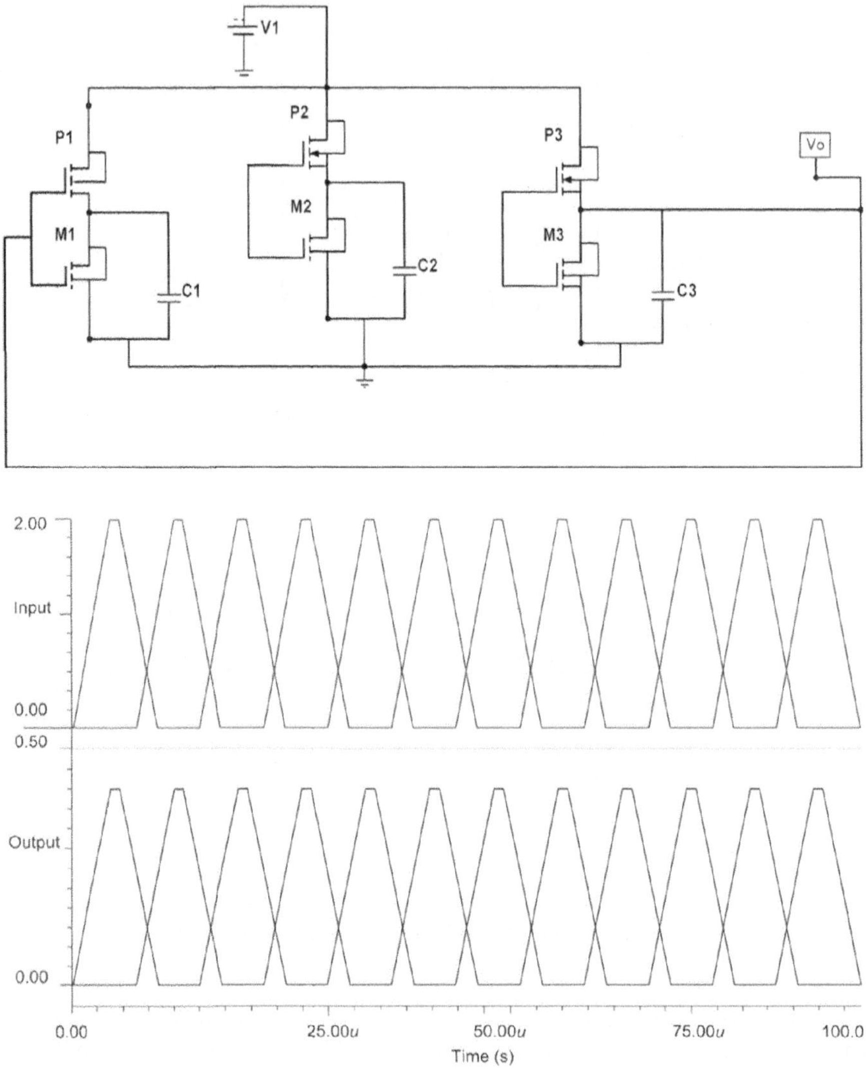

Figure 10.15 Simulated diagram and output waveform of the ring oscillator.

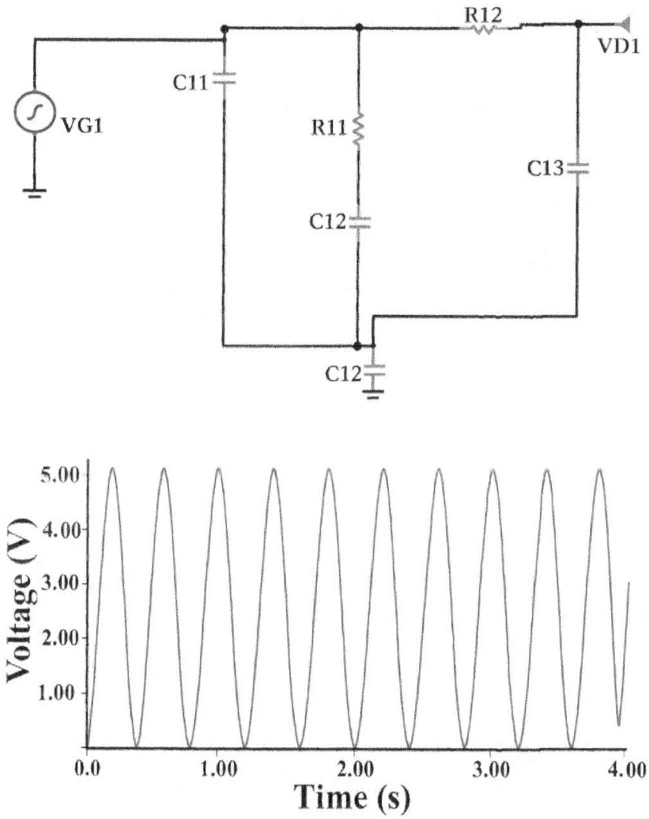

Figure 10.16 Circuit diagram and output waveform of low pass filter.

Table 10.4 The performance analysis of different parameters

Parameters	*Proposed design*	*[19]*
Technology	120 nm	130 nm
Supply voltage	1v	1v
Power consumption	1.08mW	2.07mW
Centre frequency	98.76 MHz	500MHz
No. of Stages	5	5
Oscillation frequency	333.4MHz	800MHz
Register R1	300kΩ	1.38KΩ
Load capacitance C1	10 pF	15pF
Glitch of PFD	1.016 ns to 1.426 ns	1.5μs
Delay time Td	20.25ns	----

CONFLICT OF INTEREST

The authors declare no conflict of interest.

REFERENCES

[1] L. Kyoohyun, "A Low-Noise Phase-Locked Loop Design by Loop Bandwidth Optimization," *IEEE Journal of Solid-state Circuits,* vol. 35, pp. 807–815, 2000.

[2] J. F. Parker and D. Ray, "A 1.6-GHz CMOS PLL with On-Chip Loop Filter", *IEEE Journal of Solid-state Circuits*, vol. 33, pp. 337–343, 1988.

[3] D. A. Ramey, "A 2–1600-MHz CMOS Clock Recovery PLL with Low-Capability," *Analog Integrated Circuits and Signal Processing Journal*, pp. 91–112, 1997.

[4] N. D. Patel and A. P. Naik, "A Low Jitter – Low Phase Noise Wideband Digital Phase Locked Loop in Nanometer CMOS Technology", *International Journal of Electronics and Communication Engineering and Technology*, vol. 9, no.3, pp. 1–12, 2018.

[5] K. Parmar, "All Digital Phase Locked Loop Design for Different Applications: A Review, "*IJIRT*, vol. 1, no.8, pp. 96–99, 2014.

[6] S. N. Dandare and A. H. Deshmukh, "Design of DPLL using Sub Micron 45 nm CMOS Technology and Implementation using Microwind 3.1 Software," *Journal of Science and Technology*, vol. 3, no.1, pp. 22–32, 2018.

[7] W. Abbas, Z. Mehmood and M. A. Seo," A V-Band Phase-Locked Loop with a Novel Phase-Frequency Detector in 65 nm CMOS," *Electronics*, vol. 9, pp. 1–12, 2020.

[8] D. Harikrushna, M. Tiwari, J. K. Singh and A. Khare," Design, Implementation and Characterization of XOR Phase Detector For DPLL in 45 nm CMOS Technology", *Advanced Computing: An International Journal (ACIJ)*, vol. 2, no.6, pp. 45–57, 2011.

[9] D. Ghai and N. Jain, All-Digital Phase Locked Loop (ADPLL) – A Review, *International Journal of Electronics and Computer Science Engineering*, vol. 2, no.1, pp. 94–101, 2012.

[10] P. K. Hanumolu, "Analysis of Charge-Pump Phase- Locked Loops," *IEEE Transactions on Circuits and Systems-I: Regular Papers,* vol. 51 no.9, pp. 1665–1674, 2004.

[11] B. Razavi, *Design of Analog CMOS Integrated Circuits,* Tata McGraw Hill Edition. 2002.

[12] U. Kumari and R. Yadav," Design and Implementation of Digital Phase Lock Loop: A Review," *International Journal of Advanced Research in Electronics and Communication Engineering,* vol. 7, no.3, pp. 197–201, 2018.

[13] P. Kumar and R. Yadav, "Design of CMOS Based FM Quadrature Demodulator using 45 nm Technology", *IJRSET*, vol. 3, no.7, pp. 12–16, 2016.

[14] S. Hathwalia and N. Grover, "A Comparative Study of Ring VCO & LC-VCO based PLL," *IJERA*, vol. 9, no.2, pp. 52–54, 2019.

[15] L. Zhang, Z. Wang, T. Zhang and X. Peng, "An Improved Fast Acquisition Phase Frequency Detector for High Speed Phase-Locked Loops," *Advances in Materials, Machinery, Electronics II AIP Conference Proceedings 1955*, vol. 1955, 2018, pp. 1–8.

[16] S. Tirumalasetti, S. K. Baburao, M. K. Kishore and E. R. K. Reddy," Design and Implementation of Optimum CMOS Phase Frequency Detector for PLL," *International Journal of Computer Application*, vol. 7, no.4, pp. 52–60, 2017.

[17] L. Bitla, V. Saraswathi and H. M. Akano, "Analysis of NAND Gate Based Phase Frequency Detector for Phase Locked Loop," *Journal of Critical Reviews*, vol. 7, no.9, pp. 172–176, 2020.

[18] K. Sivaraman," Design of CMOS Phase Locked Loop," *IJSRST*, vol. 4, pp. 1130–1136, 2018.

[19] N. Ahmad, N. A. B. Ishaimi and M. H. Jabbar," Charge Pump and Loop Filter for Low Power PLL Using 130nm CMOS Technology," *International Post Graduate Conference on Applied Science & Physics, IOP Publishing IOP Conference Series 1049*, 2018, pp. 1–10.

[20] G. Bhargav, G. Prasad, S. D. Canchi, B. Chanikya," Design and Analysis of Phase Locked Loop in 90nm CMOS," *IEEE*, 2016.

Chapter 11

Effective Design of a High-Speed, Low-Power Hybrid Adder for Computing Systems Using XOR-XNOR Cells

Satyajeet Sahoo and Aswini Kumar Samantaray

11.1 INTRODUCTION

For every VLSI designer in the present day, creating portable electronics devices with high speed and low power consumption—such as smartphones, tablets, PDAs, and gadgets—is a difficult undertaking. Nearly all designers will be working very hard to create gadgets with the qualities of being more dependable, high-speed, and tiny in size. To achieve this the hybrid logic style came into use. The current era has seen a rise in the demand for reliable, fast, and highly efficient electronic equipment in terms of power dissipation and speed. Digital Signal Processing (DSP)-based processors are utilized in electronic devices for video processing and filtering [1–4]. They perform a range of mathematical functions. Since addition is the most fundamental function and is utilized by nearly all devices, creating an effective adder boosts the device's overall effectiveness [5].

Adders come in a variety of design styles. They can be broadly divided into two designs: hybrid logic design and conventional logic design. The most significant logics in the conventional style include transmission gate full adder, complementary pass transistor logic, standard static CMOS, and dynamic CMOS. Different logic styles offer advantages to different performance factors. In contrast, more than one logic is involved in the construction of hybrid circuits. This type of style enhances the target circuits' performance by utilizing specific elements of the logic styles listed above [6].

In the majority of circuits, including adders, comparators, compressors, and so forth, the XOR-XNOR cells are essential. The circuit's overall performance can be raised by improving these cells' functionality and design. The design of a new XOR-XNOR circuit with full swing output and high speed and low power consumption constitutes the chapter's primary contribution. The structure of the chapter is as follows. An introduction is provided in Section 11.1, and related work is covered in Section 11.2. The concept of pass network is covered in Section 11.3. Section 11.5 wraps up Section 11.4's discussion of simulation results and the proposed work.

 DOI: 10.1201/9781003533535-11

11.2 RELATED WORK

Engineers aim to strike a balance in the PPA trade-off based on the specific requirements of a given application, taking into account factors such as power constraints, performance goals, and available silicon area [7]. Advanced semiconductor technologies and design methodologies continue to evolve to address these challenges and optimize PPA for various applications [8].

Power consumption in a digital circuit is the amount of energy it consumes during operation. It's influenced by factors such as switching activity, voltage, and frequency [9]. High-performance circuits tend to consume more power due to increased switching activity and faster clock speeds. Performance in this context refers to the speed or throughput of a digital circuit. Faster clock speeds and reduced delays generally lead to better performance [10].

However, achieving higher performance often comes at the cost of increased power consumption. Area, or physical size, is the amount of silicon area that a digital circuit occupies on a chip [6]. Smaller circuits generally consume less power and can be faster, but they may be limited in functionality [11]. Larger circuits may offer more features but could consume more power and have longer delays. To achieve higher performance, designers may increase clock speeds or employ aggressive circuit architectures. This typically results in higher power consumption and larger area requirements [12].

The circuit of the full adder mainly concentrates on the XOR-XNOR cell which is further extended to the generation of the sum and carry. Different strategies have been implied on the FA design to get the optimum results. Our strategy is to get the full output voltage swing and get the minimum PDP. With the aim of getting such an optimum result, we have designed the full adder in a hybrid style.

The transmission gate logic is a very reliable logic style in which we can obtain the full swing and also less delay to get the required output. A circuit design that combines these logic gates by incorporating CMOS and emerging technologies aims to achieve a balance between speed, power, etc. This integration of diverse technologies enables improved performance and functionality of the circuit.

Several logic types have been employed by researchers in the past [13–32] to develop complete adder circuits. The typical static CMOS full adder is one instance of a traditional design [33]. Its foundation is pull-up and pull-down typical PMOS and NMOS transistors. The excited capacitances cause a reduction in speed for static CMOS as well. Complementary pass transistor logic, which offers complete swing operation, is the other traditional approach. However, the large number of internal nodes causes it to dissipate a lot of power. Transmission gates serve as the basis for some other designs. These don't have the ability to drive, but they do use less power. Thus, scientists create hybrid designs that combine the best elements of each logic approach. Every module in the hybrid style can be implemented using a different logic style. Here, lowering the total power dissipation by minimizing the number of transistors is the primary goal. However, by doing so, the design is more viable in terms of things like noise immunity, driving capabilities, and layout complexity.

Many hybrid-style FA designs have been put out that provide the best possible performance at the output without any deterioration. Three blocks comprise a hybrid CMOS adder [5, 8]. The output signals of Block I are either an XOR, an XNOR, or both. Block II and III receive the outputs from this block as inputs to create sum and carry, respectively, as outputs, as shown in Figure 11.1. These blocks and circuits have been the subject of numerous investigations employing a variety of logics.

When evaluating the FA design's performance, the XOR-XNOR circuit's performance would be crucial. Zhang [34] presented a solution where the carry module is designed using CMOS logic and PTL to generate the XNOR-XOR outputs. In order to realize the sum and carry modules, Vastrabacka [9] suggested the XNOR and XOR circuits, which uses PTL logic and a 2-1 multiplexer logic. In recent years, various designs for XOR-XNOR circuits have been developed. One method is to use a static inverter to create the XNOR output after designing the XOR circuit. The primary limitation of this technique is that it causes output glitches and increases delay. The other method involves designing circuits with XOR-XNOR outputs generated simultaneously. This minimizes the time disparity within the signals and, to some extent, the aforementioned negatives.

In order to decrease power dissipation, Goel removed the two transistors from the XOR-XNOR circuit [5]. A short circuit current will be transmitted in this XOR-XNOR circuit design when the input shifts to AB=00 from AB = 01. This would have an impact on the circuit's general operation.

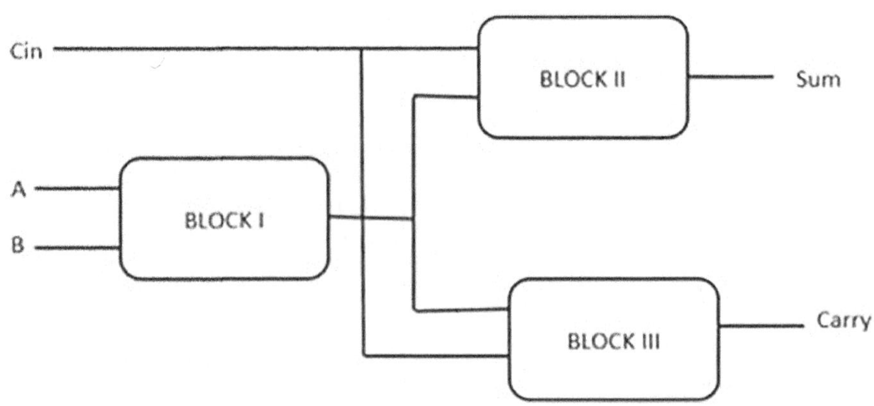

Figure 11.1 Hybrid FA block diagram.

Furthermore, the cross-coupled construction of the circuit contributes to its continued high latency.

A 12 transistor XOR-XNOR circuit architecture has been developed by Naseri and Timarchi [35]. Better delay and low power performance have been demonstrated by the design. However, the circuit has a feedback path and an external inverter. The employment of feedback paths and inverters increases output capacitance and delay, which in turn increases energy consumption.

A new ten transistor XOR-XNOR circuit has been proposed by Kandpal [8]. The ten-transistor XOR-XNOR circuit, depicted in Figure 11.2, leverages a differential pass-transistor logic (DPL) construction with a cross-coupled configuration. The XOR output side uses two PMOS, named P1 and P2, and three NMOS, named N3, N4, and N5. Conversely, the XNOR side employs three PMOS, named P3, P4, and P5, alongside two NMOS, named N1 and N2.

Within the XOR outcome section, parallel connected P1 and P2 function as a pass-transistor logic (PTL) circuit. Transistors NMOS4 and NMOS5 serve as a restoration mechanism that guarantees a rail-to-rail swing result, while NMOS3 acts by reason of a feedback transistor. Similarly, the output of XNOR utilizes N2 and N1 in a parallel PTL configuration, with P5 and P4 functioning like a restoration unit for a full-swing output. Transistor PMOS3 provides feedback within the present segment. This design offers simultaneous generation of full-swing Exclusive-OR and Exclusive-NOR results.

Upon receiving the input combination of A=0 and B=1, P4, P1, and N1 become active. PMOS2 and N1 function as pass transistors, enabling binary "1" for reaching the Exclusive-OR result and binary "0" for reaching the

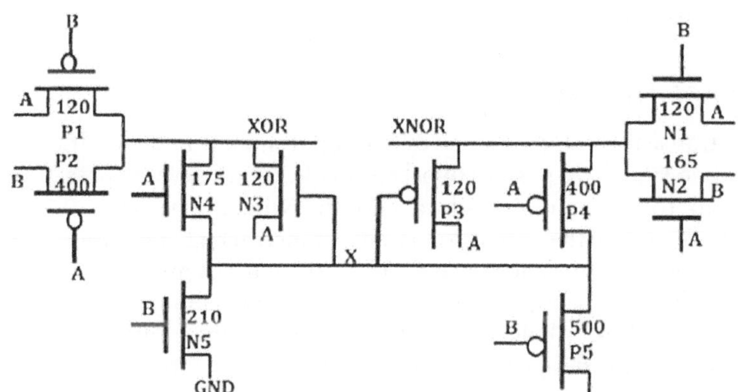

Figure 11.2 CMOS-CPL Based XOR-XNOR cell [20].

Exclusive-NOR result, respectively. Simultaneously, P4 activates transistor P3, allowing the weak logic "0" ($-V_{thp}$) to be passed to the Exclusive-NOR outcome. Likewise, according to the input state combination (A=1 and B= 0) called N4, P1, and N2 are in ON condition. Switching devices called N2 and P1 again act as pass transistors, directing digital-1,0 to the Exclusive-OR, Exclusive-NOR outcomes, independently. In this scenario, N4 activates N3, which then transmits the weak binary "1" (power supply to threshold voltage of NMOS transistor) to the Exclusive-OR outcome. It's important to note that for these specific data combinations (A=0, B=1 and A=1, B=0), the faulty logic outcome won't assume the overall outcome rail-to-rail swing because the established method already enables a rail-to-rail output, effectively resulting in a full voltage swing across the output range.

Due to the data combination (A=1, B=1) N4, N5, N1, and N2 are in ON condition. N2, N1 provide a faulty binary as "1" (power supply – threshold voltage of NMOS transistor) to the Exclusive-NOR outcome. Meanwhile, the Exclusive-OR fully turns off both the N5 and N4 transistors. Additionally, binary "0" is transmitted to the branch node called "X", which triggers P3 to activate and deliver a strong binary "1" to the Exclusive-NOR result.

With this data combination (A and B) set to "00," P5, P4, P2, and P1 become active. The P2, P1 deliver faulty logic "0" (negative threshold voltage of PMOS) to that of the Exclusive-OR outcome, Although P4, P5 transmits a straight logic as "1" to that of the Exclusive-NOR outcome and the branch node which is represented as "X", this binary "1" at junction X activates N3, resulting in a robust logic called "0" being proceed to the Exclusive-OR. The circuit shown in [35] has been enhanced by removing the external inverter. The feedback path has been used to produce full swing outputs. The size of the transistor has an impact on this circuit. Transistors that are sized incorrectly increase power consumption and delay, which have negative effects.

11.3 PASS NETWORKS

A pass transistor's source receives the input signal, and the drain receives the output signal. A pass network is made up of several pass transistors connected together to provide a certain switching function. The signal that passes across the transistor is determined using the applied signal to the gate [36]. When logic 0 is present at the gate of an NMOS transistor's instance, the path is opened and a path is established for the input to drain from source. PMOS behaves similarly when given a logic level 0 control signal. Signal X is designated as X(Y) if it is attached to the NMOS/PMOS transistor's gate and signal Y, respectively, to the source. A circuit with PMOS and NMOS transistors parallelly connected to send the signal Y is

Figure 11.3 Symbols of pass transistors: (a) nMOS transistor; (b) pMOS transistor.

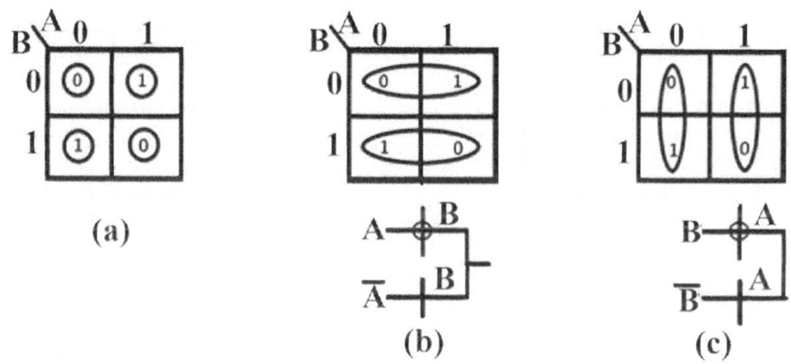

Figure 11.4 PTL implementation.

called a CMOS transmission gate. The expressions and the logic symbols for the pass gates are displayed in Figure 11.3.

An nMOS transistor produces a weak "1" and a strong "0". A pMOS transistor produces a weak "0" and strong "1". Strong "0" and strong "1" are output via the transmission gate. The moment all control signals become high, the input is connected to the output through nMOS transistors that are series connected. In contrast, when control signals are low, pMOS transistors that are series connected transfer the input to the output.

The primary distinction between the designs of a gate network and a pass network is that, in the pass network scenario, the function's 0s and 1s must be taken into account during the reduction method [37]. The pass network design process first chooses every significant pass prime implicant for the function, and after that it chooses a limited number to cover the function as whole using pass prime implicants as shown in Figure 11.4.

a. A'(B')(0) + A'(B)(1)+ A(B')(1) + AB(0), b. B'(A) + B(A'), c. A'(B) + A (B')

In pass network architecture, if the value of the pass variable is 1, a true pMOS transistor sends the input without causing any signal deterioration, and if 0 is the value of the pass variable, a true nMOS transistor sends

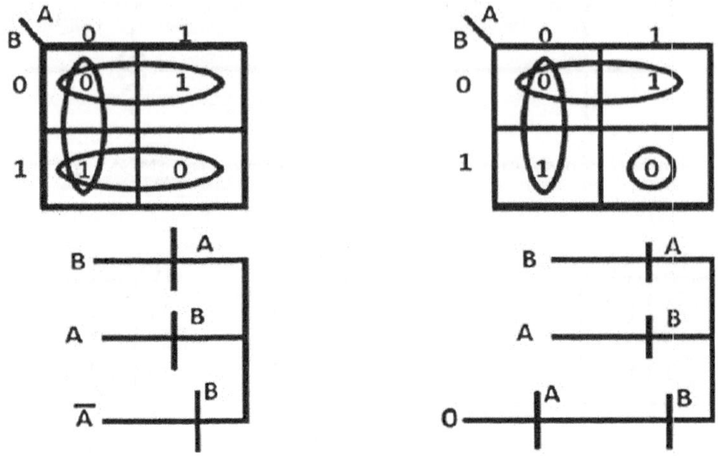

Figure 11.5 XOR logic analysis {a. *A*′(B)+ AB (1) + B(*A*′), b. *B*′(A) + *A*′B (1) +A (*B*′), c.A (*B*′)+ *A*′(*B*′)(0) + *B*(*A*′) d. *A*′(B) + B(*A*′)+ *B*′(A), e. *A*′(B)+ AB (0) + *B*′(A)}

the input without causing any signal deterioration. The output signal is deteriorated when a variable is sent apart from 1 or 0 across a pure PMOS or NMOS chain. The XOR circuits can be executed with the use of PTL logic as shown in Figure 11.5. Due to this limitation, transmission gates can be used in place of pass transistors to obtain a complete swing, however, extra inverters are needed to balance the control signals. When the drive requirements of the outputs are crucial, a CMOS pass network is employed.

11.4 PROPOSED WORK

11.4.1 Proposed XOR-XNOR Circuit

As seen in Figure 11.6, this innovative XOR-XNOR circuit design has ten transistors. The design combines CMOS logic, cross-coupled structure, and pass transistor logic. It is composed of a CMOS logic-based inverter. The XOR output employs one NMOS (N3) and two PMOS (P1 and P2), while its XNOR counterpart uses one PMOS (P3) and two NMOS (N1 and N2). There is also a cross connected PMOS (P4 and N4) and NMOS construction. The circuit's two inputs are referred to as A and B, and the following describes how the circuit operates.

The transistors P1, P2, and P3 are activated when the input combination AB=00:P1 is received. The transistors P1 and P2 pass a low logic "0" as an output for XOR, whereas the transistor P3 passes a high logic "1" as

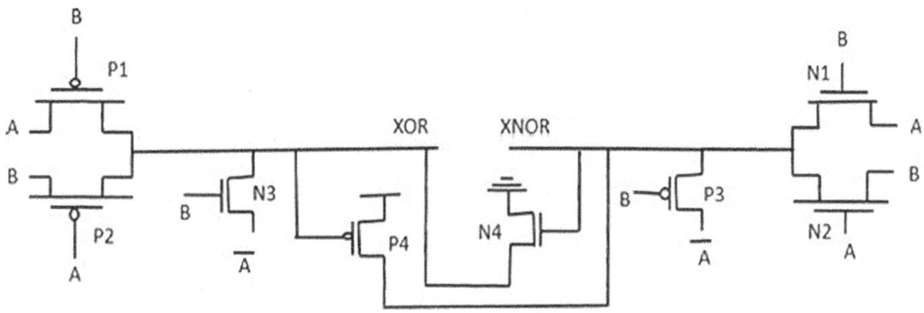

Figure 11.6 Proposed circuit design for XOR-XNOR.

an output for XNOR circuit. When the transistor N4 transistor is eventually activated, the XOR exhibits a high logic "0", thus producing the combination's full swing voltage. When AB=01, the P2 transistor switches on and outputs a strong logic of "1" at XOR, while the N1 transistor turns on and outputs a strong logic of "0" at XNOR. When AB = 10, the P1 transistor outputs a high logic "1" signal at XOR, while the transistor N2 outputs logic "0" at XNOR. Similarly, the transistor N3 goes on and sends a logic "0" as output for the XOR at AB=11, causing the P4 transistor to send logic "1" as the XNOR output.

11.4.2 The Proposed Circuit for SUM Generation

The input for the circuit are C_{in} and the previously generated XOR and XNOR signals. The pass transistor and transmission gate logic (TGL) are used in its design. The signals XNOR and XOR are provided as inputs for sources in PTL and as inputs for gates in TGL. As seen in Figure 11.7, C_{in} is similarly provided as inputs for gates in PTL and as input for sources at TGL.

11.4.3 The Proposed Circuit for Carry Generation

One of the adder circuit's outputs is Carry Out (C_{out}). With XNOR and XOR signals A as inputs to the sources and as inputs to the gate and C_{in}, it is intended as a 2:1 multiplexer as shown in Figure 11.8. Using TGL, the carry generating circuit is created with the following expression.

$$C_{out} = (A \wedge B)' A + (A \wedge B) C_{in} \tag{1}$$

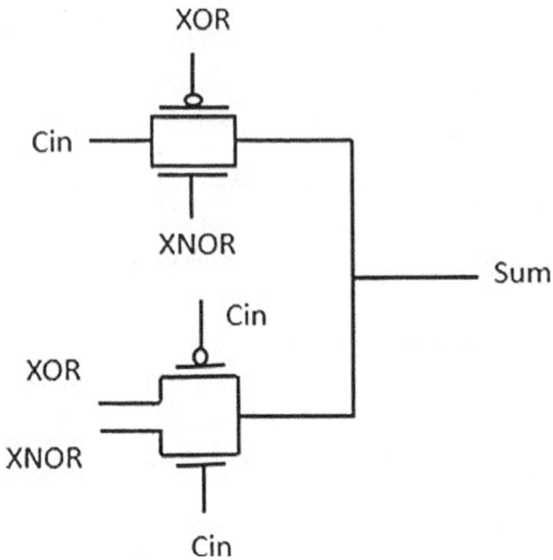

Figure 11.7 Proposed circuit for sum generation.

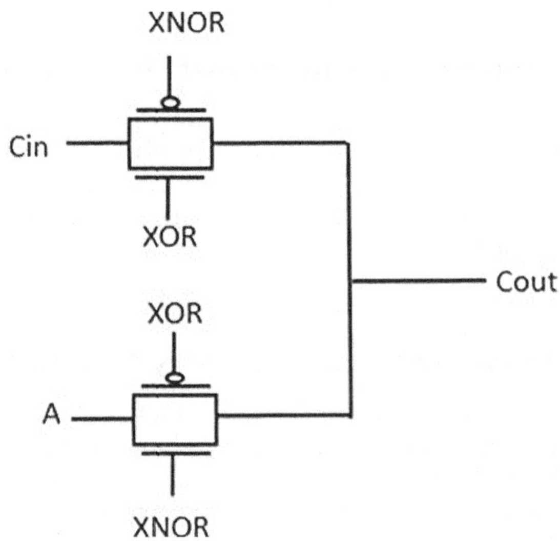

Figure 11.8 Circuit for carry generation.

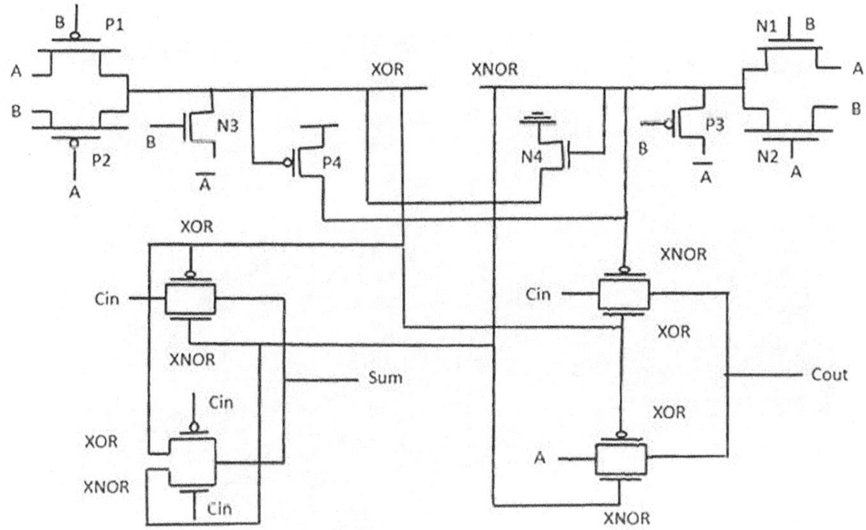

Figure 11.9 Proposed 1-bit hybrid FA design.

11.4.4 Proposed Full Adder

The three suggested circuits mentioned above are amalgamated to create a novel hybrid full adder. As seen in Figure 11.9, the circuit's outputs are sum (S) and C_{out} where its inputs are A, B, and C_{in}. The next section discusses the power, delay, and simulation findings.

11.5 SIMULATION RESULTS

The 1-bit hybrid full adder and the planned XOR-XNOR circuit are simulated in 90nm GPDK CMOS technology and examined in the Cadence Virtuoso environment at 1.2 V of voltage. The circuit characteristics, including latency, power consumption, and PDP, are measured and compared to the circuits that are currently in use.

11.5.1 The Results of XOR-XNOR Cell Simulation

The simulation results show that the entire swing voltage and the corresponding outputs for XOR and XNOR are obtained for every combination of inputs A and B as shown in Figure 11.10.

Table 11.1 compares the XOR-XNOR circuit's performance in terms of power delay product (PDP), power dissipation, and latency.

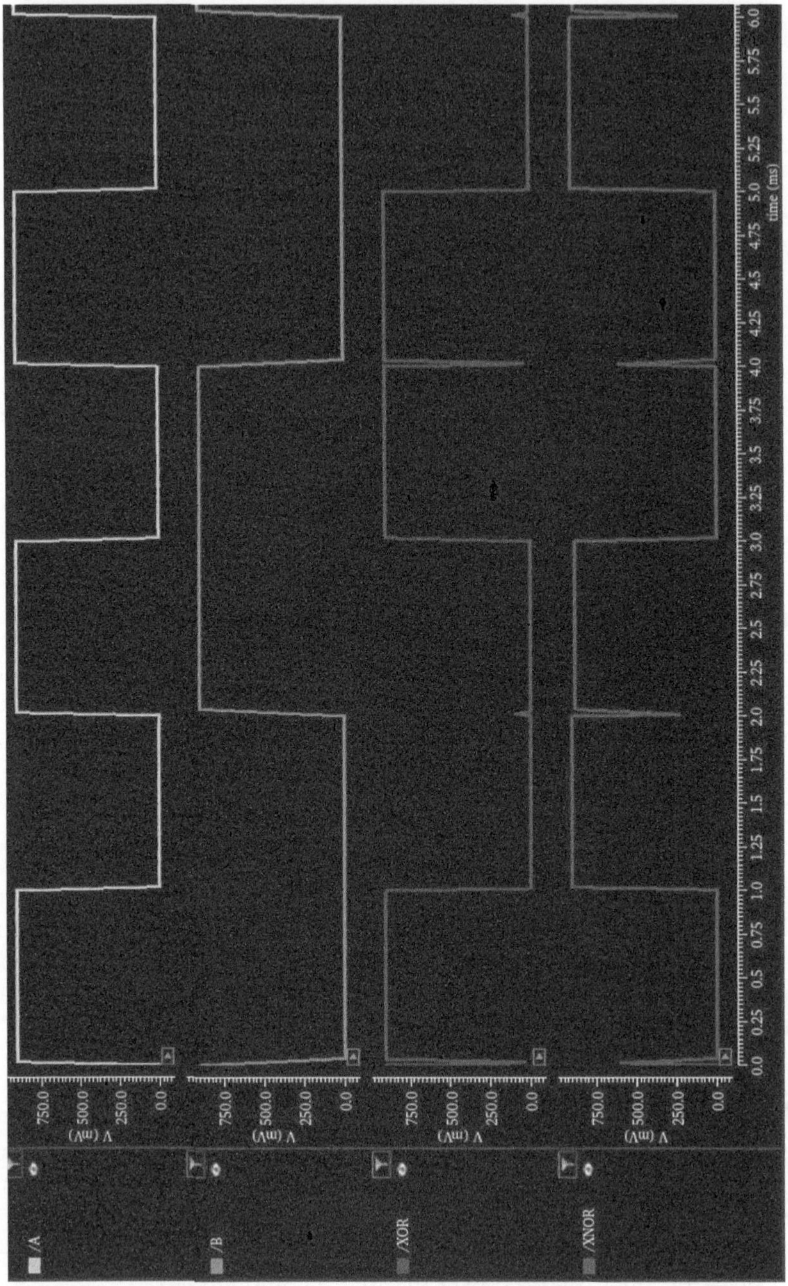

Figure 11.10 XOR-XNOR cell input-output waveforms.

Table 11.1 Comparison study of parameters of XNOR-XOR circuit with literature

XOR-XNOR Circuit	# Transistors	Delay in ps XOR	Delay in ps XNOR	Power in μW	PDP in aJ
Aguirre [38]	12	59.6	58.9	8.01	479.79
Goel [5]	8	140.9	141.8	10.37	1470.46
Radhakrishnan [39]	6	106	96.5	10.9	1144.4
Chang [40]	10	98.8	70.8	10.8	1057.16
Valshani [41, 42	10	73.7	52.2	10.7	788.59
Naseri [35]	12	52.9	49	8.98	475.04
Kandpal [8]	10	48.3	49.4	8.89	439.16
Proposed	10	14.41	15.63	1.03	15.4706

11.5.2 Simulation Results of the 1-bit Hybrid Full Adder Cell

Observations from Figure 11.11 indicate that the C_{out} = 0 and output S = 0 are obtained for the inputs A=0, B=0, and C_{in}=0. The results are S = 1 and Cout = 1 for A = 1, B = 1, and C = 1. Likewise, we obtain precise output values for the sum and carry out functions with a full swing voltage of 1.2 V for each of the eight possible input combinations.

Table 11.2 represents a parameter metric testing of the proposed adder circuit using 90 nm technology with literature.

11.6 CONCLUSION

This study proposed a new XOR-XNOR circuit architecture based on CMOS logic and pass-transistor. The Cadence Virtuoso tool was utilized to assess the circuit performance in 90 nm CMOS technology. The results of the current design were compared with other recent methods. It was found from the performance comparison that the proposed design had a reduction of delay by more than 50% than the other recent full adders and power delay product (PDP) was also reduced to 42.36%. In addition to having a strong driving capability, the circuit offered full swing at the outputs for all possible input combinations. Hence, the proposed full adder can be used for low power applications and computations.

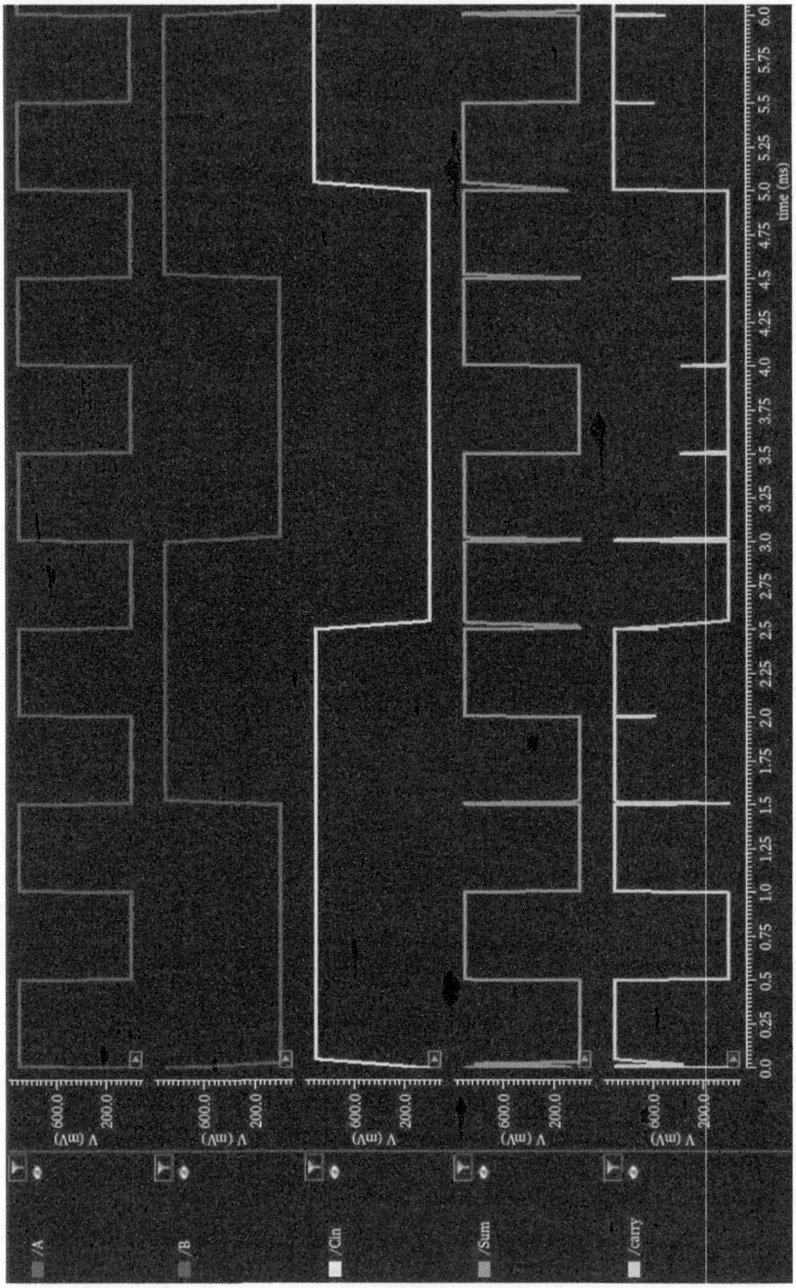

Figure 11.11 1-bit Hybrid FA cell input-output waveforms.

Table 11.2 Comparison study of different FAs with literature

Full adder Circuit	*# transistors*	*Delay in ps*	*Power in μW*	*PDP in aJ*
Zavarei [43]	26	72.3	27	1952.1
Valshani [41]	18	71.3	25	1782.5
Bhattacharyya [6]	16	65.59	22.6	1482.33
Naseri [35]	22	55.8	26.7	1489.86
Kandpal [8]	20	41.5	25.8	1070.7
Proposed	18	14.513	2.936	42.6101

REFERENCES

[1] J. Vineetha, K. Vamsi Krishna, P. Gayathri, M. Taj, and S. Sahoo, "High speed low voltage hybrid full adder for computing systems," in Robert J. Howlett, John Littlewood, *and* Lakhmi C. Jain *(Eds), Advances in Sustainability Science and Technology*, Springer Nature Singapore Pte Ltd., 2023.

[2] K. Soni, S. Sahoo, "A review on different memristor modelling and applications," in *IEEE-MECON. IEEE Conference Publication*, pp. 688–695, Apr 2022.

[3] S. Sahoo, G. Raman Murthy, S. Ramesh, and G. Anitha, "Hybrid CMOS-Memristor based operational transconductance amplifier for high frequency applications," *Sustain. Energy Technol. Assess.*, vol. 53, pp. 102506, 2022.

[4] S. Ramesh, P. Nirmala, G. Ramkumar, S. Sahoo, Anitha G., A. K. Gnanasekhar, and J. I. JoshuaRamesh, "Simulation process of injection molding and optimization for automobile instrument parameter in embedded system," *Adv. Mater. Sci. Eng.*, vol. 2021, pp. 1–10, 2021.

[5] S. Goel, A. Kumar, and M. A. Bayoumi, "Design of robust, energy efficient full adders for deep-sub micrometer design using hybrid-CMOS logic style," *IEEE Trans. Very Large Scale Integr. (VLSI) Syst.*, vol. 14, no. 12, pp. 1309–1321, Dec. 2006.

[6] P. Bhattacharyya, B. Kundu, S. Ghosh, V. Kumar, and A. Dandapat, "Performance analysis of a low-power high-speed hybrid 1-bit full adder circuit," *IEEE Trans. Very Large Scale Integr. (VLSI) Syst.*, vol. 23, no. 10, pp. 2001–2008, Oct. 2015.

[7] M. Hasan, Md. J. Hossein, M. Hossain, H. U. Zaman, and S. Islam, "Design of a scalable low power 1 bit hybrid full adder for fast computation," *IEEE Trans. Circ. Syst.*, vol. 67, no. 8, pp. 1464–1468, Aug. 2020.

[8] J. Kandpal, A. Tomar, M. Agarwal, and K. K. Sharma, "High-speed hybrid-logic full adder using high-performance 10-T XOR–XNOR cell," *IEEE Trans. Very Large Scale Integr. (VLSI) Syst.*, vol. 28, no. 6, pp. 1413–1422, June 2020.

[9] M. Vesterbacka, "A 14-transistor CMOS full adder with full voltage swing nodes," in *Proc. IEEE Workshop Signal Process. Systems. Design Implement. (SiPS)*, Oct. 1999, pp. 713–722.

[10] K. Sanapala and R. Sakthivel, "Ultra-low-voltage GDI- based hybrid full adder design for the area and energy-efficient systems," *IET Circ. Devices Syst.*, vol. 13, no. 4, pp. 465–470, May 2019.

[11] S. Roy Chowdhury, A. Banerjee, A. Roy, and H. Saha, "A high Speed 8 Transistor Full Adder Design using Novel 3 Transistor XOR Gates," *Int. J. Electron. Commun. Eng.*, vol. 2, no. 10, 2008.

[12] H. Reddy Basireddy, K. Challa, and T. Nikoubin-" Hybrid logical effort for hybrid logic style full adders in multistage structures," *IEEE Trans. Very Large Scale Integr. (VLSI) Syst.*, vol. 27, no. 5, pp. 1138–1147, May 2019.

[13] P. Chandrakasan, S. Sheng, and R. W. Brodersen, "Low-power CMOS digital design," *IEICE Trans. Electron.*, vol. 75, no. 4, pp. 371–382, 1992.

[14] N. Zhuang and H. Wu, "A new design of the CMOS full adder," *IEEE J. Solid-State Circuits*, vol. 27, no. 5, pp. 840–844, May 1992.

[15] V. Foroutan, M. Taheri, K. Navi, and A. A. Mazreah, "Design of two low-power full adder cells using GDI structure and hybrid CMOS logic style," *Integration*, vol. 47, no. 1, pp. 48–61, Jan. 2014.

[16] M. Agarwal, N. Agrawal, and M. A. Alam, "A new design of low power high speed hybrid CMOS full adder," in *Proc. Int. Conf. Signal Process. Integr. Netw. (SPIN)*, Feb. 2014, pp. 448–452.

[17] C.-K. Tung, S.-H. Shieh, and C.-H. Cheng, "Low-power high-speed full adder for portable electronic applications," *Electron. Lett.*, vol. 49, no. 17, pp. 1063–1064, Aug. 2013.

[18] H. Tien Bui, Y. Wang, and Y. Jiang, "Design and analysis of low-power 10-transistor full adders using novel XOR-XNOR gates," *IEEE Trans. Circuits Syst. II, Analog Digit. Signal Process.*, vol. 49, no. 1, pp. 25–30, 2002.

[19] M. A. Akbar, B. Wang, and A. Bermak, "Self-repairing hybrid Adder with hot-standby topology using fault-localization," *IEEE Access*, vol. 8, pp. 150051–150058, 2020.

[20] M. A. Khan and M. Hasan, "Design of high speed, energy, and area efficient spin-based hybrid MTJ/CMOS and CMOS only approximate adders," *IEEE Trans. Magn.*, vol. 58, no. 5, pp. 1–8, May 2022.

[21] L. B. Soares, M. M. A. da Rosa, C. M. Diniz, E. A. C. da Costa, and S. Bampi, "Design methodology to explore hybrid approximate adders for energy-efficient image and video processing accelerators," *IEEE Trans. Circuits Syst. I: Regular Papers*, vol. 66, no. 6, pp. 2137–2150, June 2019.

[22] M. Pashaeifar, M. Kamal, A. Afzali-Kusha, and M. Pedram, "Approximate reverse carry propagate adder for energy-efficient DSP applications," *IEEE Trans. Very Large Scale Integr. (VLSI) Syst.*, vol. 26, no. 11, pp. 2530–2541, Nov. 2018.

[23] G. Liu, L. Zheng, G. Wang, Y. Shen, and Y. Liang, "A Carry Lookahead Adder based on hybrid CMOS-Memristor logic circuit," *IEEE Access*, vol. 7, pp. 43691–43696, 2019.

[24] A. Najafi, M. Weißbrich, G. Payá-Vayá, and A. Garcia-Ortiz, "Coherent design of hybrid approximate adders: Unified design framework and metrics," *IEEE J. Emerg. Selected Top Circ. Syst.*, vol. 8, no. 4, pp. 736–745, Dec. 2018.

[25] P. Shukla, P. Kumar, and P. K. Misra, "An energy efficient, mismatch tolerant offset compensating hybrid MTJ/CMOS magnetic full adder," *IEEE Trans. Circ. Syst. II: Express Brief*, vol. 69, no. 11, pp. 4548–4552, Nov. 2022.

[26] P. Shukla, P. Kumar and P. K. Misra, "A highly reliable, dynamic logic-based hybrid MTJ/CMOS magnetic full adder for high-performance and low-power application," *IEEE Trans. Magn.*, vol. 58, no. 5, pp. 1–8, May 2022.

[27] S. M. Tabassum, M. V. Kavya Sri, D. B. Navya Sri, S. K. Muskaan, S. Sahoo," Design of low power and high speed 1-bit full adder for DSP applications," in *International Conference on Advancements in Power, Communication and Intelligent Systems (APCI)*, June 2024, pp. 1–4.

[28] B. Aiswarya, Y. S. Kumar, C. H. B. N. Jyothi, S. Sahoo, and A. K. Samantaray, "Designing of robust, energy efficient hybrid approach for implementing a 1-Bit Adder," in *2nd International Conference on Networking, Embedded and Wireless Systems (ICNEWS-2024)*, Oct 2024, pp. 1–6.

[29] B. Devika, N. Dhathri, D. Priya Darshini, T. Sarayu, S. Sahoo, and A. K. Samantaray, "Performance analysis of an energy efficient 1-Bit Hybrid Full Adder," in *2nd International Conference on Networking, Embedded and Wireless Systems (ICNEWS-2024)*, Oct 2024, pp. 1–6.

[30] V. S. Manchala, S. Sahoo, and G. Ramana Murthy "Design of energy efficient high-speed 1-bit hybrid full Adder for IoT-applications," in *IEEE ICRTCST*, May 2022, pp. 445–450.

[31] V. S. Manchala, S. Sahoo, and G. Ramana Murthy, "Design and analysis of 16-bit spares Kogge stone adder with proposed 6T-XOR cell, 1-bit HFA design for high-speed arithmetic operations," in *IEEE – 2022 International Conference on Communication, Computing and Internet of Things (IC3IoT)*, May 2022, pp. 1–6.

[32] V. G. Nissi and S. Sahoo, "Design of energy efficient high speed 1-bit Hybrid Full adder for fast computation," in Jagdish Chand Bansal, Joong Hoon Kim, and Atulya K. Nagar (Eds.), *Innovations in Sustainable Technologies and Computing*", Springer Nature Singapore Pte Ltd., 2024.

[33] R. Zimmermann and W. Fichtner, "Low-power logic styles: CMOS versus pass-transistor logic," *IEEE J. Solid-State Circuits*, vol. 32, no. 7, pp. 1079–1090, Jul. 1997.

[34] M. Zhang, J. Gu, and C.-H. Chang, "A novel hybrid pass logic with static CMOS output drive full-adder cell," *Proc. Int. Symp. Circuits Syst. (ISCAS)*, vol. 5, May 2003, p. 5.

[35] H. Naseri and S. Timarchi, "Low-power and fast full adder by exploring new XOR and XNOR gates," *IEEE Trans. Very Large Scale Integr. (VLSI) Syst.*, vol. 26, no. 8, pp. 1481–1493, Aug. 2018.

[36] S. Kishore and R. Sakthivel, "Ultra low voltage GDI based hybrid full adder design for area and energy efficient computing systems," *IET Circ. Dev. Syst.*, vol. 13, pp. 465–470, 2019.

[37] S. Lakshmi, R. C. Meenu, and D. Krishnadas, "Optimization of hybrid CMOS designs using a new energy efficient 1-bit hybrid full adder," *IEEE Trans. Very Large Scale Integr. (VLSI) Syst.*, vol. 16, pp. 905–908, 2018.

[38] M. Aguirre-Hernandez and M. Linares-Aranda, “CMOS full-adders for energy-efficient arithmetic applications,” *IEEE Trans. Very Large Scale Integr. (VLSI) Syst.*, vol. 19, no. 4, pp. 718–721, Apr. 2011.

[39] D. Radhakrishnan, “Low-voltage low-power CMOS full adder,” *IEE Proc.-Circuits, Devices Syst.*, vol. 148, no. 1, pp. 19–24, Feb. 2001.

[40] C.-H. Chang, J. Gu, and M. Zhang, “A review of 0.18-μm full adder performances for tree structured arithmetic circuits,” *IEEE Trans. Very Large Scale Integr. (VLSI) Syst.*, vol. 13, no. 6, pp. 686–695, Jun. 2005.

[41] M. A. Valashani and S. Mirzakuchaki, “A novel fast, low-power and high-performance XOR-XNOR cell,” in *Proc. IEEE Int. Symp. Circuits Syst. (ISCAS)*, May 2016, pp. 694–697.

[42] M. Amini-Valashani, M. Ayat, and S. Mirzakuchaki, “Design and analysis of a novel low-power and energy-efficient 18T hybrid full adder,” *Microelectron. J.*, vol. 74, pp. 49–59, Apr. 2018.

[43] M. J. Zavarei, M. R. Baghbanmanesh, E. Kargaran, H. Nabovati, and A. Golmakani, “Design of new full adder cell using hybrid-CMOS logic style,” in *Proc. 18th IEEE Int. Conf. Electron., Circuits, Syst.*, Dec. 2011, pp. 451–454.

Chapter 12

Low Power and High Speed Comparator of SAR ADC

Harshita Kushwah and R. S. Gamad

12.1 INTRODUCTION

Successive-approximation (SAR) analog-to-digital converters (ADC) have received attention due to their wide range of applications in various particular field, such as potable electronic system and wireless networks. In terms of energy efficiency, SAR ADC performs better than other ADCs at medium to very high speeds, low power applications, and moderate resolution. Increasing the device speed can lead to judgment errors in the SAR ADC, thereby degrading its resolution performance. It is challenging to balance the SAR ADC's speed, resolution, and power. This ADC is becoming common in ultra-low power applications, such as wireless sensor network chips and implantable biomedical devices [1-10].

SAR ADC basically is used to boost the performance of the device. A switching scheme based on a split capacitor has been proposed to achieve the goal. The function of ADC in input circuitry is to sample and convert the sensed and amplified signal into a digital form. ADC is an essential component which is the main connection between natural or analog signal in a digital processing system. There are several types of ADC, e.g., approximation ADC, flash ADC, Sigma-delta ADC, and pipeline ADC. SAR is selected for low power and high speed because it gives a maximum resolution which is enough for our design. SAR-ADC is known for its energy efficient performance, its requirement for less chip area, and its simple architecture. It has a conversion rate of 10 M samples per seconds which is compatible with our design. The most important thing about a SAR ADC is that it has excellent technology scalability. Therefore SAR-ADC has become popular among all the other ADC's [6–8, 10, 11, 12–14].

The main objective is to design a SAR ADC with an enhanced parameter which make it compatible for our design. This chapter aims to study the simulated gain and theoretical and practical knowledge of the SAR-ADC. During the research work, there were many problems which occurred and provided us with the knowledge about the behavior of the SAR-ADC. Since the research work was done on the simulating tool cadence, it provided us

DOI: 10.1201/9781003533535-12

with the ease of observing the behavior of the parameters and how they vary [8, 10, 11].

12.2 HISTORICAL CONTEXT OF ADC

12.2.1 What is an ADC?

Data converters are always used in electronic systems for making the links between analog and digital signals.

Real-time signals are analog signals. Analog to digital conversion appears as a need for processing. A data converter is also a very crucial part in a wireless receiver. Every system has different requirements for their ADCs (shown in Figure 12.1) [7].

Figure 12.2 shows simple ADC block diagram with an input and output.

The architecture of an ADC will depend on its application. ADCs can be broadly classified as follows:

SAR ADC
Flash ADC

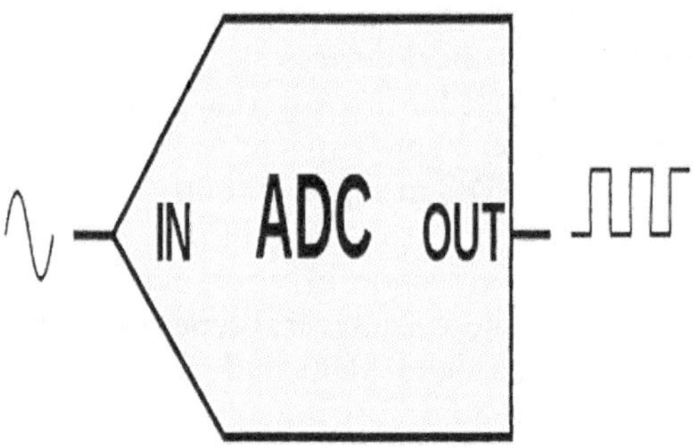

Figure 12.1 ADC diagram.

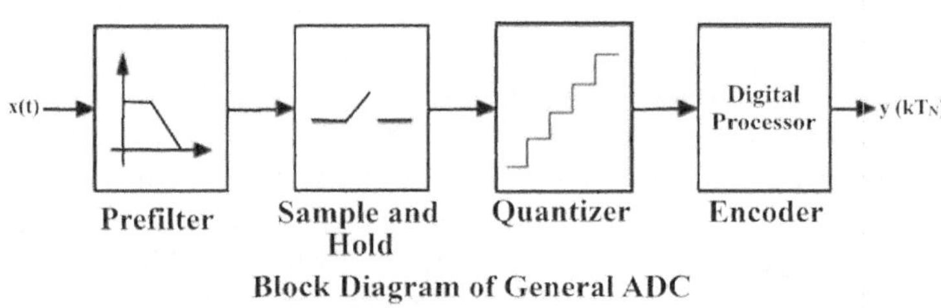

Figure 12.2 ADCs general block diagram.

Sigma-delta ADC
Folding ADC
Pipeline ADC

12.2.2 General Principle of ADC

Figure 12.2 shows an ADC block diagram with various blocks such as pre-filter, sample and hold, quantizer, and encoder.

Anti-aliasing filter or pre-filter is used to remove the aliasing or overlapping of a higher frequency component with a modulating frequency component of ADC. Their output is not certain as all inputs are sampled at discrete levels. Then the middle value of the two sampled values is considered as the value of the signal. An anti-aliasing filter is implemented by the band limiting characteristic of the ADC itself [15].

Sample and hold circuit:
The first block is used to take a time-based sample of the analog input signal. The samples are discrete in terms of time, they are taken in continuous amplitude and have original values. The conversion process is done significantly by the sampling frequency. The system's requirements will adjust the sampling rate. The "Hold" block is the block of the sample and hold circuit utilized in ADC. It simply retains the amplitude of the sample until the next sample is obtained and the hold value remains unchanged until the next value is taken out from the next sample [16–21].

Quantizer: The purpose of this block is to quantize the value. It creates a discrete amplitude from an analog or continuous sample. After passing through the "quantize" block, the continuous amplitude value becomes a discrete amplitude. The signal is converted in a digital pattern as it contains a discrete amplitude and discrete time [22].

Encoder: This is for converting the different pre-defined voltage, i.e., quantized states to digital numbers. The **encoder** block's function is to convert the digital signal into **binary form** or to convert it into bits. Given that digital devices have been shown to function on binary signals, an encoder must be used to transfer the digital signal into binary form. This is the complete procedure for applying an analog to digital converter to transform an analog signal into digital. A microsecond passes during this entire conversion. [17–23]

12.2.3 ADC Characteristics

ADC is well known by its characteristics. The characteristics curve of an ADC defines many parameter. Data converters have limitations in performance because of some phenomena. The performance parameters used in the characterization of ADC are divided into two categories: 1) dynamic and 2) static [24].

The Static Performance

An ADC experiences static errors when its transfer characteristic curve deviates from the ideal ADC transfer characteristics. Static performance of an ADC includes these terms like resolution, quantization error, full-scale error conversion time, offset error, missing code integral nonlinearity, differential non-linearity and non-monotonicity [25].

Resolution: specifies the variety of digital bits the ADC has for output. The number of bits that an ADC uses to represent the amplitude of a digital signal is called resolution.

The analog signal's amplitude is continuous. It can have any value one may dream up, including real and floating values, which are unlimited. The given signal is in digital form and has a limited and discrete set of values. Binary numbers, also called bits, are used to represent these distinct values [26].

Sampling Rate: Sampling rate, often called sampling frequency is the quantity of samples collected in a single second. It is important to adjust the sampling rate based on the signal given as an input. It should not be excessively high or low [27].

Non-monotonocity: If an increase in input signal shows no difference in the output then the converter is said to be monotonic [27].

Conversion time: This is the time required for an ADC to convert the analog input to a digital form including the setup timer of the input [27].

Quantization Error: This occurs when an analog input is quantized to a finite number of steps. The highest quantization error is $V_{LSB}/2$ [27].

Figure 12.3 shows a type of quantization error that is generated because of a difference in levels.

Full Scale Error: This is the difference between the optimal ADC transition point and the code prior to that transition point [28].

Gain Error: This is an actual output voltage differs from the ideal output voltage when the converter input voltage is at full scale [28].

Figure 12.4a and 12.4b shows graphs of error having offset error and gain error of ADC.

Differential Non-Linearity (DNL): This is a decrement of step size of ADC from step size of ideal (original) ADC. Deviation occurs

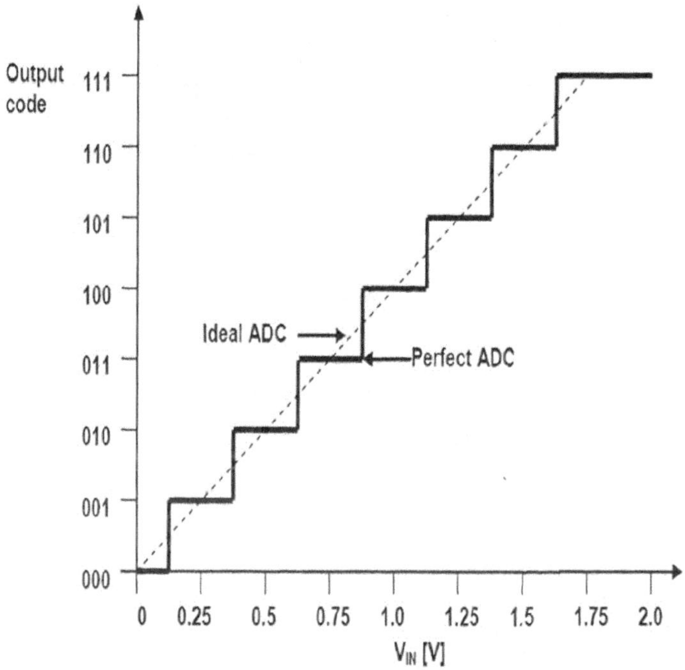

Figure 12.3 Quantization error.

because of circuit limitation, known as DNL. It is positive with a wide code and negative with a narrow code. DNL is always zero in an ideal ADC when the width of code is equal to 1LSB [24].

Integral Non-Linearity (INL): INL is the maximum difference in the center points of step size of both actual characteristics and the ideal characteristic. It is measured from the center of the step. It can be calculated as the sum of DNLs [23].

THE DYNAMIC PERFORMANCE

Static error is evaluated by applying a DC voltage. During this test, the ADC's performance related to high-frequency components and noise effects is not assessed [29].

Spurious Free Dynamic Range (SFDR): This is the peak of the input signal divided by the strongest harmonic or spur signal at output of ADC. SFDR measures of feasibility of ADC.

$$SFDR(dB) = Signal(dB) - Largest\ spur(dB) \tag{12.1}$$

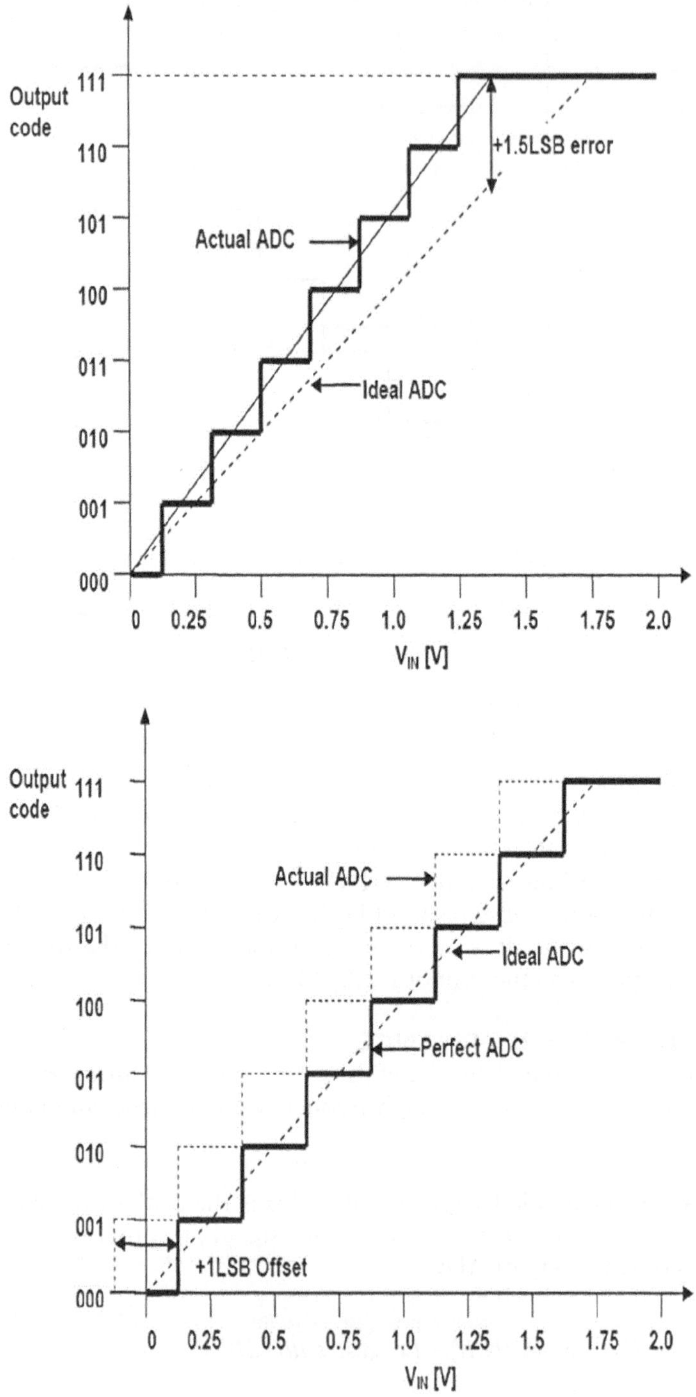

Figure 12.4 Offset error and gain error of ADC.

Signal to Noise Ratio (SNR): As its names indicates, this is the power ratio of the input signal to basic frequency to the power associated with each frequency in the spectrum, unaltered, including system noise. It is always defined in dB. Calculated as [30],

$$SNR(dB) = 10log_{10}\left(\frac{P_S}{P_N}\right) \tag{12.2}$$

where P_S is signal power, P_N is signal power including noise.
SNR of ideal ADC is also defined by number of bits as,

$$SNR(dB) = 6.02N + 1.78 \tag{12.3}$$

Signal to Noise and Distortion Ratio (SINAD): When an harmonic component is involved in the SNR calculation it is termed as SINAD. It is the ratio of the input signal divided by the summation of all harmonic and noise components present in the circuit. SINAD is always represented in dB and expressed as [31]

$$SINAD(dB) = 10log_{10}\left(\frac{P_S}{P_S} + P_d\right) \tag{12.4}$$

where P_d is the power of the distortion signal.

Effective Number of Bits (ENOB): ENOB defines the actual resolution of ADC. It represent the number of bits. Actual resolution is decreased due to non-linear errors. It is defined as,

$$ENOB(bit) = SINAD(dB) - \frac{1.76}{6.02} \tag{12.5}$$

Dynamic Range: This represents the amplitude of input signal for which the referenced Signal to Noise Ratio reaches 0 dB. It is the ratio of the lowest amplitude to the maximum amplitude of the signal that may be discriminated without overloading the converter [31].

Figure of Merit: This is defined for comparison of various ADCs performance. ADC can differ in power, resolution, and speed. To establish a valid comparison FOM is introduced. FOM can tell the difference in the performance if the resolution and bandwidth is changed for the particular application [31].

$$FOM = \left(\frac{POWER}{2^{ENOB} f_S} \right) \tag{12.6}$$

where BW denotes the ADCs bandwidth, P is power consumption, and ENOB is known as effective number of bits. The unit of FOM is measured in pico joules per conversion step, i.e., pJ/conv [19, 20].

12.2.4 Types of ADC

Table 12.1 compares the different architecture of ADC based on speed, resolution, and their application [21].

12.2.5 ADC Architecture Comparison

Application-specific factors such as resolution, power, size, sampling frequency, and performance are taken into account while selecting an ADC. Every ADC performs best in a certain application and poorly in others [17].

12.3 SUCCESSIVE APPROXIMATION ADC

This chapter provides a comprehensive explanation of the Successive Approximation Register (SAR) ADC. The rationale behind choosing this type of ADC is discussed, along with a detailed description of its working principle and the functionality of its individual sub-blocks [6].

Figure 12.5 shows the circuit design of an SAR ADC with the components of sample and hold block, comparator, delay, SAR, and DAC.

Table 12.1 Existing ADC Architecture and their application

Architecture	*Resolution*	*Speed*	*Application*
Integrating	High	Low	Measurements and Instrumentation
Cycle	Medium	Medium	Control, Microelectronics
SAR	Medium	Low to Medium	Control & Microcontrollers, Biomedical
Two – step	Medium	Medium to Low	Video, RF & Microwave,
Pipelined	Medium	High	Video, RF & Microwave,
Flash	Low to Medium	High	Telecommunication & Biomedical, RF & Microwave
Sigma – Delta	Medium to High	Low to Medium	Biomedical & Telecommunication, Audio
Incremental	High	Low	Measurement and Instrumentation

Table 12.2 Comparison of ADCS

Feature	*SAR ADC*	*Flash ADC*	*Pipeline ADC*	*Integrating ADC*	*Sigma – Delta ADC*
Frequency of Conversion	Up to 1 MHz	Up to 1 GHz	Up to 100 MHz	Up to 100 Hz	Up to 1 MHz
Resolution	8 to 18 bits	4 to 8 bits	8 to 12 bits	10 to 18 bits	12 to 24 bits
Size	Linear increase with resolution	Exponential increase with resolution	Linear increase with resolution	Does not much change with resolution	Does not much change with resolution
Cost	Cheap	Expensive	Expensive	Cheap	Cheap

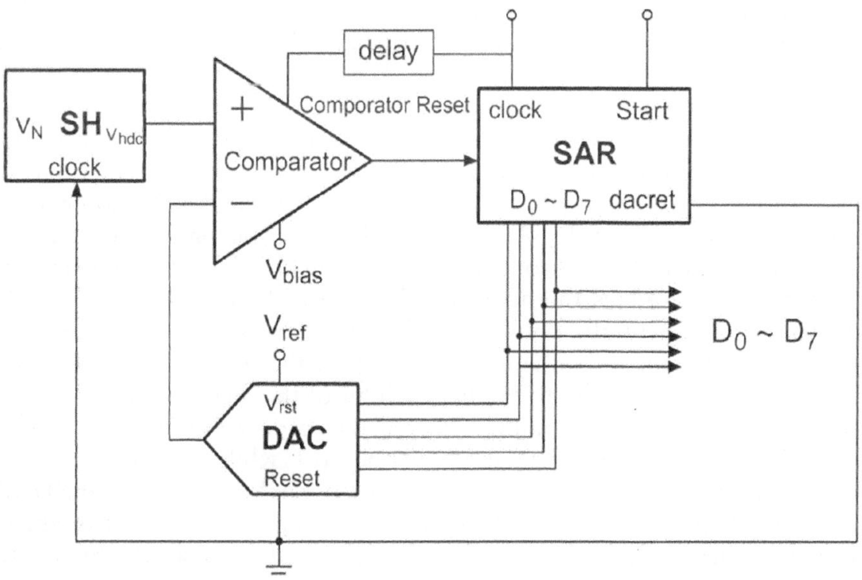

Figure 12.5 Block representation of SAR ADC.

12.4 SELECTION OF SUCCESSIVE APPROXIMATION ANALOG TO DIGITAL CONVERTER

The architecture of SAR ADC is not so complex to design. CMOS process technology can also be scaled. SAR ADCs are well-suited for many real-time applications, offering a resolution range of 8 to 12 bits and sampling frequencies up to 1 MSPS. They exhibit minimal jitter and latency issues and require fewer comparators compared to other ADC architectures. Additionally, their conversion time is minimal and remains consistent across multiple processes. Considering all these advantages, SAR ADCs are often preferred over other types of ADCs [17–26].

12.4.1 What is SAR ADC?

An analog input, or continuous waveform, is converted into a digital bit, or discrete representation, using an ADC. It sequentially steps through each potential level until it ultimately reaches the final digital output, as each conversion progressively approaches completion [17].

12.4.2 Working of SAR-ADC

SAR-ADC is mainly known for converting and storing the output in digital from. The complete Successive Approximation Register (SAR) ADC unit consists of four main blocks. It contains blocks of S/H circuit, SAR, DAC, and comparator. SAR is the main block, also known as a digital block. SAH is an important block in this SAR ADC to hold the output on the last executed value during the conversion. The SAH output is fed to the comparator to the positive terminal input. The input terminal of the comparator is fed by the DAC block output. The function of the comparator is to make decisions after making a comparison between two signals. The comparator gives two output, 0 or 1. SAR blocks collect the comparator output and store. In a SAR ADC, the control unit operates based on the comparator's output and determines the final digital output. These steps can affect the actual resolution of the converter. The SAR (Successive Approximation Register) forms the digital control circuitry of the ADC. It operates using a binary search algorithm to predict each bit of the final output. With each step, the remaining search range is halved. The process begins by setting the most significant bit (MSB) of the SAR output to 1, while all other bits are set to 0. This bit combination is converted into an analog voltage by the DAC, corresponding to the midpoint of the ADC's input range. This predicted voltage is then compared with the input signal to determine whether the bit should remain 1 or be reset to 0. The output of DAC is supplied to the comparator. The comparator then contrasts the sampled input from the SAH circuit with the DAC output. If the DAC's output is greater then the following bit is set to 1. The subsequent bit search now uses the second half of the current search. The next phase in the search for an upcoming bit is to use the bottom half of the current search interval and set the forthcoming bit to 0 if the DAC's output value is tiny. As soon as the MSB value is determined correctly, 1 is saved in another bit, and the procedure is restarted using a new search interval [10, 12–26].

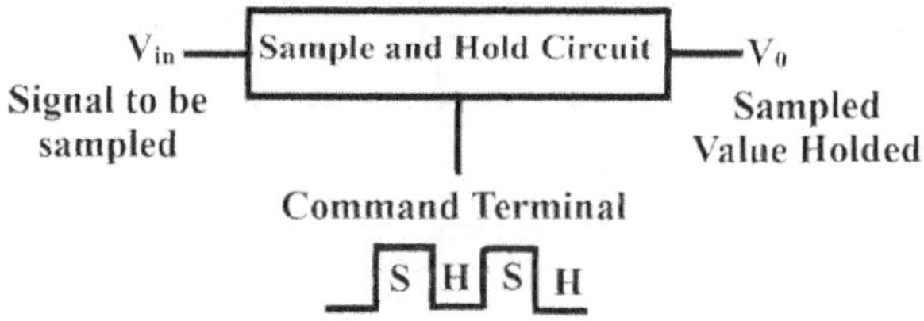

Figure 12.6 Sample and hold circuit.

12.4.3 Sample and Hold Circuit

An analog to digital converter is basically used in conjunction with an S/H circuit (Sample and Hold Circuit). It is applied for storing the sampled signal while sampling the analog input signal. The minimum period during which the S/H circuit samples the analog signal is between 10 and 1μs. After that, the sampled output remains in place until the subsequent input signal that has to be sampled comes in. On average, the sample will be held for a few milliseconds to many seconds [14–28].

Figure 12.6 shows a circuit design of a sample and hold block.

Parameters to characterize the performance of SAH circuits:

Voltage Drop

This is the phenomenon in which the voltage across the holding capacitor drops down because of leakage currents [2].

Feed Through

The switch has parasitic capacitances which may produce AC coupling between VO and Vi in hold mode and this leads to changes in the input voltage that make variations in the output voltage [3].

12.4.4 Comparator

When voltages are compared, the circuit of the comparator determines greater value by the output 1 or 0 where 0 represents the value at the negative side and 1 reflects the voltage at the positive side, or VDD in the example. Comparators are used to figure out if an input has or has not attained an initial value. An op-amp is a useful alternative to the related comparator integrated circuit (IC), that is employed in the making of a comparator. The op-amp symbol and comparator diagram are the same [19–30].

Figure 12.7 shows circuit design of a comparator with an input and output.

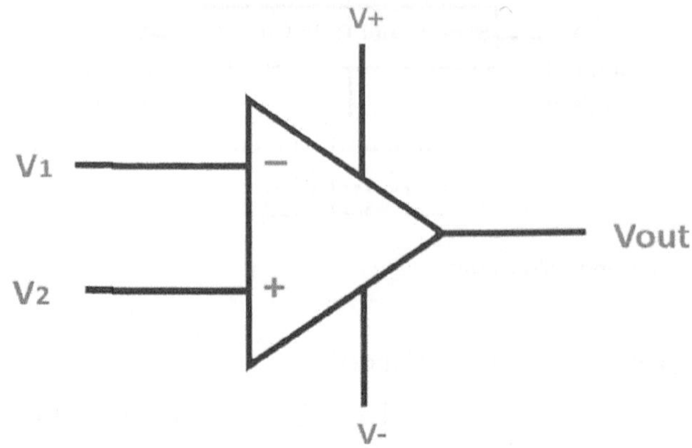

Figure 12.7 Comparator circuit.

The output of a comparator is:

V_{Out} = 1 if inp > inn

V_{Out} = 0 if inp < inn

12.4.5 Characteristics of Comparator

Static Characteristic of Comparator

Static characteristic of comparator is defined by following parameter [26]:

Gain

Gain of comparator expressed as:

$$GAIN = A_v = V_{OH} - V_{OL} / \Delta V \tag{12.7}$$

where VOH is high output voltage, ΔV is input voltage change, and VOL is low output voltage.

Input Offset Voltage

Offset in the comparator arises due to mismatches in the input transistors. It represents the minimum input voltage required to balance this mismatch and bring the comparator to its switching threshold [19].

Resolution

Resolution refers to the minimum change in input voltage required to cause a change in the output to the next corresponding binary state [20].

Dynamic Characteristic of Comparator

Dynamic characteristic of comparator include following:

Propagation Delay

It is the amount of time delay between the output signal and the input, and calculated as:

$$t_p = t_pLH + t_pHL)/2 \tag{12.8}$$

where T_p is propagation delay time, t_pLH is high to low rise time, and t_pHL is low to high rise time

The speed of the comparator is defined by the inverse of propagation delay time:

$$SPEED = 1/t_p \tag{12.9}$$

Power Dissipation

The comparator's Power Dissipation can be defined as

$$\boldsymbol{P = F\,C\,V_{dd}^2} \tag{12.10}$$

where f is frequency, C is capacitance, and V_{dd} is input voltage.

Sampling Frequency

Inverse of time interval T is termed as sampling frequency "f_s" as:

$$f_s = 1/T \tag{12.10}$$

12.4.6 Architecture of Comparator

Open loop and regenerative are two types of comparator. An op-amp based comparator is known as an open loop comparator with any feedback and compensation, whereas a regenerative comparator uses some sort of positive feedback to do comparison tasks between the two signals. A new type of comparator combines both regenerative and open-loop architectures, resulting in improved speed and performance [19–30].

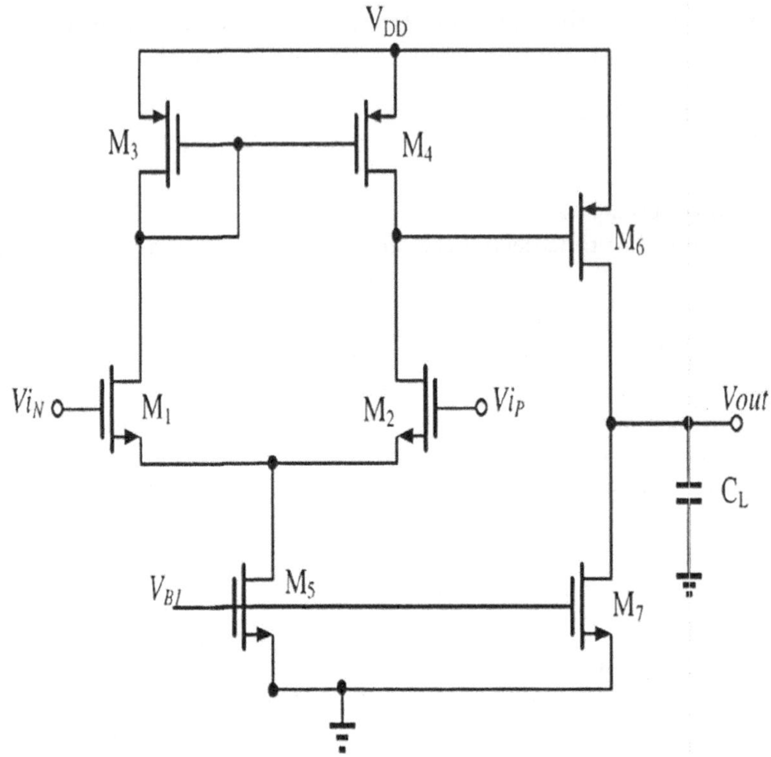

Figure 12.8 Open loop comparator.

Open Loop Comparator

Comparators with a high gain stage driving their outputs between VOH and VOL for very small input voltage changes are an important class of comparators. For an open loop, a high gain comparator is the uncompensated two stage operational amplifier. The comparator's speed is reduced in a single stage amplifier with a single pole on the gain bandwidth. Although lowering the gain in every stage that increases speed, it also introduces more input referred offset voltage. The comparator is incompatible with both high speed and low power A/D converters due to its higher power consumption [30,31].

Figure 12.8 shows the circuit design of a transistor and capacitor.

Dynamic Latched Comparator

Clocked comparators are comparators that employ a latch. A latch is used to amplify the signal and provide positive feedback, ensuring fast and stable decision-making. The comparator operates in two stages: evaluation and reset. The comparator's architecture determines whether to connect the output node to the supply to charge or to the ground to discharge. The reset phase determines the output node's charging and discharging [7, 8, 10, 11, 12–14].

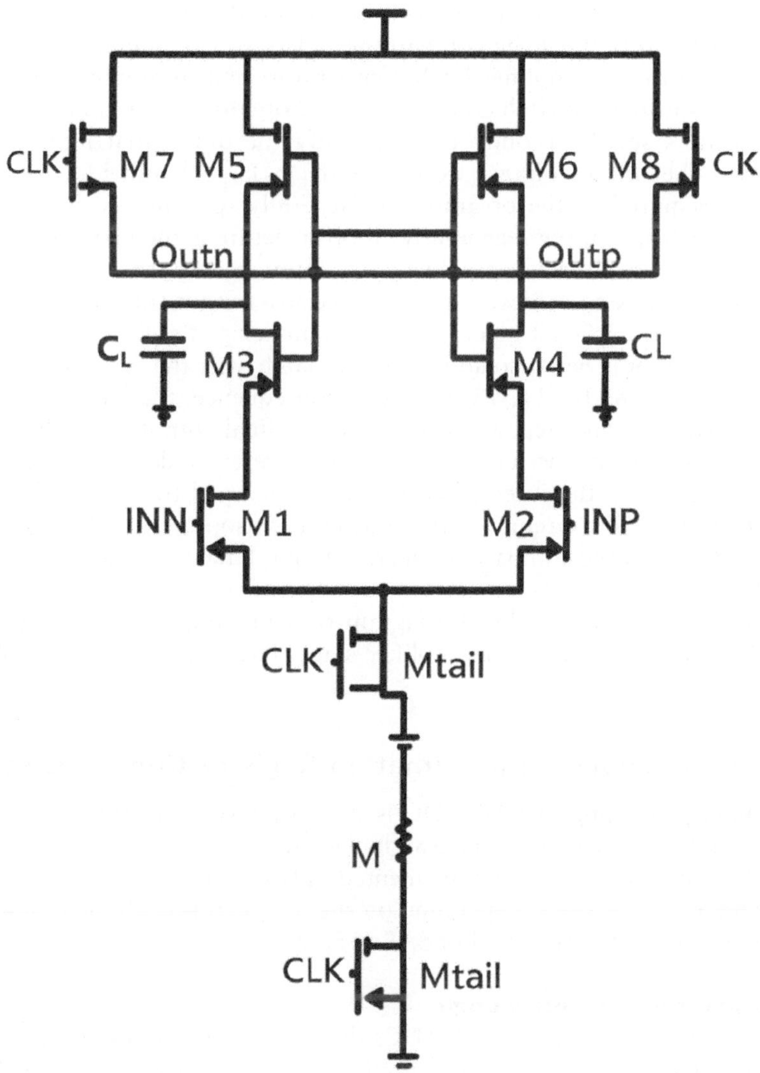

Figure 12.9 Dynamic latch comparator.

Comparator tracking occurs during the reset phase. Positive assessment feedback provides the digital value at the comparator's output

Figure 12.9 shows the circuit design of a transistor and capacitor.

Pre-amplified Based Comparator

The purpose of the preamplifier is to amplify analog input signals. Subsequently, the preamplifier's larger signals are compared by the regenerative latch comparator. The latch output, the final stage, processes the output of regenerative latch comparator and produces the output either to 1 or 0 as the output of the comparators. Here the preamplifier plays two key roles in the preamplifier latch comparator: first, it amplifies the input signals to shorten the latch comparator time of comparison, improving the comparator's speed; second, it can amplify the differential input signals to lessen the offset voltage. Consequently, a large bandwidth and high gain are required of the preamplifier. Generally speaking, to increase the comparator's speed, the regenerative loop structure is the first option for a latch comparator. In a regenerative latch, input signals are compared using positive feedback to achieve rapid and decisive output transitions. For the improving of the speed it quickly amplified input signals. To achieve a stable digital output, it is better to add an output latch after the regenerative latch comparator [7, 8, 10, 11, 12–15]. As a consequence, the latch's output is the comparator's last step, and it will offer the final output when the output latch persists on the last clock period. In the reset mode of a regenerative latch comparator, the internal nodes are precharged to a defined voltage level to prepare the circuit for the next comparison cycle. When the latch comparator is in the compare mode, the output latch will generate the final conclusion.

Figure 12.10 shows a block diagram of a pre-amplifier with blocks of input signal, clock pulse, preamplifier stage, latch stage, refrence voltage, and output.

12.4.7 Successive Approximation Register Control Circuit

The main processing of SAR ADC is done by SAR control logic. Control logic is used to design a binary search algorithm.

SAR control logic can be implemented in two ways: redundant and non-redundant architectures, depending on the design approach. Minimum 2N flip flops are used in this kind of SAR [19, 20].

Redundant SAR Control Logic

SAR architecture adopted SAR ADCs due to its simple design. It is known that a shift register can be configured to operate as a ring counter. Every conversion in the clock cycle zero causes the EOC signal to be high, resetting all flip flop outputs to 0. The EOC signal remains low for the remainder of the cycle. One of the ring counters sets a FF in the code register on every

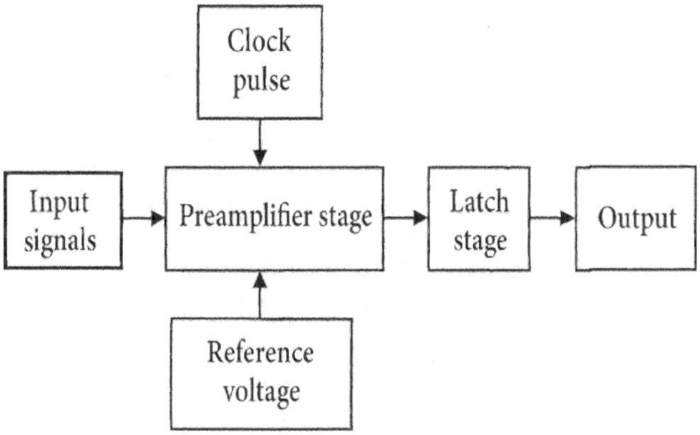

Figure 12.10 Pre-amplifier based comparator.

cycle of clock. The clock signal for the prior flip flop is obtained from the output of this flip flop, which is determined by the ring counter. At the clock's rising edge flip flop having the result of the comparator [19–30].

Figure 12.11 shows the circuit design of a 12-bit DAC. It consists of 12 CMOS switches and 12 NMOS. This NMOS sizes are binary in ratio.

Non-Redundant SAR Logic

Digital control in an N-bit SAR ADC requires at least N flip-flops to perform the conversion. The non-redundant SAR architecture utilizes the minimum number of flip-flops necessary for this operation. There are three possibilities for each bit:

1. Memorization mode
2. Shifting right
3. Comparator results

An OR chain and N bit shift register are components of the non-redundant successive approximation control logic. Every flip flop (FF) has the ability to accept OR gate value in each step, the output of the preceding FF, or the comparator's result. In the initialization stage, the leftmost flip flop is set to 1, while the remaining flip flops are reset to 0. The start signal is an external control signal that provides this situation. The following step involves selecting one of the three input options for each flip flop. The need for a multiplexer arises from the presence of three possible inputs, with its control signals, A and B, integrated within the flip-flop. By providing the OR chain with a high voltage, the conversion is eliminated. As a result, the successive approximation register enters the code storage phase. At the end of the conversion, SAR stores the converted result. This is performed by

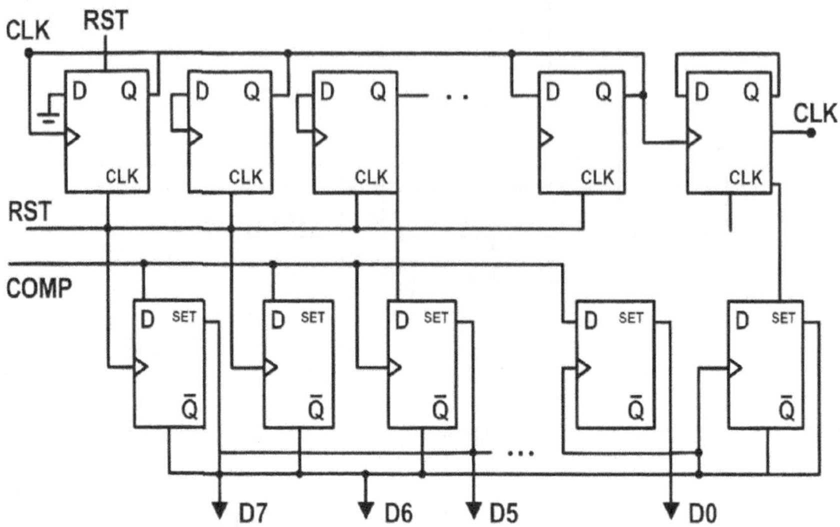

Figure 12.11 Redundant SAR architecture.

joining the output of the last FF to the OR chain. Hence, the least significant flip flop on the right defines the end of conversion [6–15].

12.4.8 Digital to Analog Converter

The DAC converts the digital output of the SAR logic into an analog voltage. This analog value is then compared with the input signal by the comparator. An N-bit DAC is designed to handle 2^n possible input combinations [31].

Types of DAC

Several types of DAC are given below. Redistribution circuit, voltage mode and current mode are some of them.

1. Summing Amplifier

Summing amplifier circuits are employed in digital to analog conversion since the conversion process is sum of the binary input. This is just an op-amp amplifier with many resistors connected to one input. The summing junction, also called virtual ground, is the point where the resistors come together. The resistors receive the binary input, while the op-amp's output provides the analog output. The resistors are what drive this circuit; to have a precise analog output, each resistor needs to be properly selected and matched. It's not usually realistic to require more varied values of resistors the more bits you have [19–24].

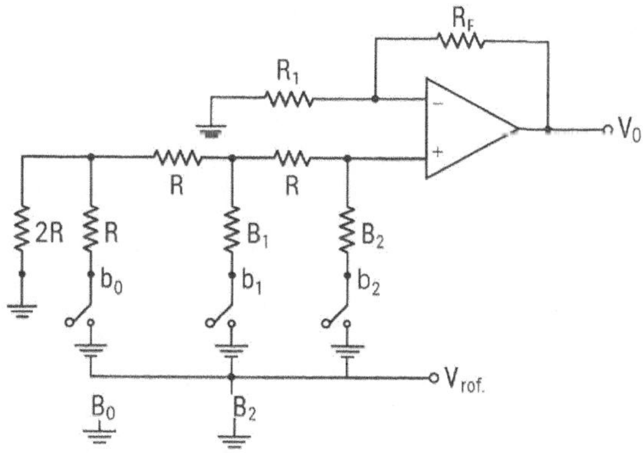

Figure 12.12 Resistive DAC.

II. R-2R Ladder

All that is required for this kind of DAC is a ladder arrangement of just two resistor values. Output is obtained at the base of the ladder and is fed into the 2R resistors from the binary input.

Figure 12.12 shows circuit design which consist of resistor and Amplifier.

III. PWM DAC

Using a PWM signal, an Arduino microcontroller can produce analog signals. Signal seems to be a binary waveform having peak high and low and a changeable duty cycle (the ratio of on time to timing). But this is meant to be used in conjunction with an RC filter to filter out the AC component and leave the DC component in order to convert the PWM signal into a voltage value. The duty cycle of the input derives the voltage output; the higher duty cycle can give us the high output voltage of the filter [26].

12.5 Circuit and Schematic Design of 12-Bit SAR ADC and Its Sub-Modules

12.5.1 Circuit Design of Inverter

Figure 12.13 shows schematic diagram of inverter having 1 NMOS AND 1 PMOS.

12.5.2 Circuit Design of S/H Circuit

Figure 12.14 shows the schematic of an SAH circuit which shows two NMOS, one capacitor and one inverter. The inverter is used to invert the clock signal. The output is taken across the capacitor.

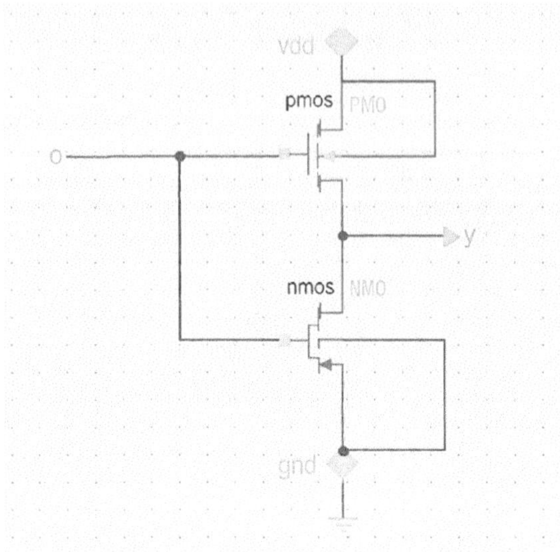

Figure 12.13 Schematic of inverter.

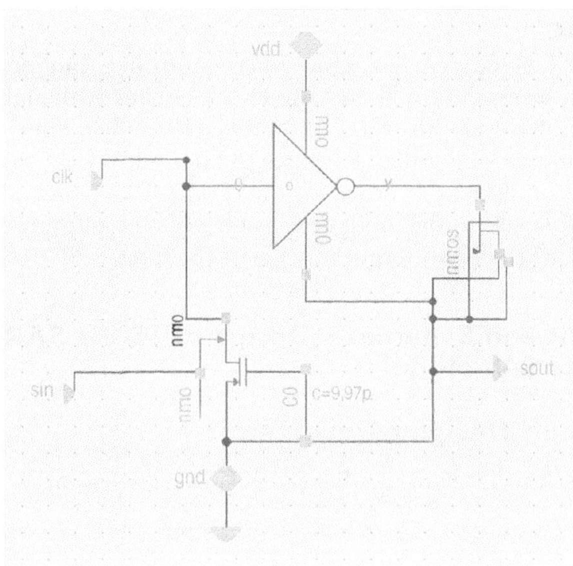

Figure 12.14 Schematic of sample and hold circuit.

12.5.3 Schematic Design of Comparator

Figure 12.15 shows a schematic of a comparator.

12.5.4 Schematic Design of Control Unit bitslice

Figure 12.16 shows a schematic diagram of bit-slice which is the control unit of a SAR controller. Here bit slice works as a control unit. It plays a major role in controlling the number of bits. The control unit for MSB bits is used in the SAR controller which sets the first bit as 1 and the control unit LSB bit used in SAR controller which resets the remaining bits as 0 [27].

12.5.5 Circuit Design of SAR Controller

Figure 12.17 shows a schematic of the SAR controller which has 12-bit slices as a control unit. Each control unit has a bit slice.

12.5.6 Circuit Design of a 12 Bit ADC

Figure 12.18 shows a circuit design of a 12-bit ADC.

12.5.7 Circuit Design of DAC

Figure 12.19 shows a circuit design of 12-bit DAC. It consists of 12 CMOS switches and 12 NMOS. This NMOS sizes are binary in ratio.

12.5.8 Circuit Design of CMOS Switch

Figure 12.20 shows a schematic design of a CMOS switch which consists of six transistors. It is a two-input, single-output circuit. The clock signal controls the selection between the inputs.

12.6 SUMMARY

This chapter describes the designing of a dynamic comparator and its simulated results to achieve fast speed and low power consumption. A dynamic regenerative latch and pre-amplifier are the foundations of a comparator design, which enables fast speed and low power consumption. 1.8 volts is the supply used to get the simulation results. Here we achieved 100 MHz which is a high sampling frequency and a low power usage of 129.8 μW. A simple sample and hold circuit is described for the sampling the input. It consists of two NMOS transistor and capacitor. The average power dissipation by the S/H circuit is 0.05 μW. A very basic topology for a comparator is used for quantization. Eight MOS are connected without

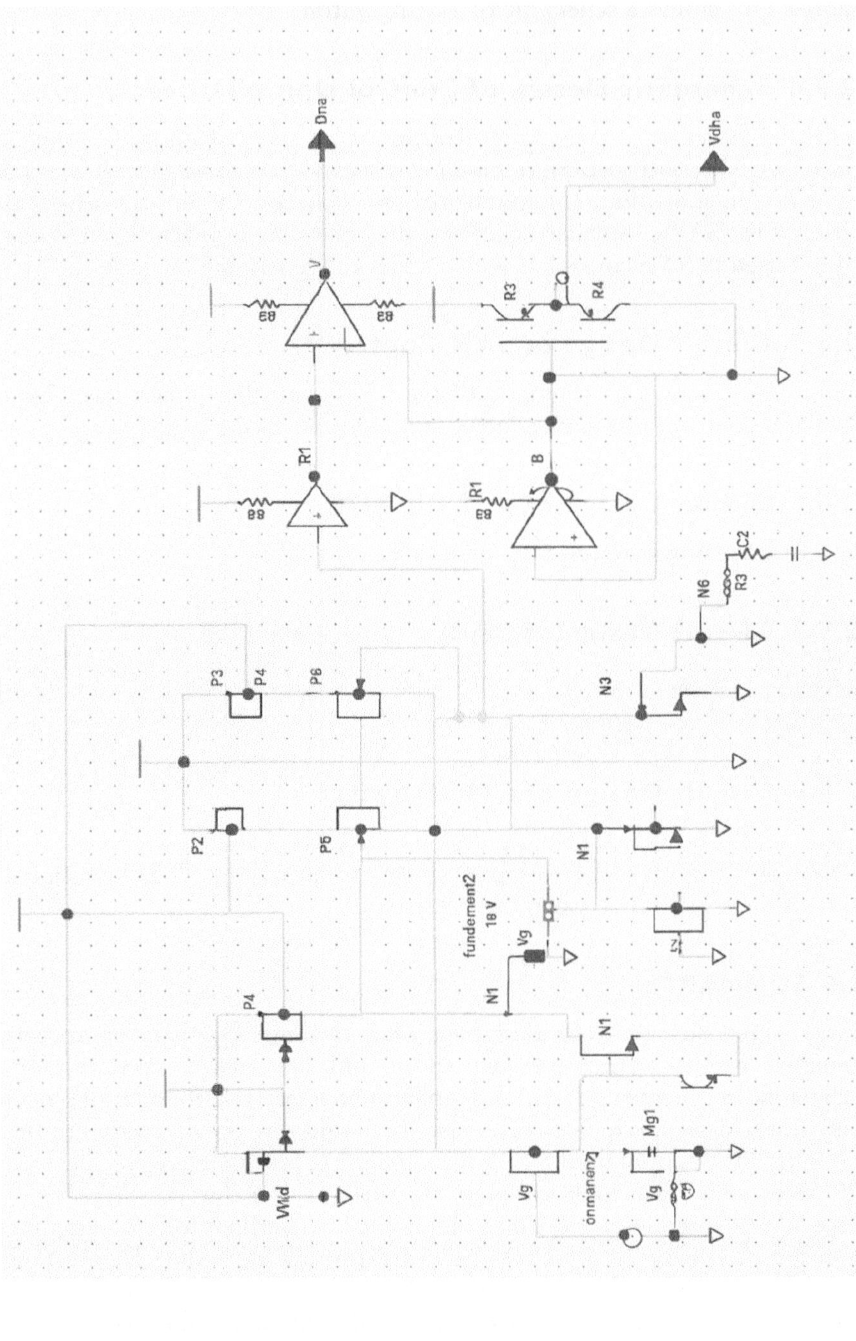

Figure 12.15 Schematic of comparator.

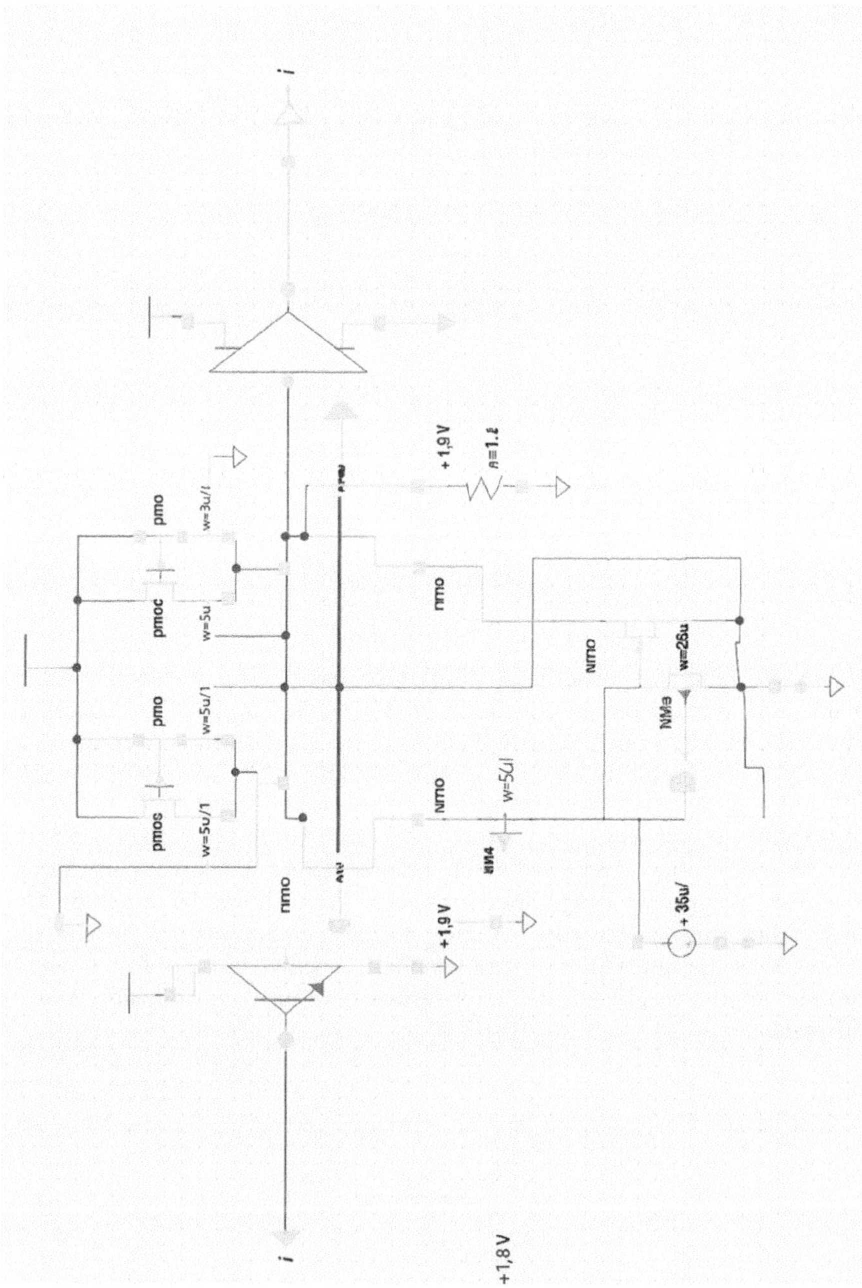

Figure 12.16 Bit slice.

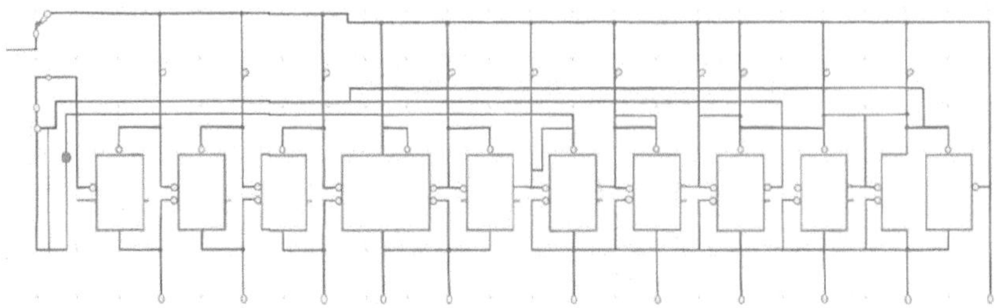

Figure 12.17 Schematic of SAR controller.

any compensating capacitor. The comparator has an input voltage offset of 5 mV and average power consumption 1.29 μW. The DAC is designed with the use of MOS only. A SAR block is designed with the use of a control unit. Bit-slice used in the SAR controller. The ADC is simulated with a sinusoidal signal of 10 kHz frequency. The maximum sampling rate is 100MHz but the sampling rate used is 100 kHz. The supply voltage used is 1.8 V. The new design can be evolved to further reduce the power consumption of ADC. Instead of a capacitive array type DAC should be targeted. By proper scaling, this architecture can alter the performance as it will make it possible for the designer to easily implement it. The invention of low power techniques can be employed for better results without compromising ADC performance.

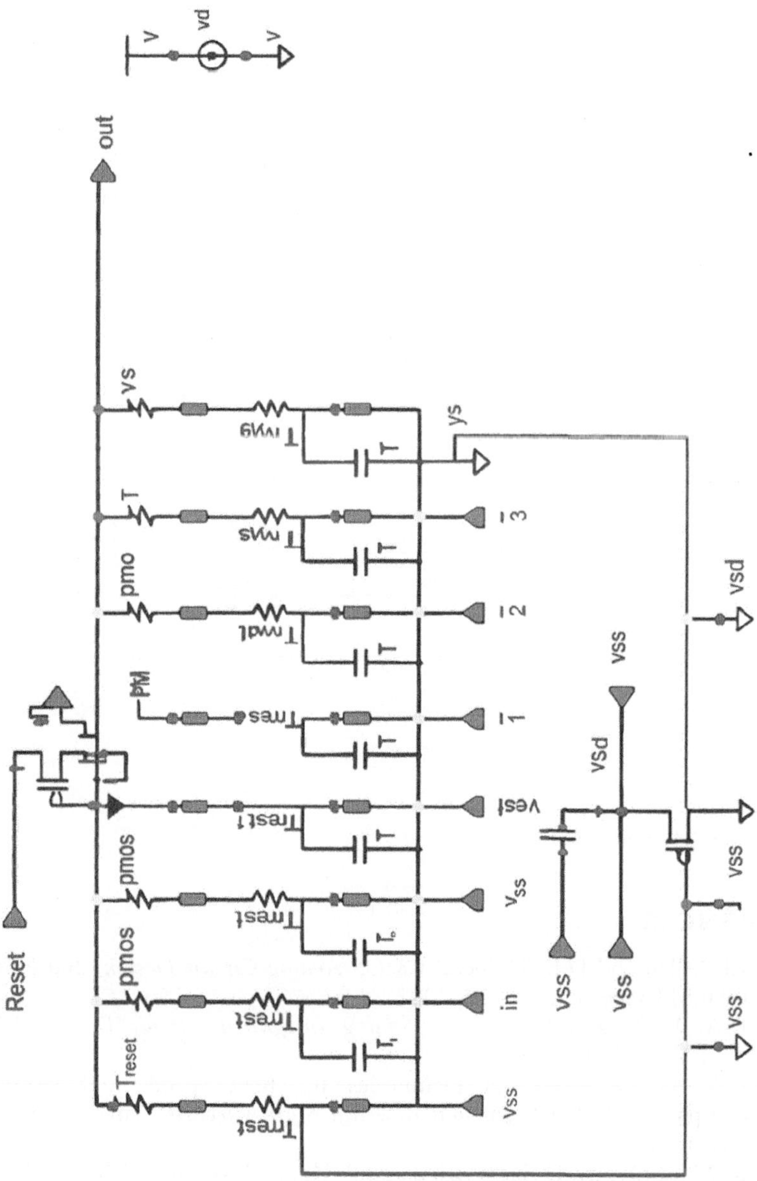

Figure 12.18 SAR ADC circuit diagram.

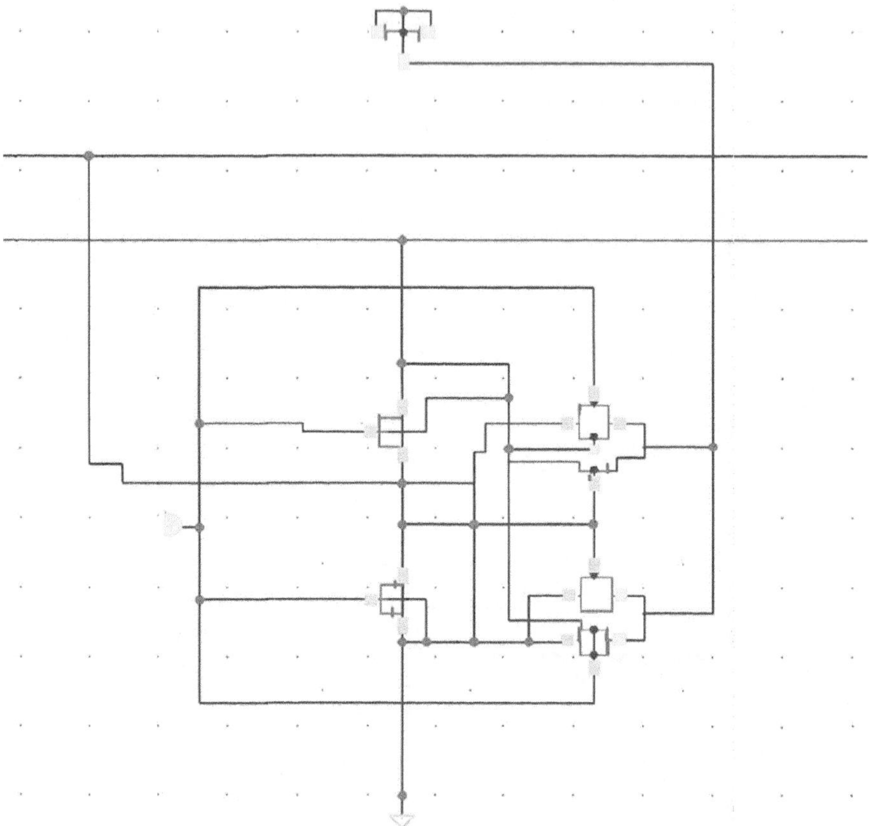

Figure 12.19 Schematic of DAC.

REFERENCES

1) P.E. Allen and D.R. Holberg, *CMOS Analog Circuit Design*, 2nd Edition, Oxford University, Oxford, 2007.
2) D.A. Johns and K. Martin, *Analog Integrated Circuit Design*, Wiley, New York, 1997.
3) B.A. Wooley, "Design techniques for high speed, high-resolution comparator", *IEEE Journal of Solid State Circuits*, Vol. 27, No. 12, Dec.1992.
4) R. Ma, L. Wang, D. Li, R. Ding and Z. Zhu, "A 10-bit 100-MS/s 5.23-mW SAR ADC in 0.18.m CMOS", *Microelectronics Journal*, 2018, pp. 68–72.
5) Y. Degerli, N. Fourchers, M. Rouger and P. Lut, "Low power autozeroed high-speed comparator for the read-out chain of a CMOS monolithic active pixel sensor based vertex detector," *IEEE Transaction on Nuclear Science*, Vol. 50, No. 5, 2003, pp. 1–21.

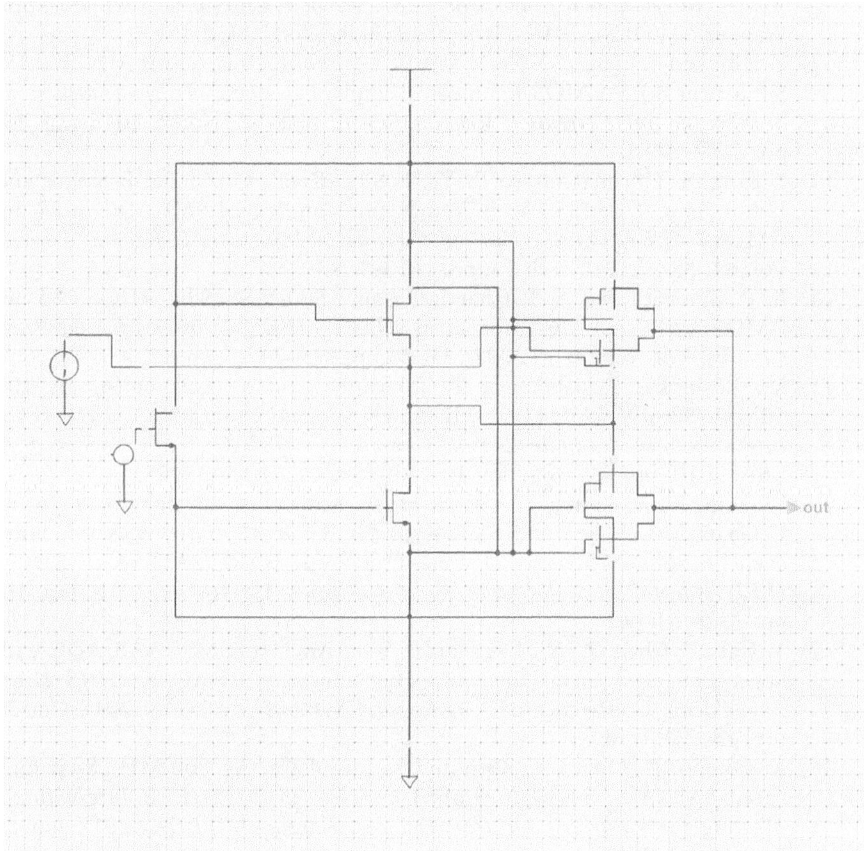

Figure 12.20 Schematic of CMOS switch.

6) M. Panchore and R. S. Gamad, "Low power and high speed cmos comparator design using 0.18 µm technology," *International Journal of Electronics Engineering Research India Publication*, Vol. 2, No. 1, 2010, pp. 71–77.
7) S. Yewale and R. S. Gamad, "Design of low power and high speed CMOS comparator for A/D convereter application," *Wireless Engineering and Technology*, Vol. 3, 2012, pp. 90–95.
8) M. Hossiennejad and H. Shamsi, "Fully differential charge-pump comparator-based pipelined ADC in 90nm CMOS," *Microelectronics Journal*, Vol. 53, 2016, pp. 8–15.
9) H. Nyquist, "Certain topics in telegraph transmission theory," *Transaction of the AIEE*, Vol. 47, No. 2, 1928, pp. 71–77.
10) P. R. Gray and R. G. Meyer, *Analysis and Design of Analog Integrated Circuits*, 3rd Edition, Wiley, Hooken, 1993.

11) C. Sargaiya and R. S. Gamad, "A low power comparator design for SAR ADC," *International Journal of VLSI system Design and Communication Systems (IJVDCS)*, 2016, Hyderabad, India, ISSN 2332-0929.
12) Z. Zhu and Y. Liang, "A 0.6-V 38-nW 9.4-ENOB 20-kS/s SAR ADC in 0.18-$\mu {\rm m} $ CMOS for medical implant devices," *IEEE Transactions on Circuits and Systems I: Regular Papers,* Vol. 62, No. 9, 2015 Aug 19, pp. 2167–2176.
13) J. Fang, S. Thirunakkarasu, X.Yu, F. Silva-Rivas, C. Zhang, F. Singor and J. Abraham, "A 5-GS/s 10-b 76-mW time-interleaved SAR ADC in 28 nm CMOS," *IEEE Transactions on Circuits and Systems I: Regular Papers,* Vol. 64, No. 7, 2017 Feb 14, pp. 1673–1683.
14) B. P. Ginsburg and A. P. Chandrakasan, "500-MS/s 5-bit ADC in 65-nm CMOS with split capacitor array DAC," *IEEE Journal of Solid-State Circuits,* Vol. 42, No. 4, 2007 Mar 26,pp. 739–747.
15) M. Yoshioka, K. Ishikawa, T. Takayama and S. Tsukamoto, "A 10b 50MS/s 820µW SAR ADC with on-chip digital calibration," *2010 IEEE International Solid-State Circuits Conference - (ISSCC)*, San Francisco, CA, USA, 2010, pp. 384-385, doi: 10.1109/ISSCC.2010.5433965..
16) W. Kim, H. K. Hong, Y. J. Roh, H. W. Kang, S. I. Hwang, D. S. Jo, D. J. Chang, M. J. Seo and S. T. Ryu, "A 0.6 V 12 b 10 MS/s low-noise asynchronous SAR-assisted time-interleaved SAR (SATI-SAR) ADC," *IEEE Journal of Solid-State Circuits,* Vol. 51, No. 8, 2016 Jun 10, pp. 1826–1839.
17) Y. Cao, Y. Chen, Z. Ni, F. Ye and J. Ren, An 11b 80MS/s SAR ADC with Speed-Enhanced SAR Logic and High-Linearity CDAC. In *2018 IEEE Asia Pacific Conference on Circuits and Systems (APCCAS)* 2018 Oct 26 (pp. 18–21). IEEE.
18) K. Okuno, K. Obata, T. Kato and K. Sushihara, An 800-MHz 8-bit high speed SAR ADC in 16nm FinFET process. In *2017 IEEE International Meeting for Future of Electron Devices, Kansai (IMFEDK)* 2017 Jun 29 (pp. 24–25). IEEE.
19) H. N. Minh, D. N. Quoc and T. Hoang, A design of 10-bit 25-MS/s SAR ADC using separated clock frequencies with high speed comparator in 180nm CMOS. In *2015 International Conference on Advanced Technologies for Communications (ATC)* 2015 Oct 14 (pp. 133–138). IEEE.
20) M. Rui, B. Wenbin and Z. Zhangming, "An energy-efficient and highly linear switching capacitor procedure for SAR ADCs," *Journal of Semiconductors,* Vol. 36, No. 5, 2015 May, pp. 055014.
21) V. S. Sooraj and G. M. Joseph, Speed & Resolution Enhancement of 12-Bit SAR ADC. In *2018 Second International Conference on Intelligent Computing and Control Systems (ICICCS)* 2018 Jun 14 (pp. 1655–1658). IEEE.
22) C. Wulff and T. Ytterdal, "A compiled 9-bit 20-MS/s 3.5-fJ/conv.step SAR ADC in 28-nm FDSOI for Bluetooth low energy receivers," *IEEE Journal of Solid-State Circuits*, Vol. 52, No. 7, 2017 July, pp. 1915–1926.
23) M. Ding et al., "A hybrid design automation tool for SAR ADCs in IoT," *IEEE Transaction of Very Large Scale Integration (VLSI) System*, Vol. 26, No. 12, 2018 December, pp. 2853–2862.

24) B.-S. Song, "Nyquist-rate data converters." In Bang-Sup Song (Eds.) *MicroCMOS Design*, Boca Raton, FL, USA: CRC Press, 2011.
25) E. Siragusa and I. Galton, "A digitally enhanced 1.8-V 15-bit 40-MSample/s CMOS pipelined ADC," *IEEE Journal of Solid-State Circuits*, Vol. 39, No. 12, 2004 December, pp. 2126–2138.
26) A. M. Abo and P. R. Gary, "A 1.5-V 10-bit 14.3 MS/s CMOS pipeline analog-to-digital converter", *IEEE Journal of Solid-State Circuits*, Vol. 34, No. 5, 1999 May, pp. 599–606.
27) N. D. Dalt, M. Harteneck, C. Sandner and A. Wiesbauer, "On the jitter requirements of the sampling clock for analog-to-digital converters," *IEEE Transaction Circuits of Systems I Fundamental Theory Application*, Vol. 49, No. 9, 2002 September, pp. 1354–1360.
28) L. Chen, A. Sanyal, J. Ma, X. Tang and N. Sun, Comparator common-mode variation effects analysis and its application in SAR ADCs. In *Proceedings IEEE International Symposium Circuits Systems (ISCAS)*, 2016 May (pp. 2014–2017).
29) S.-E. Hsieh and C.-C. Hsieh, "A 0.44-fJ/conversion-step 11-bit 600-kS/s SAR ADC with semi-resting DAC", *IEEE Journal of Solid-State Circuits*, vol. 53, no. 9, pp. 2595–2603, Sep. 2018.
30) X. Tang, L. Chen, J. Song and N. Sun, A 1.5fJ/conv-step 10b 100kS/s SAR ADC with gain-boosted dynamic comparator. In *Proceedings IEEE Asian Solid-State Circuits Conference (A-SSCC)* 2017 Nov (pp. 229–232).
31) H. S. Bindra, C. E. Lokin, D. Schinkel, A.-J. Annema and B. Nauta, "A 1.2-V dynamic bias latch-type comparator in 65-nm CMOS with 0.4-mV input noise," *IEEE Journal Solid-State Circuits*, Vol. 53, No. 7, 2018 July, pp. 1902–1912.

Chapter 13

Future Perspectives in FETs

Angshuman Khan, Abheek Gupta, Soumya Sen, Ashish Raman, and Prabhat Singh

13.1 INTRODUCTION TO FET EVOLUTION

Field-Effect Transistors (FETs) have been the cornerstone of modern electronics, shaping the semiconductor industry for over five decades. As the fundamental building blocks of integrated circuits, FETs have evolved to meet the ever-increasing demands for performance, miniaturization, and energy efficiency. This section provides a detailed overview of the historical development of FET technology and the emerging need for novel solutions to overcome current technological limitations. The future of FETs lies in both scaling existing designs and developing new materials and architectures that can address the physical and electrical challenges of next-generation electronics.

13.1.1 Historical Overview of FET Development

This subsection explores the key milestones in the evolution of FET technology as shown in Figure 13.1, beginning with the introduction of the Metal-Oxide-Semiconductor Field-Effect Transistor (MOSFET) in the 1960s. It highlights the significance of MOSFETs in driving Moore's Law, enabling the continuous miniaturization of transistors and the integration of billions of devices on a single chip. The transition from planar MOSFETs to advanced architectures like FinFETs and Silicon-on-Insulator (SOI) FETs is discussed, showcasing how these innovations addressed early challenges such as short-channel effects and leakage currents. This historical perspective sets the stage for understanding the ongoing need for further advancements in FET technology [1]-[5].

13.1.2 Motivation for Future FET Advancements

As the limits of silicon-based scaling approach, this subsection delves into the driving factors behind the search for alternative solutions. The

DOI: 10.1201/9781003533535-13

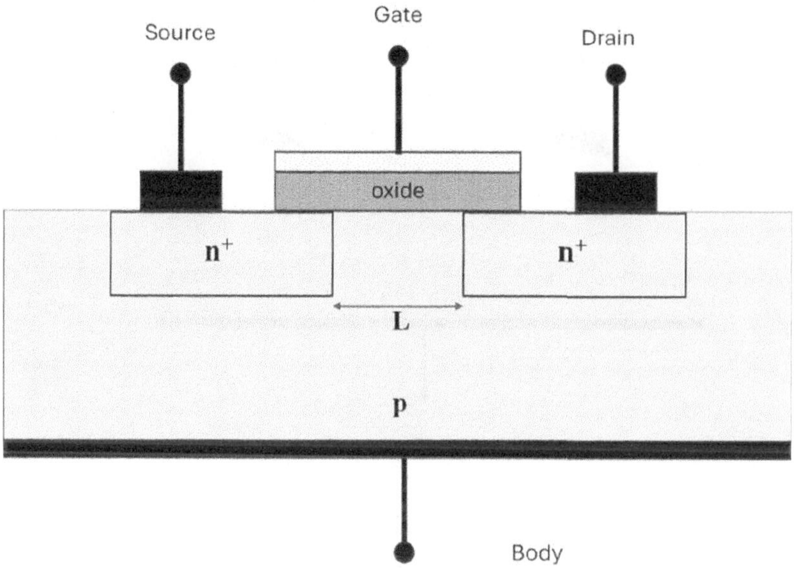

Figure 13.1 Schematic view of a field effect transistor (FET).

continued demand for high-performance, low-power electronics, especially in applications such as mobile devices, high-speed computing, and Internet of Things (IoT) systems, necessitates innovations beyond traditional FETs. The discussion focuses on the key challenges faced by current FET technologies, including increased leakage, thermal management issues, and the diminishing returns of shrinking transistor size. This motivates the exploration of novel materials, such as 2D semiconductors, and new device architectures, such as Tunnel FETs (TFETs), to meet future industry requirements.

This introduction paves the way for understanding the critical role of material science and device engineering in the future of FET evolution [2,5].

13.2 CHALLENGES IN SCALING DOWN

As transistors have continued to shrink, they have faced a series of challenges that undermine their performance and reliability. This section delves into the critical issues that arise when scaling down FETs, focusing on the physical and electrical limits that hamper further miniaturization. The goal is to provide a comprehensive understanding of the obstacles encountered

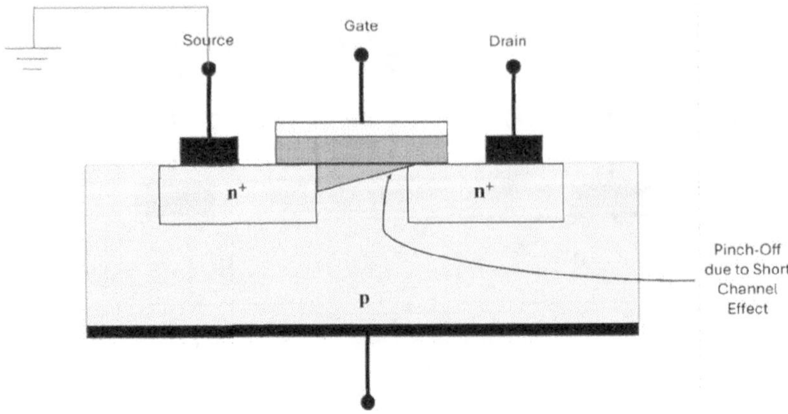

Figure 13.2 Schematic diagram for short channel effects.

at the nanoscale and the ways in which researchers are addressing them. Mathematical models and expressions are introduced to quantify these effects and highlight their impact on FET performance.

13.2.1 Short-Channel Effects (SCEs)

One of the primary challenges when scaling FETs to nanoscale dimensions is the manifestation of **short-channel effects** (SCEs), as shown in Figure 13.2. As the channel length between the source and drain becomes comparable to the depletion regions around the junctions, the ability to control the channel with the gate diminishes. This leads to issues such as threshold voltage reduction, drain-induced barrier lowering (DIBL), and increased subthreshold leakage currents [6–10].

13.2.1.1 Threshold Voltage Roll-Off

The threshold voltage, V_{th}, tends to decrease as the channel length, L, is reduced. The relationship can be approximated as in equation 13.1:

$$\Delta V_{th} \propto \frac{t_{ox}}{L} \tag{13.1}$$

where t_{ox} is the oxide thickness. A thinner gate oxide can partially mitigate this effect, but there are limits due to gate leakage [1,3,6].

13.2.1.2 Drain-Induced Barrier Lowering (DIBL)

DIBL occurs when the drain voltage V_D influences the potential barrier at the source, reducing the control of the gate. The barrier height ϕ_b can be illustrated by equation 13.2:

$$\phi_b \propto V_D \left(\frac{L}{t_{ox}} \right) \tag{13.2}$$

As L becomes shorter, the gate loses its ability to effectively modulate the barrier, leading to increased off-state current [1,3,6].

13.2.1.3 Subthreshold Leakage Current

In short-channel FETs, the subthreshold slope S degrades, causing an increase in leakage current I_{leak} in the off-state, as stated by equation 13.3:

$$I_{leak} = exp\left(\frac{V_G - V_{th}}{S} \right) \tag{13.3}$$

where V_G is the gate voltage and S is the subthreshold swing, ideally 60 mV/decade for MOSFETs. Reducing the gate length increases leakage exponentially [1,3,6].

13.2.2 Leakage Currents and Power Dissipation

Power dissipation becomes a crucial concern as devices scale down, driven largely by increased leakage currents and static power consumption. There are several sources of leakage in nanoscale FETs:

13.2.2.1 Gate Oxide Leakage

Thinner gate oxides lead to significant gate leakage due to **quantum tunneling.** The tunneling current I_{ox} through the gate oxide is given by equation 13.4 [1,3,6]:

$$I_{ox} \propto \exp\left(-\frac{q.t_{ox}}{\hbar \cdot \sqrt{2m^* V_{ox}}} \right) \tag{13.4}$$

where q is the charge of an electron, t_{ox} is the oxide thickness, m^* is the effective electron mass, and V_{ox} is the oxide voltage. As the oxide thickness

t_{ox} decreases, I_{ox} increases exponentially, contributing to significant power losses.

13.2.2.2 Subthreshold Leakage

As noted earlier, short-channel effects result in increased subthreshold leakage, which leads to higher static power consumption, even when the device is in the off-state. The total static power P_{static} is given by equation 13.5:

$$P_{static} = V_{DD} \cdot I_{leak} \tag{13.5}$$

where V_{DD} is the supply voltage. Reducing V_{DD} can help mitigate power dissipation, but it also negatively impacts performance and switching speed [1,3,6].

13.2.2.3 Junction Leakage

Leakage through the source/drain junctions increases as the junction depth becomes shallow. This can be approximated using Shockley-Read-Hall (SRH) recombination mechanisms, which are exacerbated by high doping concentrations at the junctions.

13.2.2.4 Hot Carrier Effects

When scaling FETs, electrons gain high kinetic energy, leading to hot carrier effects, where carriers generate electron-hole pairs due to impact ionization, causing reliability issues like gate oxide degradation.

13.2.3 Power-Performance Trade-Offs

Scaling down transistors has traditionally improved performance by enabling faster switching and higher densities. However, the simultaneous increase in power dissipation has led to the so-called **power-performance trade-off.** This subsection explores how reducing the supply voltage V_{DD} and adjusting the threshold voltage V_{th} impacts both performance and power consumption

13.2.3.1 Performance Metrics

The switching speed of a FET is determined by the gate capacitance C_G and the on-state current I_{ON}. The delay time τ can be given by equation 13.6 [1,3,6]:

$$\tau \propto \frac{C_G \cdot V_{DD}}{I_{on}} \tag{13.6}$$

Reducing V_{DD} improves power efficiency, but it also reduces I_{on}, increasing the delay τ.

13.2.4 Power Dissipation

The total power consumption P in a FET includes both dynamic and static components, given by equation 13.7:

$$P = P_{dynamic} + P_{static} \tag{13.7}$$

where the dynamic power $P_{dynamic}$ is approximated by equation 13.8:

$$P_{dynamic} = \alpha \cdot C_G \cdot V_{DD}^2 \cdot f \tag{13.8}$$

with α representing the switching activity and the operating frequency. Reducing V_{DD} decreases dynamic power quadratically but increases delay, creating a fundamental trade-off [1,3,6,11].

13.2.5 Reliability and Manufacturing Challenges

As transistors scale down, maintaining reliability and high manufacturing yield becomes increasingly difficult. This subsection focuses on challenges such as process variation, electromigration, and gate oxide breakdown

13.2.5.1 Process Variability

Variations in doping concentrations, channel lengths, and oxide thicknesses during manufacturing lead to device-to-device variability, which becomes more pronounced at smaller geometries. This variability causes inconsistencies in performance, power, and threshold voltages

13.2.5.2 Electromigration

High current densities in nanoscale interconnects lead to electromigration, a phenomenon where metal atoms are displaced by the flow of electrons, eventually causing failure. The time to failure (TTF) due to electromigration can be modeled by Black's equation depicted in 13.9.

$$TTF \propto J^{-n} \cdot exp\left(\frac{E_a}{kT}\right) \tag{13.9}$$

where J is the current density, E_a is the activation energy, k is the Boltzmann constant, and T is the temperature [1,3,6].

13.2.5.3 Gate Oxide Breakdown

Thin oxides are prone to breakdown due to high electric fields and the accumulation of defects over time. Breakdown models like the time-dependent dielectric breakdown (TDDB) are used to predict the oxide lifetime under stress conditions.

In summary, this section identifies the key challenges faced when scaling down FETs, including short-channel effects, leakage currents, power-performance trade-offs, and reliability concerns. By quantifying these issues through mathematical expressions, the section highlights the pressing need for innovative solutions in the design of future FETs.

13.3 EMERGING MATERIALS IN FETS

As traditional silicon-based FETs approach their physical limits, researchers are exploring alternative materials to overcome the challenges of miniaturization and performance enhancement. This section focuses on the potential of new materials, including 2D semiconductors, III–V compounds, and ferroelectric materials, to revolutionize FET design. Each of these materials offers unique properties that can address issues like short-channel effects, mobility degradation, and energy efficiency. Mathematical models and expressions are used to describe their impact on device performance.

13.3.1 Two-Dimensional (2D) Materials for FETs

Two-dimensional (2D) materials, such as graphene, transition metal dichalcogenides (TMDs), and black phosphorus, have gained significant attention due to their atomically thin nature and excellent electronic properties. Their unique characteristics make them ideal candidates for next-generation FETs, especially in reducing short-channel effects and improving gate control.

13.3.1.1 Graphene FETs (GFETs)

Graphene, a single layer of carbon atoms arranged in a honeycomb lattice, exhibits high carrier mobility and excellent thermal conductivity. However, its zero-bandgap nature limits its use in digital logic applications. To model carrier transport in graphene, the current-voltage characteristics of a graphene FET (GFET) can be expressed as in equation 13.10:

$$I_D = \frac{q}{h}\int_{E_{c1}}^{E_{c2}} T(E)\left[f_s(E) - f_d(E)\right]dE \tag{13.10}$$

where I_D is the drain current, q is the electron charge, h is Planck's constant, $T(E)$ is the transmission probability at energy E, and $f_s(E)$, $f_d(E)$ are the Fermi-Dirac distributions at the source and drain, respectively.

13.3.1.2 Transition Metal Dichalcogenides (TMDs)

TMDs, such as MOS_2 and WS_2, have gained attention due to their inherent bandgaps and high on/off current ratios, making them suitable for digital applications. The mobility μ of carriers in TMD-based FETs can be modeled by equation 13.11 [1,3,6]:

$$\mu = \frac{q\tau}{m^*} \tag{13.11}$$

where τ is the relaxation time, and m^* is the effective mass of carriers. The current in TMD-based FETs, assuming a ballistic transport regime, can be expressed as in equation 13.12:

$$I_D = \frac{qn_s v_{sat}}{L} \tag{13.12}$$

where n_s is the charge density, v_{sat} is the saturation velocity, and L is the channel length [1,3,6,12-15].

13.3.1.3 Black Phosphorus (BP) FETs

Black phosphorus is a 2D material with a tunable direct bandgap, offering a good balance between mobility and on/off current ratio. The direct bandgap energy E_g of black phosphorus can be modeled as a function of its thickness, illustrated by equation 13.13:

$$E_g = E_g^0 - \frac{h^2\pi^2}{2m^*t^2} \tag{13.13}$$

where E_g^0 is the bulk bandgap, t is the thickness, and m^* is the effective mass. This tunable bandgap makes BP highly suitable for use in FETs with optimized electronic properties [1,3,6].

13.3.2 III–V Compounds for High-Speed FETs

III–V compound semiconductors, such as GaAs, InP, and InGaAs, offer high electron mobility and superior saturation velocity, making them ideal for high-speed and low-power applications. These materials can significantly enhance the performance of FETs by reducing carrier scattering and improving overall device efficiency.

13.3.2.1 Electron Mobility in III–V Materials

The mobility μ of electrons in III–V semiconductors is much higher than in silicon, which can be modeled using the following expression shown in equation 13.14:

$$\mu = \frac{q\tau_c}{m^*} \tag{13.14}$$

where τ_C is the carrier relaxation time, and m^* is the effective mass of the electron. In high-speed FETs, the higher mobility leads to faster switching and lower power dissipation [1,3,6].

13.3.2.2 Velocity Saturation and Ballistic Transport

III–V semiconductors exhibit high saturation velocities v_{sat}, which can be modeled as in equation 13.15:

$$v_{sat} \approx \frac{2\times10^7}{1+\left(\frac{E}{E_{sat}}\right)} \tag{13.15}$$

where E is the electric field, and E_{sat} is the saturation electric field. The high saturation velocity in III–V materials allows for rapid carrier transport in short-channel devices [1,3,6].

13.3.2.3 Quantum Capacitance in III–V FETs

Quantum capacitance C_q becomes significant in III–V based FETs, particularly at nanoscale dimensions. It is given by equation 13.16:

$$C_q = \frac{q^2 g(E_F)}{A} \tag{13.16}$$

where $g(E_F)$ is the density of states at the Fermi level, and A is the area of the FET. The quantum capacitance limits the gate control and contributes to short-channel effects, particularly as devices approach the quantum mechanical regime [1,3,6,16–20].

13.3.3 Ferroelectric Materials for Negative Capacitance FETs (NCFETs)

Ferroelectric materials offer the potential to significantly reduce power consumption in FETs by utilizing negative capacitance, which amplifies the gate voltage and reduces the supply voltage required for switching. This is achieved through the unique properties of ferroelectric materials, which exhibit a non-linear polarization response.

13.3.3.1 Negative Capacitance Effect

The concept of negative capacitance arises from the Landau-Khalatnikov equation, which models the free energy of a ferroelectric material as a function of polarization P:

$$F(P) = \frac{\alpha}{2}P^2 + \frac{\beta}{2}P^4 \tag{13.17}$$

Equation 13.17 shows above, α and β as material-specific coefficients. Negative capacitance occurs when the device operates in the unstable region of the polarization curve, where the slope of the $P - V$ curve is negative [1,3,6].

13.3.3.2 Energy-Efficient Switching in NCFETs

The negative capacitance amplifies the input gate voltage, reducing the power required for switching. The voltage amplification factor γ can be expressed as in equation 13.18:

$$\gamma = 1 + \frac{C_f}{C_{ox}} \tag{13.18}$$

where C_f is the ferroelectric capacitance, and C_{ox} is the gate oxide capacitance. The effective reduction in supply voltage V_{DD} due to negative capacitance enables energy-efficient operation without compromising on performance [1,3,6].

13.3.3.3 NCFET Device Modeling

The total capacitance of an NCFET is given by the combination of the ferroelectric capacitance C_f and the traditional oxide capacitance C_{ox}, shown in equation 13.19,

$$\frac{1}{C_{total}} = \frac{1}{C_f} + \frac{1}{C_{ox}} \tag{13.19}$$

when C_f is negative, the total capacitance can be made larger than the oxide capacitance alone, leading to improved gate control and subthreshold swing below the thermionic limit of 60 mV/decade, enabling faster and more efficient switching [1,3,6].

13.3.4 Impact of Material Properties on Device Scaling and Performance

The performance of FETs is deeply influenced by the choice of materials. In this subsection, the impact of material properties such as bandgap, mobility, and dielectric constant on the scaling and overall performance of FETs is discussed.

13.3.4.1 Bandgap Engineering

The bandgap of a semiconductor plays a critical role in determining the on/off ratio, subthreshold slope, and leakage current in FETs. Narrow bandgap materials (like InGaAs) offer higher electron mobility but suffer from higher leakage currents, while wide bandgap materials (like GaN) provide better thermal stability and reduced leakage.

13.3.4.2 Dielectric Constant and Capacitance

The dielectric constant εrεr of the gate oxide material affects the gate capacitance C_{ox}, which in turn controls the switching speed and threshold voltage of the FET. The gate capacitance can be expressed as in equation 13.20:

$$C_{ox} = \frac{\varepsilon_0 \varepsilon_r A}{t_{ox}} \tag{13.20}$$

where ε_0 is the permittivity of free space, A is the gate area, and t_{ox} is the oxide thickness. High-κ materials like HfO_2 have been employed to maintain high gate capacitance while minimizing leakage currents [1,3,6].

13.4 DEVICE ARCHITECTURE INNOVATIONS FOR FUTURE FETS

With the ongoing miniaturization and performance demands of transistors, it is evident that traditional planar FET architectures are reaching their limits. This section discusses how innovations in device architecture—such as multigate devices, tunnel FETs, and vertical FETs—are shaping the future of field-effect transistors (FETs). These novel designs provide better control over channel electrostatics, reduce power consumption, and enhance scalability for next-generation electronics.

13.4.1 Multigate FETs

Multigate architectures, such as FinFETs, Gate-All-Around FETs (GAAFETs), and Nanowire FETs, have emerged as effective solutions to the short-channel effects that hinder traditional planar transistors. These architectures provide improved gate control, suppress leakage currents, and enable further scaling of transistor dimensions

13.4.1.1 FinFET Architecture

The **FinFET** structure uses a thin silicon fin as the channel, with the gate wrapping around three sides of the fin. This design enhances the electrostatic control over the channel, reducing short-channel effects and improving the device's subthreshold slope. The drain current in a FinFET can be modeled as in equation 13.21:

$$I_D = \mu \cdot C_{ox} \cdot \frac{W_{eff}}{L} \cdot \left(V_{GS} - V_{th}\right) \cdot V_{DS} \tag{13.21}$$

where μ is the carrier mobility, C_{ox} is the gate oxide capacitance, W_{eff} is the effective channel width, L is the channel length, V_{GS} is the gate-source voltage, V_{th} is the threshold voltage, and V_{DS} is the drain-source voltage [1,3,6].

13.4.1.2 Gate-All-Around FETs (GAAFETs)

GAAFETs take the multigate concept further by surrounding the entire channel with the gate, providing the best possible electrostatic control. This allows for even smaller channel lengths and better scalability. The electrostatic potential $\phi(x,y)$ in a GAAFET can be described by solving the 2D Poisson's equation, illustrated in 13.22:

$$\nabla^2\phi(x,y) = -\frac{\rho(x,y)}{\varepsilon} \tag{13.22}$$

where $\rho(x,y)$ is the charge density, and ε is the dielectric constant of the gate oxide. GAAFETs exhibit superior electrostatic integrity compared to FinFETs, enabling further scaling beyond the 5 nm technology node [1,3,6].

13.4.1.3 Nanowire FETs

Nanowire FETs are an extension of the GAAFET concept as illustrated in Figure 13.3, where the channel is reduced to a nanometer-scale wire. The gate surrounds the nanowire, allowing for 360-degree control over the channel. The current in a nanowire FET, assuming ballistic transport, is given by equation 13.23:

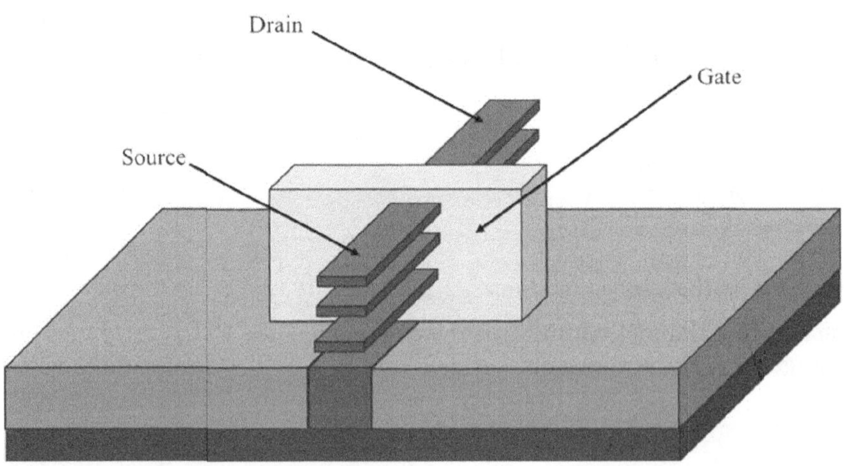

Figure 13.3 Gate all around structure of FET.

$$I_D = \frac{q^2}{h} \cdot \frac{V_{DS}}{L} \tag{13.23}$$

where L is the channel length and h is Planck's constant. This expression highlights the direct relation between current and the applied voltage V_{DS}, and the fact that shorter channel lengths enhance device performance in terms of speed and power consumption [1,3,6].

13.4.2 Tunnel Field-Effect Transistors (TFETs)

Tunnel FETs (TFETs), as depicted in Figure 13.4, leverage quantum mechanical tunneling to achieve extremely low subthreshold swings, potentially below the 60 mV/decade limit of traditional MOSFETs. TFETs have gained attention for their ability to reduce power consumption in future ultra-low-power applications.

13.4.2.1 Operating Principle of TFETs

Unlike conventional MOSFETs, which rely on thermionic emission for carrier injection, TFETs use band-to-band tunneling to inject carriers into the channel. When the gate voltage is applied, the energy bands of the source and channel align in such a way that electrons can tunnel from the valence band

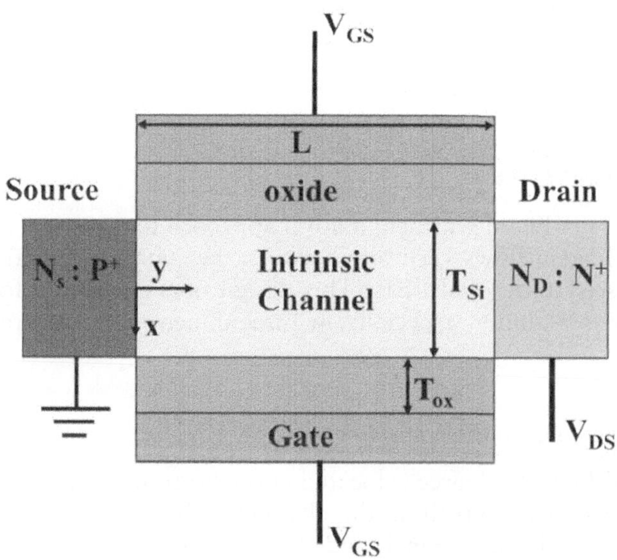

Figure 13.4 Cross sectional view of tunnel field effect transistor (TFET).

of the source to the conduction band of the channel. This tunneling process results in an exponential increase in current, even at low gate voltages.

The tunneling current IT*IT* in a TFET can be described by the following expression in 13.24:

$$I_T \propto exp\left(-\frac{2\pi\sqrt{2m^* E_g}}{3q\hbar E}\right) \tag{13.24}$$

where m^* is the effective mass of the carrier, E_g is the bandgap of the material, E is the electric field, q is the elementary charge, and $\hbar$ is the reduced Planck's constant. The strong dependence of the tunneling current on the electric field and the material's bandgap makes TFETs extremely sensitive to device geometry and material choice [1,3,6].

13.4.2.2 Advantages of TFETs

TFETs offer several advantages over traditional MOSFETs, including:

- Lower Power Consumption: Since TFETs can achieve a subthreshold swing below the 60 mV/decade limit, they can switch at much lower voltages, reducing overall power consumption.
- Higher Energy Efficiency: Due to their low power requirements, TFETs are ideal for energy-efficient applications, especially in battery-powered devices.
- Reduced Leakage Current: The use of band-to-band tunneling for current injection naturally reduces leakage currents, further enhancing energy efficiency.

13.4.3 Vertical FETs

Vertical FETs (VFETs) represent a novel approach to transistor architecture, where the current flows vertically between the source and drain instead of laterally, as in traditional FETs. This design offers potential for improved density and scalability, especially in three-dimensional integrated circuits (3D ICs).

13.4.3.1 Operating Principle of VFETs

In a vertical FET, the source is located at the bottom, and the drain is at the top, with the gate controlling the channel in between. The current flows vertically through the channel, and the gate wraps around the channel to control the carrier flow. This vertical design significantly reduces the device footprint and enhances transistor density.

The vertical current flow can be modeled as in 13.24:

$$I_D = \mu \cdot C_{ox} \cdot \frac{W}{L} \cdot \left(V_{GS} - V_{th}\right) \cdot V_{DS} \tag{13.24}$$

where W is the width of the vertical channel, and L is the length of the channel (the height in this case). The small footprint of the vertical FET allows for more devices to be integrated into a smaller area, enabling greater scalability [1,3,6, 21–27].

13.4.3.2 Applications and Advantages of VFETs

Vertical FETs are particularly suited for 3D integration, where multiple layers of transistors are stacked on top of each other to improve device density. Some of the key advantages of VFETs include:

- Reduced Area: Since VFETs are built vertically, they occupy less area on the chip, allowing for higher transistor density in integrated circuits.
- Improved Performance: The short vertical channel length allows for higher drive currents and faster switching times, making VFETs ideal for high-performance applications.
- Better Heat Dissipation: The vertical design facilitates improved heat dissipation compared to traditional lateral FETs, helping manage thermal issues in densely packed circuits.

13.4.3.3 Challenges in VFET Fabrication

The vertical FET (VFET) architecture offers substantial advantages in terms of device density, performance, and scalability for advanced technology nodes. However, the fabrication of VFETs presents a range of technical challenges that arise from their vertical structure and the need for precision in nanoscale manufacturing. Here are the major challenges associated with VFET fabrication:

13.4.3.4 Complex Etching Processes

In VFETs, the channel is vertically oriented, which requires precise etching techniques to form the vertical structure. Unlike traditional planar FETs, where the channel is created through a straightforward lateral etch, VFETs involve deep vertical etching of the semiconductor material, often silicon or III–V materials. Some of the difficulties include:

- High Aspect Ratio Etching: Creating tall, narrow vertical channels with high aspect ratios can be challenging. Achieving the desired aspect ratio while maintaining sidewall smoothness and uniformity across the wafer is difficult, as imperfections can severely affect device performance.
- Anisotropic Etching: To avoid lateral etching and preserve the integrity of the vertical structure, highly anisotropic etching techniques (like reactive-ion etching, RIE) are required. However, this can introduce sidewall roughness or tapering, which degrade the electrical properties of the channel.

13.4.3.5 Gate Deposition and Control

The gate in VFETs needs to wrap around the vertical channel, providing complete control over the current flow. This requires careful deposition of the gate material and gate oxide:

- Conformal Gate Deposition: A key challenge is depositing the gate material conformally around the vertical channel. Atomic layer deposition (ALD) is often used for this purpose due to its ability to create uniform thin films on 3D surfaces. However, controlling the thickness of the gate oxide and metal layer across the entire vertical structure is difficult.
- Gate Alignment: Precise alignment of the gate with the vertical channel is essential to ensure proper electrostatic control. Misalignment can lead to uneven gate coverage, leakage, or degraded device performance.

13.4.3.6 Doping and Junction Formation

Doping the source, drain, and channel regions in VFETs requires high precision to ensure proper carrier injection and device operation:

- Vertical Doping Profiles: In traditional FETs, doping is done laterally using ion implantation. In VFETs, doping needs to be controlled vertically, which is more complex. Achieving the desired doping concentration and profile along the height of the vertical structure, particularly in the channel and source/drain junctions, is challenging.
- Junction Sharpness: The formation of sharp p-n junctions in the vertical direction is critical to minimize leakage currents and ensure high on-off ratios. Maintaining sharp doping profiles across the entire

height of the device is difficult due to ion scattering and diffusion effects.

13.4.3.7 *Material and Interface Quality*

The quality of the materials used in VFETs, including the semiconductor channel, gate dielectric, and metal contacts, is critical to device performance:

- Surface Roughness: Since the channel is vertical, the surface quality of the sidewalls plays a major role in determining mobility and overall performance. Any roughness or irregularities in the etched channel can scatter carriers and degrade device performance.
- Interface Defects: The interface between the semiconductor and gate dielectric is crucial for VFETs to achieve high performance. Any defects or traps at this interface can lead to increased leakage currents, threshold voltage shifts, or degraded reliability. Achieving a high-quality interface, especially on the sidewalls of the vertical channel, requires advanced surface passivation techniques and high-quality deposition methods.

13.4.3.8 *Scaling and Uniformity Across Wafers*

As the dimensions of VFETs continue to shrink, maintaining uniformity across a wafer and ensuring consistent performance across all devices becomes more difficult:

- Wafer-Level Uniformity: Ensuring that each VFET on a wafer has identical vertical dimensions and doping profiles is a significant challenge. Variations in the etching, doping, or gate deposition processes across the wafer can lead to variations in device performance, yield loss, and reliability issues.
- Scaling to Smaller Nodes: As the technology scales to sub-10 nm nodes, the precision required in each step of the fabrication process increases dramatically. Even small variations in vertical dimensions or gate oxide thickness can lead to significant performance deviations, making the fabrication of such small-scale vertical devices increasingly complex.

13.4.3.9 *Integration with 3D ICs*

One of the main advantages of VFETs is their potential for integration into three-dimensional integrated circuits (3D ICs), illustrated in Figure 13.5. However, integrating VFETs into multilayer stacks presents additional challenges:

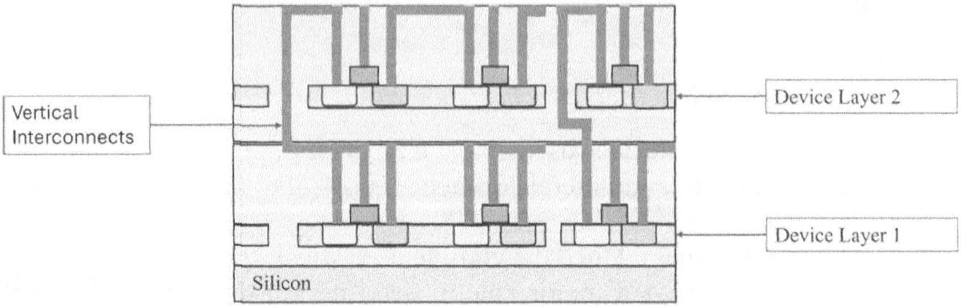

Figure 13.5 3-D integrated circuits.

- Interconnects and Routing: In a 3D IC, multiple layers of transistors need to be connected by interlayer vias and interconnects. Managing the vertical and lateral routing of these connections without introducing excessive parasitic resistance or capacitance is challenging.
- Thermal Management: VFETs integrated into 3D ICs face increased thermal management challenges. Heat generated in the lower layers of the stack must be dissipated effectively to avoid thermal degradation of the upper layers. Ensuring effective heat dissipation across the stacked layers is difficult, particularly as the device density increases.

13.4.3.10 Reliability and Defect Management

VFETs, like all advanced semiconductor devices, must be highly reliable for commercial applications. However, the complexity of their vertical structure introduces several reliability challenges:

- Gate Oxide Reliability: The gate oxide in VFETs, which wraps around the vertical channel, is susceptible to breakdown due to the high electric fields present in scaled devices. Ensuring the long-term reliability of the gate oxide, particularly at high operating voltages, is critical.
- Defects and Yield: The fabrication process for VFETs involves many complex steps, each of which can introduce defects. Managing these defects and ensuring high yield across a wafer is challenging, especially as device dimensions continue to shrink.

While VFETs hold tremendous promise for future transistor technologies, their fabrication presents significant challenges that must be overcome. Innovations in etching, doping, gate deposition, and material quality are essential to successfully fabricate VFETs at nanoscale dimensions. Additionally, ensuring uniformity, reliability, and effective integration with 3D ICs will be critical to their adoption in next-generation semiconductor devices. As fabrication techniques improve and these challenges are addressed, VFETs have the potential to enable continued scaling and performance improvements in semiconductor technology.

13.5 ADVANCED FABRICATION TECHNIQUES AND INTEGRATION FOR FUTURE FETS

The evolution of Field-Effect Transistors (FETs) demands not only advancements in device architecture but also corresponding innovations in fabrication techniques and integration methods. As technology nodes scale down to sub-5 nm, traditional fabrication methods face significant limitations, necessitating new approaches. This section explores the intersection of advanced fabrication techniques and their role in integrating next-generation FETs, such as Vertical FETs (VFETs) and Nanowire FETs (NWFETs), into three-dimensional integrated circuits (3D ICs).

13.5.1 Advanced Fabrication Techniques

As device dimensions shrink, precision in the fabrication process becomes more crucial. Advanced fabrication techniques address the limitations of conventional methods like planar etching and doping by introducing new processes tailored to vertical and three-dimensional structures. Key advancements include.

13.5.1.1 High-Aspect-Ratio Etching

High-aspect-ratio structures, essential for VFETs and NWFETs, require advanced etching techniques like Reactive Ion Etching (RIE) and Deep Reactive Ion Etching (DRIE). These methods ensure vertical profiles with smooth sidewalls and high precision. However, challenges such as controlling etch depth uniformity and sidewall roughness persist.

The mathematical formulation of etch depth uniformity can be modeled by defining the desired etch profile, $D(x,y)$, across the wafer, where x and y are lateral coordinates on the wafer, and D is the etch depth. Uniformity U can be quantified as given by equation 13.25:

$$U = \frac{\max(D) - \min(D)}{\max(D)} \tag{13.25}$$

Achieving low values of U is crucial for consistent device performance across a wafer [1,3,6].

13.5.1.2 Conformal Deposition Technique

For VFETs and nanowire devices, the gate oxide and metal gate need to be conformally deposited along the vertical structures. Techniques such as Atomic Layer Deposition (ALD) and Chemical Vapor Deposition (CVD) are widely used for their ability to coat complex 3D surfaces with thin, uniform films. ALD, in particular, offers atomic-level control over the thickness of the deposited material, which is essential for gate oxide scaling.

Mathematically, the conformality C of a deposition process can be expressed as in equation 13.26:

$$C = \frac{T_{sidewall}}{T_{top}} \times 100\% \tag{13.26}$$

where $T_{sidewall}$ and T_{top} are the thicknesses of the film on the sidewall and top surface, respectively. Achieving close to 100% conformality is crucial for ensuring uniform gate control over the vertical channel [1,3,6].

13.5.1.3 Selective Doping and Junction Formation

Doping in 3D structures like VFETs requires selective techniques to control the vertical doping profile. Traditional ion implantation methods struggle with vertical profiles, leading to techniques such as plasma immersion ion implantation (PIII) and laser doping. These methods enable precise control over doping concentration and depth, critical for forming sharp p-n junctions in vertical devices.

13.5.2 Integration in 3D ICs

The integration of FETs into 3D ICs offers numerous advantages, including higher device density, improved performance, and lower power consumption. However, integrating VFETs and other advanced FET architectures into 3D stacks presents several challenges:

13.5.2.1 Interconnect and Routing Challenges

In 3D ICs, inter-layer interconnects play a critical role in connecting the stacked devices. The vertical nature of VFETs and nanowire FETs complicates routing, as it increases the parasitic resistance and capacitance associated with the interconnects. Through-Silicon Vias (TSVs) are used for vertical interconnects, but ensuring their electrical performance without introducing excessive heat or signal delay is challenging

The total resistance R_{total} of interconnects in 3D ICs can be expressed as in equation 13.27:

$$R_{total} = R_{via} + R_{wire} + R_{contact} \tag{13.27}$$

where R_{via}, R_{wire}, and $R_{contact}$ are the resistances of the TSV, metal wire, and contact regions, respectively. Minimizing R_{total} is critical for high-performance 3D ICs [1,3,6].

13.5.2.2 Thermal Management

Heat dissipation is a major challenge in 3D ICs due to the increased power density in stacked layers. VFETs and nanowire FETs, when integrated into 3D ICs, exacerbate this issue as vertical structures generate more localized heat. Innovative thermal management techniques, such as using thermal vias and heat spreaders, are required to prevent thermal runaway and ensure device reliability.

The heat generation in a device layer Q can be modeled as in equation 13.28:

$$Q = I^2 R \tag{13.28}$$

where I is the current flowing through the device and R is the total resistance. Managing Q across multiple layers require effective thermal conduction paths to dissipate the generated heat [1,3,6].

13.5.2.3 Water Bonding and Alignment

Aligning multiple wafers in 3D ICs requires precise wafer bonding techniques such as Cu-to-Cu bonding and fusion bonding. These methods ensure minimal misalignment and create strong electrical connections between layers. The success of wafer bonding is critical for integrating VFETs into 3D stacks without affecting their performance.

The misalignment tolerance Δx in wafer bonding can be modeled as shown in 13.29:

$$\Delta x = \frac{W_{device} - W_{bond}}{2} \tag{13.29}$$

where W_{device} and W_{bond} are the widths of the device and bond pad, respectively. Minimizing Δx is essential for reducing defects and ensuring proper device functioning [1,3,6].

Advancements in fabrication techniques and integration strategies are essential for realizing the potential of next-generation FET architectures like VFETs and NWFETs. The challenges of high-aspect-ratio etching, conformal deposition, and selective doping must be addressed to ensure these devices can be manufactured at scale. Additionally, successful integration into 3D ICs requires solving complex interconnect, thermal, and alignment challenges. Overcoming these hurdles will be key to enabling the continued scaling and performance improvements of future semiconductor technologies [28–30].

13.6 CONCLUSION

The evolution of Field-Effect Transistors (FETs) represents one of the most significant technological advancements in the semiconductor industry, driving the progression of Moore's Law and fostering the rapid development of modern electronics. As traditional planar FET architectures reach their fundamental physical limits, the exploration of novel transistor geometries, materials, and fabrication techniques has become paramount for continued device scaling and performance improvements.

In this chapter, we explored the future perspectives of FET technology by examining advanced architectures like Vertical FETs (VFETs), Nanowire FETs (NWFETs), and 2D material-based FETs. These architectures offer substantial improvements in key performance metrics such as current density, electrostatic control, and scalability. Their vertical or quasi-1D geometries are poised to address the short-channel effects and power consumption limitations faced by conventional planar devices. The chapter highlighted how materials like III–V compounds, transition metal dichalcogenides (TMDs), and other 2D materials enable superior carrier mobility, lower leakage currents, and improved device sensitivity, making them promising candidates for the next generation of FETs.

Key to the successful implementation of these novel FET architectures is the development of advanced fabrication techniques. High-aspect-ratio etching, conformal deposition methods, and selective doping processes were discussed as critical enablers for realizing vertical and nanostructured devices. Furthermore, the integration of these FETs into 3D ICs offers new avenues for performance scaling, despite presenting challenges in terms of interconnects, thermal management, and wafer alignment. Through-silicon vias (TSVs), thermal vias, and precise wafer bonding techniques will be pivotal for ensuring the efficient functioning of 3D-stacked devices.

While these advancements promise significant performance gains, they also bring unique challenges. Issues such as increased fabrication complexity, scalability of new materials, and cost-effectiveness must be carefully addressed to ensure the widespread adoption of future FET architectures.

Furthermore, the integration of new device types into existing manufacturing ecosystems requires concerted efforts across material science, device engineering, and system-level design.

In conclusion, the future of FET technology is at a crossroads, where innovation in both device architecture and fabrication methods will dictate the trajectory of the semiconductor industry. As we continue to push the boundaries of what is possible, the successful integration of next-generation FETs into commercial technologies will define the pace of progress in fields ranging from consumer electronics to high-performance computing and artificial intelligence. The collaboration between academia, industry, and research institutions will be essential in overcoming the technical and economic challenges, ensuring that FETs remain at the forefront of technological innovation in the coming decades.

REFERENCES

[1] Tsividis, Y. (2002). *Mixed analog-digital VLSI devices and technology.* World Scientific.

[2] Cao, Wei, Huiming Bu, Maud Vinet, Min Cao, Shinichi Takagi, Sungwoo Hwang, Tahir Ghani, and Kaustav Banerjee. "The future transistors." *Nature* 620, no. 7974 (2023): 501–515.

[3] Taur, Yuan, and Tak H. Ning. *Fundamentals of modern VLSI devices.* Cambridge University Press, 2021.

[4] Taur, Yuan, Y-J. Mii, David J. Frank, H-S. Wong, Douglas A. Buchanan, Shalom J. Wind, Stephen A. Rishton, G. A. Sai-Halasz, and Edward J. Nowak. "CMOS scaling into the 21st century: 0.1 μm and beyond." *IBM Journal of Research and Development* 39, no. 1.2 (1995): 245–260.

[5] Ratnesh, Ratneshwar K., A. Goel, G. Kaushik, H. Garg, M. Singh, and B. Prasad. "Advancement and challenges in MOSFET scaling." *Materials Science in Semiconductor Processing* 134 (2021): 106002.

[6] Sze, Simon M., Yiming Li, and Kwok K. Ng. *Physics of semiconductor devices.* Wiley, 2021.

[7] Wong, H. "On the CMOS device downsizing, more Moore, more than Moore, and more-than-Moore for more Moore." In *2021 IEEE 32nd International Conference on Microelectronics (MIEL)*, pp. 9–15. IEEE, 2021.

[8] Equbal, Md Salim, and Shubham Sahay. "Scaling the MOSFET: Detrimental short channel effects and mitigation techniques." In *Nanoelectronics: Physics, Materials and Devices*, pp. 11–37. Elsevier, 2023.

[9] Rawat, Udit, Jackson D. Anderson, and Dana Weinstein. "Design and applications of integrated transducers in commercial CMOS technology." *Frontiers in Mechanical Engineering* 8 (2022): 902421.

[10] Kumari, N. Aruna, V. Bharath Sreenivasulu, and P. Prithvi. "Impact of scaling on nanosheet FET and CMOS circuit applications." *ECS Journal of Solid State Science and Technology* 12, no. 3 (2023): 033001.

[11] Gatermann, C., and Ralf Sommer. "Teaching the MOSFET: A Circuit Designer's View." In *2022 18th International Conference on Synthesis, Modeling, Analysis and Simulation Methods and Applications to Circuit Design (SMACD)*, pp. 1–4. IEEE, 2022.

[12] Zhao, Yuda, Marco Gobbi, Luis E. Hueso, and Paolo Samorì. "Molecular approach to engineer two-dimensional devices for CMOS and beyond-CMOS applications." *Chemical Reviews* 122, no. 1 (2021): 50–131.

[13] Kuo, Yi-Shan, Shen-Yang Lee, Chia-Chin Lee, Shou-Wei Li, and Tien-Sheng Chao. "CMOS-compatible fabrication of low-power ferroelectric tunnel junction for neural network applications." *IEEE Transactions on Electron Devices* 68, no. 2 (2021): 879–884.

[14] Kaushal, Priya, and Gargi Khanna. "The role of 2-Dimensional materials for electronic devices." *Materials Science in Semiconductor Processing* 143 (2022): 106546.

[15] Zheng, Wei, Xianghong Liu, Jiayue Xie, Guocai Lu, and Jun Zhang. "Emerging van der Waals junctions based on TMDs materials for advanced gas sensors." *Coordination Chemistry Reviews* 447 (2021): 214151.

[16] Cheng, Keh-Yung. *III–V Compound semiconductors and devices.* 2020: Springer.

[17] Liu, T. C., & Hu, C. (2011). *MOS transistor physics and simulation.* Wiley.

[18] Agarwal, Arun, Gourav Misra, and Kabita Agarwal. "Semiconductor III–V nanowires: Synthesis, fabrication and characterization of nanodevices." *Journal of The Institution of Engineers (India): Series B* 103, no. 2 (2022): 699–709, Springer.s

[19] Netchitailo, Vladimir S. "Cosmology and classical physics." *Journal of High Energy Physics, Gravitation and Cosmology* 8, no. 4 (2022): 1037–1072.

[20] Sen, Soumya, Mamta Khosla, and Ashish Raman. "Design and investigation of electrostatic doped heterostructure vertical Si (1-x) Gex/Si nanotube TFET." *Microelectronics Journal* 153, (2024): 106417. Elsevier.

[21] Sen, Soumya, Mamta Khosla, and Ashish Raman. "Design and challenges in TFET." In Dharmendra Singh Yadav, Shiromani Balmukund Rahi, Sukeshni Tirkey (Eds.), *Advanced Field-Effect Transistors*, pp. 23–45. CRC Press, 2023.

[22] Sen, Soumya, Ashish Raman, and Mamta Khosla. "A literature survey on tunnel field effect transistors." *AIJR Proceedings* (2021): 506–512. AIJR Publisher.

[23] Rahi, Shiromani Balmukund, Shubham Tayal, and Abhishek Kumar Upadhyay. "A review on emerging negative capacitance field effect transistor for low power electronics." *Microelectronics Journal* 116 (2021): 105242.

[24] Singh, Avtar, and Chandan Kumar Pandey. "Improved DC performances of gate-all-around si-nanotube tunnel FETs using gate-source overlap." *Silicon* 14, no. 4 (2022): 1463–1470.

[25] Bhol, Krutideepa, Biswajit Jena, and Umakanta Nanda. "Journey of Mosfet from planar to gate all around: A review." *Recent Patents on Nanotechnology* 16, no. 4 (2022): 326–332.

[26] Sen, Soumya, Mamta Khosla, and Ashish Raman. "Deciphering phenomenal performance with gate-all-around strained heterostructure nanotube

tunneling FET (TFET) for better drive currents and steep subthreshold slope." In *2024 International Conference on Integrated Circuits, Communication, and Computing Systems (ICIC3S)*, vol. 1, pp. 1–4. IEEE, 2024.

[27] Zurzolo, Chiara. "Tunneling nanotubes: Reshaping connectivity." *Current Opinion in Cell Biology* 71 (2021): 139–147.

[28] Zhao, Gang, Cen Song, and Binghan Wu. "3D Integrated Circuit (3D IC) technology and its applications." *Journal of Industrial Engineering and Applied Science* 2, no. 4 (2024): 60–65.

[29] Abdullah, M. F., and H. W. Lee. "Technology review of CNTs TSV in 3D IC and 2.5 D packaging: Progress and challenges from an electrical viewpoint." *Microelectronic Engineering* 290 (2024): 112189.

[30] Dhananjay, Krithika, Prachi Shukla, Vasilis F. Pavlidis, Ayse Coskun, and Emre Salman. "Monolithic 3D integrated circuits: Recent trends and future prospects." *IEEE Transactions on Circuits and Systems II: Express Briefs* 68, no. 3 (2021): 837–843.

Chapter 14

Web Monitoring and Speed Control of BLDC Motor with IoT

Delshi Howsalya Devi. R., R. Naresh, C.N.S.Vinoth Kumar, and Sharmila. S

14.1 INTRODUCTION

The Internet of Things (IoT) lets you do wireless ON/OFF control of the brushless DC motor. Using an android device, you can toggle the BLDC motor on and off by linking both of their IP addresses to a certain web site. By using an internet-based domain, the user can turn the BLDC motor ON or OFF anywhere they are. Using the Ethernet shield, an Arduino controller can be utilized for downloading and broadcasting signals from the internet [1]. Configuring a function with a specific IP address for a web domain was done in the Arduino controller. The Ethernet shield was used to receive signals from the World Wide Web domain. A specified website address was used to ON/OFF the Brushless DC motor. The phrase "Internet of things" describes the internet-based networking of physical objects such as residences, automobiles (also called "interconnected in order devices" or "smart devices"), furniture, and other objects that have sensors, actuators, software, internet connectivity, and other parts that make data collection and exchange easier [2–6].

The Global Standards Group on Internet of Things (IoT-GSI) defines IoT as "the architecture of information-rich society" that may be constantly monitored or managed through a well-established network architecture [7]. This presents opportunities for a faster connection between computer-powered systems and the real world. Additionally, this technology enhances efficiency, accuracy, and economic advantage while lowering the requirement for human involvement. Once IoT is further developed with sensors and actuators, it is anticipated to be included in the broader category of computerized systems with physical components, which also includes the smart grid, smart house, intelligent public transit, and smart city technologies. Even while they are able to interact with one another using the present-day internet infrastructures each object has an embedded computing system that contributes to making it achievable for it to be identified by its unique features [8,9]. According to researchers, the IoT could consist of up to 50 billion gadgets by 2020. It is often anticipated that IoT will offer a range

DOI: 10.1201/9781003533535-14

of guidelines, domains, and applications, as well as stronger machine-to-machine (M2M) interconnections for systems, devices, goods, and services. It is projected that the smooth integration of these embedded technologies, which together constitute smart things, would facilitate the automation of nearly every industry and promote the growth of innovative projects such as smart neighborhoods and a smart grid [10–12].

The term "things" in the context of the Internet of Things can refer to a broad variety of devices, including implants, farm animal biochip transponders, sensors incorporated in cars, DNA analysis equipment for food, environmental, and pathogen monitoring, or outdoor electronic devices designed to help emergency workers during search and rescue operations. Legal experts recommend thinking of "Things" as a "inextricable mixture of hardware, software, data, and service" [13,14]. These devices receive relevant data and then seamlessly transmit it to various other devices using a range of contemporary technology. One contemporaneous market trend is automation of homes, which is commonly known as technological advancements in smart homes. Examples include equipment like dishwashers and washers, air purifiers, artificial vacuums, stoves and ovens, and refrigerators and freezers with Wi-Fi for remote surveillance, as well as lighting, heating (with a smart thermostat), ventilation, and air conditioning [15]. It is anticipated that IoT would generate enormous quantities of data from many sources, requiring rapid data consolidation and an increase in capabilities to index, store, and manage such data more effectively in various ways. Additionally, a variety of fresh fields of application have lately been made possible by Internet-connected automation [16].

14.2 LITERATURE SURVEY

It is now necessary for electrical installations to use less carbon dioxide and run more effectively. Since brushless DC (BLDC) motors run entirely on electricity and are more efficient than traditional induction motors that require air conditioning, they can meet these requirements. By monitoring the position of the rotor, sensors that measure the hall effect control the motor. The control method is a trapezoidal pulse width modulation (PWM) wave, which is additionally referred to as trapezoidal control. An API controller can be used to change the rotation speed. It has been proven through a number of investigations that PI control produces a motor speed that is more precise as well as consistent than control without PI control [17–24].

Bilal Akin, Manish Bhardwaj, and John Warriner proposed the use of Hall effect sensors for trapezoidal control of BLDC motors.

In this application, brushless DC motors are programmed by TMS320F2803x microcontrollers. The C2000 series of microcontrollers, which includes the TMS320F280x devices, enables the efficient and economical development of advanced controllers for three phase motors.

The implementation of regulation algorithms that are considerably more accurate is made possible by these advances in technology. The ref [25] is a comprehensive solution suggestion that includes assessments on the effectiveness of energy transformation, organizational structures for management, power hardware configuration, and management hardware [25].

Jose Carlose suggested employing sensorless methods and application guidelines to regulate the speed and location of brushless DC motors. In addition to background information on the usage of sensors and their benefits and drawbacks, this project offers a technical investigation of sensorless approaches to brushless direct current (BLDC) motor drive control [26].

Thanks to the development of sensorless technology, BLDC motor controllers have better performance and reliability compared to traditional control and sensing methods. This is followed by a discussion of recent developments in the field, their advantages and limitations, and the analysis summarizes real-world implementation challenges and applications. This study comprehensively examines the latest backward electromagnetic field methods for detection, including PWM, terminal current detection, third harmonic voltage integration, terminal voltage detection, and back EMF integration. A brief explanation is also provided of the most significant estimate and model-based methods, such as neural networks with artificial intelligence, full-order and pseudo-reduced-order adaptive observers, the extended Kalman filter, and the adaptive mathematical reference system [27].

The BLDC motor's wide-range speed control system used Dong Hee Lee's suggested Hall Sensor Signal as a basis. For brushless DC motors with different low- and high-speed controllers, this study offers a Hall sensor-based wide-range speed management strategy. Utilizing the Hall sensor data to determine the motor's velocity in the low-speed region is insufficient due to the low sensor resolution and time delay [12]. The recommended technique describes a micro-stepping current management strategy based on the variation in torque angle. The reference speed, not the actual velocity supplied by the hallway sensor, controls the motor's frequency and rotation angle in this mode. A motor speed control method employs distinct high-speed and low-speed controllers for wide-range operation. A brushless DC (BLDC) motor wide-range speed management technique is presented in this study and is based on distinctive low- and high-speed controllers [18]. Due to the low sensor resolution and time delay, the usage of the Hall sensor signal is insufficient to determine the motor's velocity in the low-speed area. The suggested method suggests using a torque angle-based micro-stepping current control methodology. Instead of using the actual velocity provided by the Hall sensor, in this mode the motor frequency and rotation angle are regulated by the speed utilized as a reference. In order to apply the vector control method based on discrete low- and high-speeds controllers, the recorded speed and Hall sensor direction are used to calculate the continuous rotor position. When the mode shifts between the low and standard speed ranges, calculations are done to identify the proper initial current command and reference location of the rotor [19, 20]. The calculated current instruction can be used to reduce the torque that causes a ripple in transient mode. The suggested method is simple but effective in extending the acceleration range of a conventional BLDC motor with a Hall sensor without the use of a high-resolution encoder. On an actual BLDC motor, several trials show the efficiency of the proposed method [28].

Control and Position Estimation of Compact BLDC Motors Using an Analog Linear Hall Effect Sensor Design by Alex Simpkins and Emanuel Todorov: Bio-inspired robotic systems demand engineering components such as compact motors that offer high torque, minimal torque ripple, and precise position control [23]. It is often difficult to build an engine that meets all these standards. Several companies have recently released incredibly well designed small brushless DC motors [25]. However, pulsation-free accurate control of asynchronous motors requires accurate position measurement. Simply installing encoders mounted on the shafts of these motors can be difficult, and a sensorless solution may not be the best for precise positioning control. We introduce a novel method for rotor position determination that measures the magnetic field of rotor magnets using analogue Hall sensors. A computationally straightforward motor control system could be created with this technology's assistance, which offers absolute power cycle monitoring. Position tracking is made possible with a very slight increase in motor size because these Hall effect sensors, which are easily placed on the motor, have a very broad bandwidth, are quite inexpensive, and are very accurate [12–19].

14.3 PROPOSED SYSTEM

We introduce a novel technique for pinpointing the precise location of an electrical circuit by tracking the electromagnetic energy of a rotor magnet using analogue Hall effect sensors. A motor control system that is computationally simple can be effectively created using this method. Directly placed Hall effect sensors on the engine have a high bandwidth, are incredibly small, inexpensive, and precise. They enable unidirectional measurements with only a modest increase in motor size. DC drive systems are frequently utilized in robotics, operations, and manipulators, among other industrial applications. This project aims to use the IoT to monitor and control the ON/OFF status of a brushless direct current (BLDC) motor. Both the continually growing network of tangible items with IP addresses that can connect to the Internet along with the communication between these objects and other Internet systems and devices have been referred to as IoT. Brushless DC motors (BLDCMs) have become commonplace in industry because of their excellent weight-to-torque ratio, trustworthiness, and efficiency. The pace can be raised using this IoT controller until the user gets what they want from it.

14.3.1 Introduction to IoT

The Internet of Things, or IoT, connects computers, motorized and digital devices, objects, animals, or people with distinctive identification numbers (UIDs) and the ability to exchange data across a network without requiring human-to-human-to-computer contact. A "thing" is any natural or artificial device that can transmit data across a network and has the ability to be assigned an IP address. Examples include people with implanted heart rate monitors, pets with biochip transmitters, cars with integrated tire pressure monitoring, and a number of other devices [16–25].

IoT is being used more and more by organizations in a variety of sectors to improve comprehension of clients, streamline operations, and make more informed decisions that raise total corporate value.

14.3.2 How IoT works

The networked smart devices that together make up the IoT ecosystem use integrated components like CPUs, cameras, sensors, and communication gadgets to collect, transmit, and consume data. IoT devices connect to an IoT gateway or an alternative edge device. Typically, the gateway delivers the data to a cloud service for analysis or processes it locally [19].

On occasion, these widgets may interact with other similar entities and respond to the information they provide by acting. Despite the fact that people can interact with electronic devices to change settings, provide directions, or receive information, the majority of work is completed without human intervention [12].

The connection, social networks, and communication habits of these web-enabled devices are greatly influenced by installed IoT applications. Additionally, by using machine learning and artificial intelligence (AI), IoT can facilitate dynamics and accelerate gathering information procedures [9].

14.3.3 Why IoT is Important

People can live more intelligent lives and take complete control of their life by using the Internet for a variety of objectives. In addition to providing smart home automation devices, the IoT is essential for businesses. Thanks to IoT, which provides information on the logistics and supply chain processes as well as device functionality, businesses can rapidly determine whether their systems are operating as intended. Businesses can lower labor

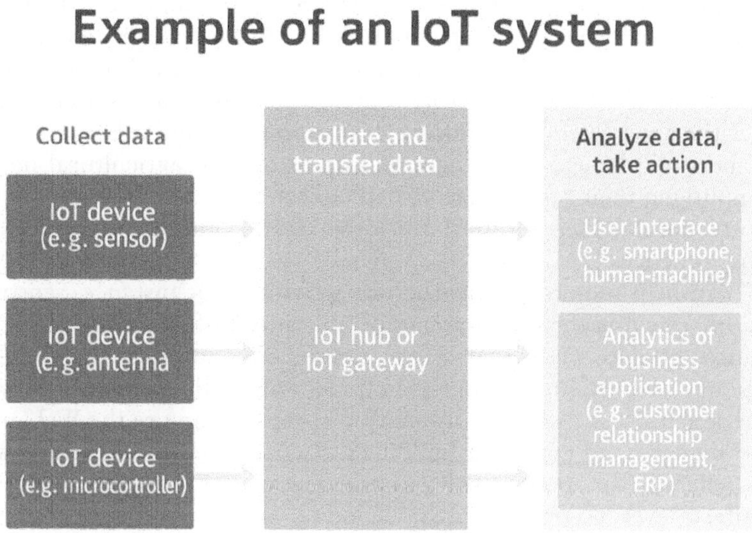

Figure 14.1 Internet of Things.

expenses and automate procedures thanks to IoT. Additionally, it lowers waste, enhances service delivery, lowers manufacturing and marketing costs, and gives insight into client events. Because of this, the Internet of Things is one of the most commonly utilized technologies in contemporary life. As additional businesses realize how connected gadgets may help businesses stay competitive, the IoT is growing increasingly popular [9–15]. An overview of the IoT is shown in Figure 14.1.

14.3.4 IoT Advantages to Organizations

Businesses can benefit tremendously from the IoT. While several benefits are exclusive to certain companies, others are applicable to a wide range of industries. The following benefits are a few examples of how IoT can benefit enterprises [19–24].

- Track your entire business
- Improve customer CX
- Save time and money
- Increase labor productivity
- Incorporate and change the business model
- Better business overviews
- Increase income

Businesses are compelled to reconsider their business practices by the IoT, which also provides consumers with the resources they need to improve their objectives. Nevertheless, technology has also been applied to construction, agriculture, and home automation, which has caused some businesses to go digital. The business sectors that typically use sensors and other IoT devices the most are manufacturing, transportation, and urban areas [23].

Farmers will benefit from the adoption of IoT for agricultural purposes by simplifying their tasks. Sensors can collect information about soil type, moisture, rainfall, temperature and other factors that can help automate farming practices [26] [27].

IoT can help monitor infrastructure activities. For instance, sensors are useful for tracking the development or alterations of structural elements of buildings, bridges, and other infrastructure. The advantages include time and money savings, improvements in quality of life through utilization, and a paperless process. A home automation company can use the Web and the IoT to manage and control a building's electrical and mechanical systems. More generally, smart cities will assist citizens cut back on energy and trash. IoT affects every economic area, including manufacturing, banking and finance, retail, and healthcare [26–28].

14.3.5 Pros and Cons of IoT

Some of the advantages of IoT include:

- Access to information on any device, at any time, and from any location.
- Improved electronic device communication; time and cost savings through transmission of information packets over a network connection.
- Task automation to lower the need for human intervention and raise the quality of services provided by the business.

Some disadvantages of IoT include the following:

- A hacker has a better chance of stealing sensitive data as there are more connected devices and more data being exchanged between them.
- Companies may eventually deal with millions of IoT devices, and gathering and managing data from all of these devices will be challenging.
- Devices from different manufacturers find it challenging to communicate with one another due to the lack of a global IoT interoperability standard; if there is an issue with the system, all linked devices may fail [23,24].

14.3.6 IoT Standards and Frameworks

There are several new IoT standards, including those listed below.

- LiteOS is an operating system (OS) for wireless networks of sensors, similar to Unix. Numerous accessories can be operated by LiteOS, such as Internet of Vehicles (IoV), home automation, mobile smartwatches, and applications for efficient manufacturing. The operating system serves as a framework for the construction of smart devices [24].
- ZigBee is a wireless network type with a low data rate and low power consumption that is primarily utilized in industrial settings. The IEEE (Institute of Electrical and Electronic Engineers) 802.15.4 standard is the foundation of ZigBee. Smart commodities can work together and safely communicate over any internet connection thanks to the Dotdot universal IoT language from the ZigBee Alliance [25].
- M2M is a service layer that facilitates machine-to-machine (M2M) communication between devices through hardware and software implementation. OneM2M, a global standardizing structure of organizations, has been created to offer reusable standards that would enable IoT applications from various sectors of the economy to connect with IoV [26].
- The Object Management Group (OMG) developed the Data Distribution Service (DDS), an Internet of Things standard for effective real-time M2M communication.
- Asynchronous communication across the wire is made possible by the free and open-source Advance Message Queuing Protocol (AMQP). Applications and businesses can be transported securely and interoperably thanks to AMQP. The network protocol is used to manage IoT devices while promoting client-server communication [27].
- The Internet Engineering Task Force created the Constrained Application Protocol (CoAP) outlines the way low-power, constrained computational services can operate in the IoT [28].
- To facilitate huge networks, such as those observed in smart cities with millions of low-power devices, a WAN protocol known as LoRaWAN was created.

IoT frameworks include the following:

- Amazon launched a cloud-based IoT computing platform called Amazon Web Services, which is occasionally referred to as AWS IoT. This design allows intelligent devices to securely connect and communicate with other connected devices in addition to the AWS cloud.

- The Arms MbedIoT platform enables the creation of Internet of Things-related applications for Arms microcontrollers. The Arms campaign Mbed Internet of Things, or IoT, platform aims to provide IoT devices with a flexible, connected, and trustworthy environment by using Mbed technology and services.
- Users can interact with one another, access data collected from linked devices, and perform a range of procedures on the data, including complex evaluations, adjustments, and celebrations, thanks to a group of services known as Microsoft's Azure IoT Suite. Users can choose the actions that best suit their organization's requirements.
- The Google Brillo/Weave platform makes it achievable to create IoT apps quickly. Brillo, an operating system based on Android for low-power applications in embedded systems, and Weave, an Internet of Things-focused communication protocol that serves as a language for communication between devices and the cloud, are the platform's two primary foundations.
- To create and administer shared software programs that facilitate device connectivity, Ericsson created Calvin, a free and open-source IoT platform. Calvin incorporates a prototype environment for developers working on new apps and additionally includes a runtime setup to administrate applications that are currently running.

IoT apps for consumers and businesses:

- The Internet of Customers (IoT) has a number of useful applications, from commercial and consumer IoT through industrial and industrial IoT (IIoT). Energy, social media, and autos are just a few of the organizations that leverage IoT applications [12].
- In the consumer market, smart homes can be controlled by remote workstations and smartphones, where heating, cooling and electronic devices are connected to the network in addition to smart devices [13].
- In order to make people's lives easier, wearable technologies connect with other human-related technology and use sensors and software to gather and analyze user data. In public safety, wearable technology is also being employed to help first responders respond to emergencies more quickly. For instance, on potentially hazardous building sites, they might keep an eye on the health of firemen or construction workers or offer the best routes to a destination [14].
- The Internet of Things has many implications for the healthcare industry, one of which is improving patient monitoring through the data analysis it produces. Hospitals are making heavy use of IoT devices to manage tasks such as their medical supplies and drug inventory [15].

- For example, smart buildings may decrease energy costs by using sensors that count the number individuals present in a room. For instance, the air conditioner can be set to turn on when sensors detect that the conference space is full, or the heat can be turned immediately when all employees have left the building [16].
- Using linked sensors, IoT-based intelligent agricultural systems can continuously track soil moisture, light, temperature, and humidity in areas used for agriculture. Irrigation system automation is another benefit of the IoT [17].
- By using IoT sensors and applications, such as intelligent lighting or meters, a smart city may improve sanitation, manage traffic, save energy, and identify and treat pollution-related problems [18].

IoT security and privacy issues:

- The Internet of Things requires the protection of billions of data points and millions of connected devices. IoT security and privacy have been identified as critical concerns because of the expanded attack surface [12].
- The Mirai botnet in 2016 was one of the most well-known IoT assaults in recent years. It was one of the largest distributed denial-of-service (DDoS) attacks ever documented by registration server provider Dyn, and it briefly disrupted several different websites. Through improperly secured IoT devices, attackers were able to enter the network [19].
- Due to the interconnectedness of IoT devices, a hacker can use a single vulnerability to corrupt and destroy all of the data. If devices are not updated regularly or not at all by manufacturers, it leaves them susceptible to hackers [13].
- Users of linked devices are frequently asked for personal information, including residences, phone numbers, titles, age ranges, and even social network profiles. Hackers may take advantage of this data [18].
- In addition to posing a threat to the Internet of Things, hackers also raise significant privacy issues for users. Companies that produce and sell IoT devices to consumers, for instance, might use these capabilities to advertise and gather user data [8].
- In addition to the potential for personal data compromises, IoT threatens critical infrastructure, particularly government, financial institutions, and infrastructure [6].

History of IoT:

- In a 1999 presentation to Procter and Gamble (PandG), Kevin Ashton, co-founder of the Auto-ID Centre at the Massachusetts Institute of Technology (MIT), first discussed the Internet of Things. Ashton titled the presentation "Internet of Things" in a reference to the popular new trend of 1999, the Internet, in addition to attracting the attention of RFID PandG executives. "When Objects Start to Think" was written by Neil Gershenfeld, a professor at MIT, in 1999. even though it did not use the exact same terminology, it clearly illustrated the development of the Internet of Things [1–5].

- Wireless, MEMS, microservices, and Internet technologies are all combined to form the Internet of Things (IoT). Unstructured machine-generated data can now be examined to find insights that propel change by crossing the silos of information technology (IT) and operational technology (OT). Although the concept of networked gadgets was initially introduced by Ashton, it has existed since the 1970s and is referred to as pervasive computing and embedding the internet. The first internet gadget was a Coke dispenser at Carnegie Mellon University in the early 1980s. When they decided to stop by, programmers could check the gadget's status online and see whether a refreshing drink was ready [8,9].
- The Internet of Things began with M2M conversations, in which machines connected to another machine through a network without human help. M2M refers to the collection and management of data from a cloud-connected device. A network of sensors comprising billions of intelligent gadgets, the IoT links individuals, systems, and other programs to collect and exchange data. It increases the bar for M2M. In essence, M2M offers the connectivity required for the Internet of Things [10,11].
- The Internet of Things' logical next step is the SCADA (supervisory control and data collecting) software category. SCADA collects real-time data on controlling remote devices and circumstances. SCADA systems comprise both software and hardware components. The physical component collects and sends data to the computer, where it will be processed by the software and shown in contemporaneous fashion. As SCADA technology advanced, first-generation IoT technologies and late-generation SCADA systems emerged [15].

14.3.7 Application

Electrical, electronic, and robotics are essentially the main applications of the proposed system [7].

- Drone
- Social Robot
- Industrial Robotics
- Operate the connected device all over the world

14.3.8 Module Explanation

This chapter offers web-based speed control and brushless direct current (BLDC) motor monitoring for solar-powered BLDC motors. Because it dominates its counterpart motors, brushless DC motors are a new trend in industrial applications. A part of the Controller board, an element of the Inverter board, the MPPT algorithm, a MOSFET driver circuit, and a communication module make up the hardware setup. With a cheap atmel 328p microcontroller, the BLDC motor speed may be managed [12–19].

14.4 MODULE NAMES

- Potentiometer
- Electronic speed controller
- Lithium polymer battery
- LED display
- Arduino Uno
- Node MCU (ESP8266)
- BLDC motoe

14.5 MODULE DESCRIPTION

14.5.1 Arduino Uno

Building blocks for interactive and digital things that can perceive and operate actual machines have been created by an open source project called Micro Controller. Microcontroller board layouts made by various manufacturers employing various microcontrollers serve as the foundation for the design architecture. Both digital and analog input/output (I/O) connections are capable of being used to interface various boards, boards for expansion (also known as shields), along with additional circuits with the system. Some of the cards allow USB connectivity, and others have serial interfaces for running computer software. The microcontroller project supports C and C for microcontroller programming and offers an integrated development framework (IDE) mainly for the CPU programming language [21–26].

The ATmega328P microprocessor serves as the foundation for the Uno microcontroller board (shown in Figure 14.2). Six analogue inputs, 14 digital I/O pins, comprising six for PWM outputs, a connector for USB 2.0, a power socket, an ICSP header, a reset button, and a 16 MHz crystal made of quartz are some of its features. All you need to do is plug it into your computer using a USB wire or power it with an AC-DC converter because it already has everything you need to operate the microcontroller. Numerous features on the Uno microcontroller allow it to interact with

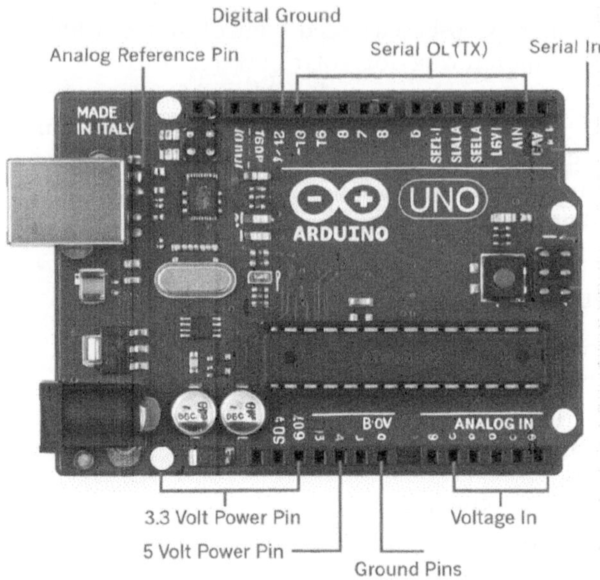

Figure 14.2 Arduino Uno.

other microcontrollers, another microprocessor board, or the device itself [23–28].

14.6 FEATURES

- Input voltage: 7–12 V
- 32 kB of flash memory
- EEPROM: 1 kB
- Microcontroller: ATmega328P
- Operating voltage: 5 V

14.7 APPLICATIONS

- Real-time biometrics
- Robotic applications
- Academic applications

The Micro Chip Controller Uno is a publicly downloadable microcontroller board made by Micro Controller and is based on the Microchip ATmega328P microprocessor. The board's alternative digital as well as analog I/O connections allow it to be attached to a variety of motherboards for expansion (shields) and other circuits [23]. PWM may be produced using six of the board's 14 digital and six analogue I/O pins. It may be programmed using a USB Type B connector and a micro and medium driver IDE (Integrated Development Environment). It can be used with a USB cable or an external 9-V battery, and its voltage range is 7–20 V. The Leonardo, Nano, and Micro consoles are similar. The actual electronic reference design is accessible without charge on the microcontroller's website under a Creative Commons Attributes Share Alike 2.5 license. Additional layouts and build files are available for different hardware versions [27].

The first microcontroller release of software was commonly referred to as "uno" (meaning "one") in the Italian language. The Uno board was the first in a line of USB-based microcontroller chips when the Micro IDE controller had initially released. Later, the Uno board and the current Micro-controllers IDE version 1.0 were upgraded. New code can be added without the use of an additional hardware reprogramming tool thanks to the bootloader on the embedded ATmega328 [27,28]. The Uno differs from the other devices in that it does not use the FTDI-USB serial controller chip. Rather, it continues to use the old STK500 protocol for connectivity. Rather, a USB-to-serial converter is used when working with the Atmega16U2 (or the Atmega8U2 version up to R2).

14.8 BACKGROUND

The Ivrea Institute (IDII), an interactive design school in Ivrea, Italy, is where the microcontroller project got its start. A microcontroller, commonly referred to as BASIC, was expensive for a lot of students at the time. In 2003, Hernando Barragán developed the electrical wiring development platform as part of his IDII master's degree with Massimo Banzi and Casey Reas, who have become recognized for their work together on process programming languages. The project's goal was to provide low-cost, user-friendly tools that people without engineering backgrounds could use to create digital creations. The Wires platform included a processor-based integrated development environment (IDE), an integrated circuit board (PCB) for an ATmega168 microcontroller, and functionality of libraries for rapid microcontroller coding. In 2003, David Cuartielles, Massimo Banzi, and David Mellis, another IDII student, modified the wiring to include support for the less expensive ATmega8 CPU. They divided the project into discrete components and dubbed it a microcontroller rather than carrying on with the wiring. The ATmega168 and the FTDI-USB serial controller chip were utilized in early microcontroller boards. The combination of the ATmega328P CPU and ATmega16U2 (up to the Atmega8U2 R2 version) in a USB serial converter set the Uno motherboard apart from all other boards [20–27].

14.8.1 Applications

As of 2013, many projects and independent vehicles that require a simple and inexpensive microcontroller generally use the ATmega328 [21]. This chip is most commonly used in the Micro, Medium, Controller Uno and Micro Controller Nano versions of the widely recognized Micro controller development architecture. The rated containment failure rate is significantly less than 1 PPM for 20 years at 85°C or 100 years at 25°C, according to reliability criteria.

The ATmega328 is the dominant microcontroller used in various projects, including autonomous vehicles, since 2013 [22].

14.8.2 Serial Programming

Table 14.1, shows the used signal, pin name, and I/O function for the serial programming. For data entering and exiting the MCU, serial data has been scheduled on the descending edge and the elevation edge, respectively [27]. VCC receives power while SCK and RESET are both set to 0. After a delay of at least 20 milliseconds, the software should initiate the transmission of the serial instructions 0xAC, 0x53, 0x00, and 0x00 through the MOSI pin. The second byte (0x53) will be reflected back by the MCU.

Serial data transmission is configured such that data leaving the MCU is aligned with the rising edge, while data entering the MCU is aligned with the falling edge. Both SCK and RESET are held low during VCC power-up. After a delay of at least 20 milliseconds, the software initiates the transmission of the serial instruction sequence: 0xAC, 0x53, 0x00, and 0x00 to the MOSI pin. The MCU is expected to echo back the second byte (0x53) in response.

14.8.3 Programming in Micro Controller

Programming in microcontrollers requires an appropriate Integrated Development Environment (IDE). To program the Electronic Controller Uno offline, use the Microcontroller Desktop IDE provided by the semiconductor manufacturer. This IDE is specifically designed to support code

Table 14.1 Serial Programming

Programming signal	*Pin Name*	*I/O*	*Function*
Symbol	**Pins**	**I/O**	**Description**
MOSI	PB3	I	Serial data in
MISO	PB4	O	Serial Data out
SCK	PB5	I	Serial Clock

development, compilation, and uploading for various Uno boards. Before starting, ensure that the Microcontroller IDE is properly installed on your system, as detailed on the official Getting Started page. The Uno board is connected to the PC using an A-B USB cable (commonly referred to as a USB printer cable), which serves as the primary interface for code uploading and serial communication between the microcontroller and the host computer [19,20].

14.9 NODE MCU (ESP8266)

Developed on Lua, NodeMCU is an open-source firmware and development platform designed exclusively for IoT applications. Its hardware is based on the ESP-12 module, and its software is driven by the ESP8266 Wi-Fi SoC from Espressif Systems [22].

A brushless Washington, DC electric motor (BLDC motor or BL motor) (depicted in Figure 14.3), also known as an electronically commutated motor (ECM or EC motor), is a synchronous motor that generates electricity in the form of alternating current (AC) to drive each phase of the drive unit through a closed loop controller. It can be powered by direct current (DC) electricity via an inverter or switching power supply. By applying short electrical charge bursts to the motor windings, the controller can adjust the torque and speed of the motor [23–26].

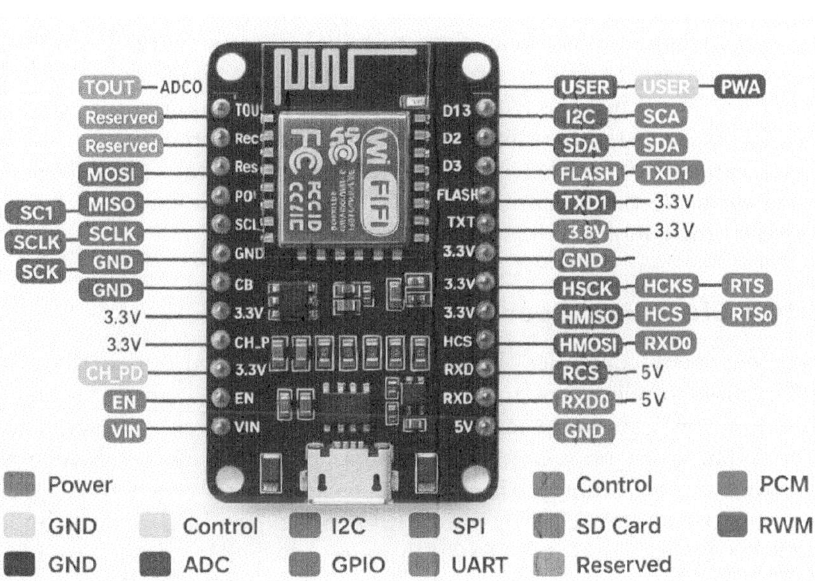

Figure 14.3 BLDC motor.

14.9.1 Brushless vs. brushed motors

Developed in the 19th century, brushed DC motors are widely utilized. In the 1960s, developments in solid state electronics made brushless DC motors feasible. Torque is produced by switching the polarity of the revolving magnets connected to the electric motor's moving rotor and the permanent magnets in the stator, which houses the rotor. One or both sets of electromagnets are encircled by a wire coil and an iron core [20–28].

14.9.2 Variations in Construction

Brushless motors can be made in a number of methods, such as: In the "conventional" (also known as inrunner) arrangement, permanent magnets constitute a crucial component of the rotor. The rotor is surrounded by three stator windings. The stator coils are the central component of the motor in the outrunner (or external-rotor) design. The outrunner additionally reverses the radial relationship between the filaments and magnets by spinning the permanent magnetic fields inside an overhanging rotor that surrounds the core. The stator and rotor plates used in the conventional or rectangular axial flux type are positioned face to face when there are limitations on space or shape [15–19].

14.10 POTENTIOMETER

A potentiometer, which is a three-terminal resistor with a sliding or revolving contact, is used to provide an adjustable voltage divider (shown in Figure 14.4). If just the wiper and one connector end are utilized, it acts as a rheostat or variable resistor. Potentiometers are rarely used to directly regulate large amounts of electricity (greater than one watt) since the power dissipated inside the potentiometer would be proportionate to the energy utilized by the controlled load [24–28].

14.10.1 Electronic Speed Controller

Electronic speed controller (ESC) refers to an electrical circuit that alters the direction, acceleration, and braking characteristics of an electric motor. These are now commonplace parts of electrically driven radio-controlled models. The most popular modification is brushless motors, which give the motor a low voltage, three-phase, technologically generated energy source in the end. While toy-grade R/C cars typically include a built-in Electronic Speed Controller (ESC), in more advanced systems, the ESC is often a separate module connected to the throttle channel of the receiver. Some R/C manufacturers design integrated electronic systems – especially

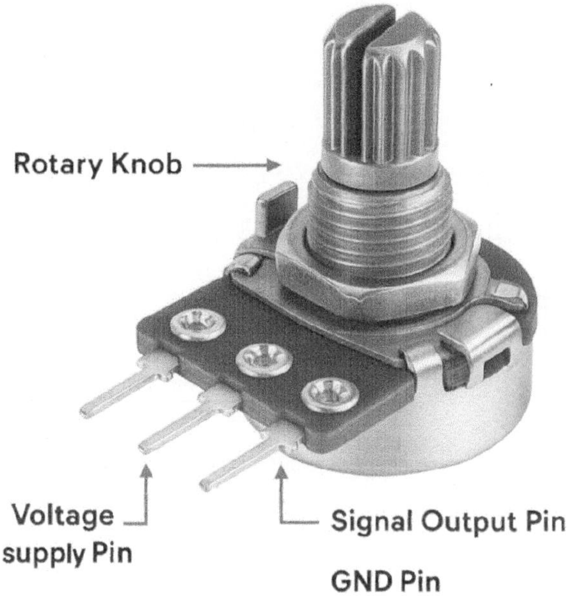

Figure 14.4 Potentiometer. (The Basics of Quadcopter Anatomy.)

for entry-level trucks, boats, or aircraft – that consolidate multiple control functions onto a single circuit board, limiting compatibility to only their own proprietary hobby-grade components (shown in Figure 14.5) [18–25].

14.10.2 Components Used in ESC

The following are the primary components used in ESC:

- Servo signal or input of the PWM signal.
- Positive (+) LIPO connections.
- Negative (-) LIPO connections.
- GND reference of the PWM signal.
- Solder jumper, for adjusting the direction of rotation (CW/CCW).
- Solder jumper, for adjusting the type of the PWM input signal State LED.

14.11 EXPERIMENTAL RESULT

The experimental circuit of the work is shown in Figure 14.6. It contains all the necessary connections to achieve the desired result/output.

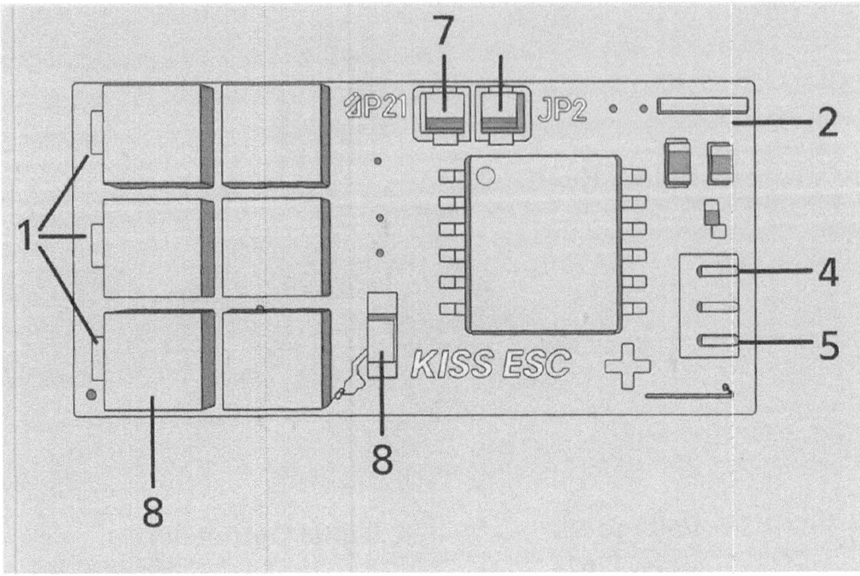

Figure 14.5 ESC.

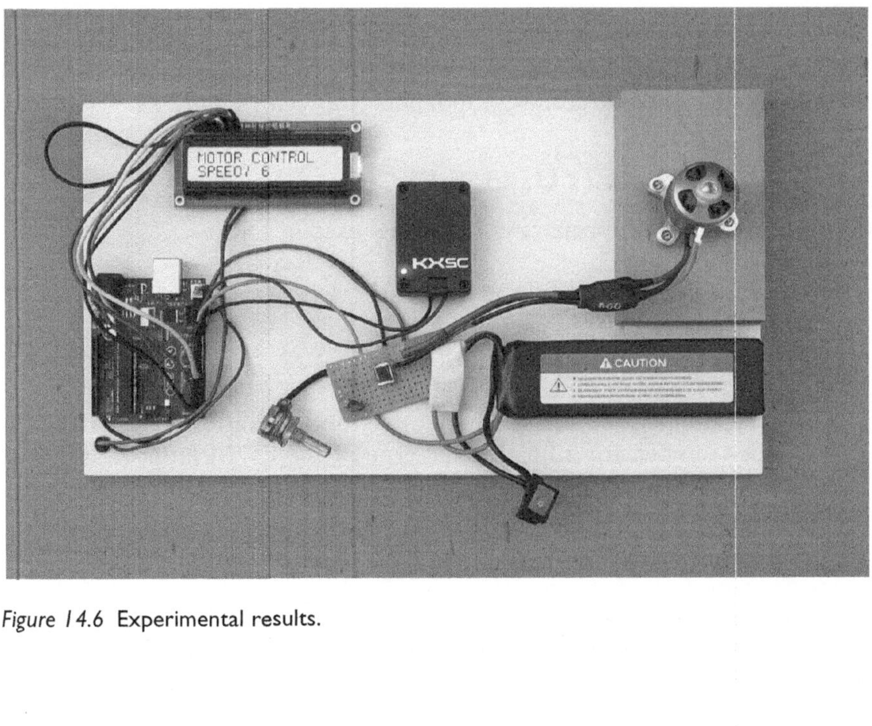

Figure 14.6 Experimental results.

14.12 CONCLUSION

Using PWM technology, the motor speed can be changed to run at the desired or necessary speed. Monitoring is done through the IoT. The Internet of Things, or IoT, is a crucial part of this system. The collaboration process is facilitated by this communication tool. Companies gain from this and have a promising future.

REFERENCES

1. K. Sreeram, "Design of Fuzzy Logic Controller for Speed Control of Sensorless BLDC Motor Drive", *IEE International Conference On Control, Power, Communication and computing Technologies*, pp. 18–24, 2018.
2. K.S.K. Veni, N.S. Kumar, and J. Gnanavadivel, "Low Cost Fuzzy Logic Based Speed Control of BLDC Motor", *IEEE International Conference On Advances In Electrical Technology For Green Energy*, pp. 1–5, 2017.
3. K. Sowmiya, and R. VishnuPriya, "Microcontroller based speed control scheme of BLDC motor using proteus", *International Journal*, vol. 5, no. III, 2017.
4. R.L. Valle, P.M. de Almeida, A.A. Ferreira, and P.G. Barbosa, "Unipolar PWM predictive current-mode control of variable-speed low inductance BLDC motor drive", *IET Electric Power Applications*, vol. 11, no. 5, pp. 688696, 2017.
5. S. Chen, G. Liu, and L. Zhu, "Sensorless control strategy of a 315K High – Speed BLDC motor based on a speed independent flux linkage function", *IEEE Transactions on Industrial Electronics*, vol. 64, no. 11, pp. 8607–8617, 2017.
6. W. Li, J. Fang, H. Li, and J. Tang, "Position sensorless control without phase shifter for high Speed BLDC motors with low inductance and nonideal back Emf", *IEEE Transactions on Power Electronics*, vol. 31, no. 2, pp. 1354–1366, 2016.
7. R. Arivalahan, S. Venkatesh, and T. Vinoth, "An effective speed regulation of brushless DC motor using hybrid approach", *Advances in Engineering Software*, vol. 174, p. 103321, 2022, ISSN 0965-9978.
8. J. Sun, Y. Chai, C. Su, Z. Zhu, and X. Luo, "BLDC motor speed control system fault diagnosis based on LRGF neural network and adaptive lifting scheme", *Applied Soft Computing*, vol. 14, Part C, pp. 609–622, 2014, ISSN 1568-4946.
9. X. Huang et al. "Design of a 5-phase brushless DC motor for a safety critical aerospace application", *IEEE Transactions on Industrial Electronics*, vol. 59, pp. 3532–3541, 2011.
10. X.Y. Zhang et al. "Development review of drive motor applied in electric vehicle", *Advanced Materials Research*, vol. 433, pp. 7333–7337, 2012.
11. F. Aghili, "Fault-tolerant torque control of BLDC motors", *IEEE Transactions on Power Electronics*, vol. 26, no. 2, pp. 355–363, 2011. IEEE.
12. T.S. Tamir, G. Xiong, Z. Shen, X. Gong, S. Liu, E. Lodhi, L. Wan, X. Dong, "Comparative study of four speed controllers of brushless DC motors for

industrial applications", *IFAC-Papers OnLine*, vol. 53, no. 5, pp. 59–64, 2020, ISSN 2405-8963.
13. M.A.H. Azman, J.M. Aris, Z. Hussain, A.A.A. Samat and A.M. Nazelan, "A comparative study of fuzzy logic controller and artificial neural network in speed control of separately excited DC motor", *2017 7th IEEE International Conference on Control System, Computing and Engineering (ICCSCE)*, pp. 336–341, 24–26 Nov. 2017.
14. H. Chaudhary, S. Khatoon and R. Singh, "ANFIS based speed control of DC motor", *2016 Second International Innovative Applications of Computational Intelligence on Power, Energy and Controls with their Impact on Humanity (CIPECH)*, pp. 63–67, 18–19 Nov. 2016.
15. K. Cheon, J. Kim, M. Hamadache, and D. Lee, "On replacing PID controller with deep learning controller for DC motor system", *Journal of Automation and Control Engineering*, vol. 3, no. 6, pp. 452–456, December 2015.
16. R. Kushwah, and S. Wadhwani, "Speed control of separately excited Dc motor using fuzzy logic controller", *International Journal of Engineering Trends and Technology (IJETT)*, vol. 4, no. 6, pp. 2518–2523, 2013.
17. G. Liu, C. Cui, K. Wang, B. Han, and S. Zheng, "Sensorless control for high-speed brushless DC motor based on the line-to-line back EMF", *IEEE Transactions on Power Electronics*, vol. 31, no. 7, pp. 4669–4683, 2016.
18. T.S. Tamir, G. Xiong, H.M. Menkir, X. Shang, Z. Shen, X. Dong, and X. Gong, "Developing SCADA systems to monitor and control liquid and detergent factories", *2020 IEEE 16th International Conference on Automation Science and Engineering (CASE)*, pp. 691–696, 20–21 Aug. 2020.
19. T.S. Tamir, G. Xiong, Y. Tian, and G. Xiong, "Passivity based control of stewart platform for trajectory tracking", *2019 14th IEEE Conference on Industrial Electronics and Applications (ICIEA)*, pp. 988–993, 19–21 June 2019.
20. S.N. Thanh, K.N. Thanh, C.N. The, P.P. Hung, and H.H. Xuan, "Development of fuzzy logic controller for DC motor using personal computer and inexpensive microcontroller", *2014 13th International Conference on Control Automation Robotics & Vision (ICARCV)*, pp. 1310–1314, 10–12 Dec. 2014.
21. G. Agrawal, H. Mohan and M. Pathak, "Improved speed sensorless control of induction motor drive using artificial neural network", *2nd International Conference on Power Electronics &IoT Applications in Renewable Energy and its Control (PARC), 2022*, pp. 1–6, 2022.
22. S. Potturi, and R.P. Mandi, "Critical survey on IoT based monitoring and control of inductionmotor", *IEEE Student Conference on Research and Development (SCOReD)*, Selangor, Malaysia, pp. 1–6, 2018.
23. W. Tsai, Y. Shih, and T. Tsai, "IoT-type electric fan: remote-controlled by smart-phone", *Third International Conference on Computing Measurement Control and Sensor Network (CMCSN)*, Matsue, pp. 12–15, 2016.
24. C.N.S. Kumar, U. Sakthivelu, R. Naresh, and S.S. Kumar, "Secured Smart Meal Delivery System for Women's Safety", In S. Ponnusamy, V. Bora, P. Daigavane, & S. Wazalwar (Eds.), *AI Tools and Applications for Women's Safety* (pp. 275–290). IGI Global Scientific Publishing. https://doi.org/10.4018/979-8-3693-1435-7.ch017

25. I. Kamble, and Y.M. Patil, "A review of parameters monitoring and controlling system for industrial motors using wireless communication", *International Journal for Research in Applied Science and Engineering Technology*, 2019.
26. L.D. Xu, W. He, and S. Li, "Internet of things in industries: A survey," *IEEE Transactions on Industrial Informatics*, vol. 10, no. 4, pp. 2233–2243, Nov. 2014.
27. D. Miorandi, S. Sicari, F. De Pellegrini, and I. Chlamtac, "Internet of things: Vision, applications and research challenges," *Ad Hoc Networks*, vol. 10, no. 7, pp. 1497–1516, 2012, ISSN 1570-8705.
28. A. Al-Fuqaha, M. Guizani, M. Mohammadi, M. Aledhari and M. Ayyash, "Internet of Things: A survey on enabling technologies, protocols, and applications", *IEEE Communications Surveys & Tutorials*, vol. 17, no. 4, pp. 2347–2376, Fourthquarter 2015.
29. G. Agrawal, H. Mohan and M. Pathak, "Improved speed sensorless control of induction motor drive using artificial neural network", *2nd International Conference on Power Electronics & IoT Applications in Renewable Energy and its Control (PARC), 2022*, pp. 1–6, 2022.

Chapter 15

A Recapitulation of Advancements in Water Treatment Using Cutting-edge Renewable Energy Technologies for a Sustainable Future

Kirti B. Zare and Manisha Tanwer

15.1 INTRODUCTION

As the global population continues to grow and industrial activities expand, the demand for clean water and sustainable energy has reached unprecedented levels. Traditional water treatment methods, while effective, often rely heavily on fossil fuels, contributing to environmental degradation and climate change. In response, there is a growing interest in leveraging renewable energy sources to power water treatment processes, aligning with the broader goals of sustainability and environmental conservation. Renewable energy technologies such as solar, wind, and bioenergy offer promising alternatives that can reduce the carbon footprint, lower operational costs, and enhance the resilience of water treatment systems.

This review aims to explore recent advancements in the integration of these cutting-edge renewable energy technologies in water treatment, assessing their potential to revolutionize the field and address the pressing challenges of resource scarcity and environmental impact. This paper reviews advancements in renewable energy-powered water treatment, highlighting current technologies, practical applications, and emerging innovations. It identifies key challenges and opportunities, providing a roadmap for future research and sustainable development in the field.

Figure 15.1 showcases a water treatment facility powered by a diverse mix of renewable energy sources like solar, wind, and hydropower. This design offers significant environmental benefits by reducing dependence on fossil fuels, promoting sustainable energy use, minimizing pollution from greenhouse gases and potential spills, and enabling efficient operation through energy storage.

This paper discusses various renewable energy sources for desalination, highlighting solar and wind energy as key players. It concludes that renewable energy can significantly reduce the operational costs and environmental impact of desalination processes [1]. This study explores the utilization of solar energy for water purification, with a focus on solar distillation and

 DOI: 10.1201/9781003533535-15

Figure 15.1 Renewable sources for different water treatment.

photocatalytic processes. It identifies the main challenges as technological and economic barriers, suggesting that further research is needed to enhance efficiency [2]. This review explores the potential of wind energy in driving water treatment processes, particularly in remote areas. It concluded that wind energy is a feasible option, but its intermittent nature requires effective storage solutions [3].

This chapter discusses the use of biogas generated from wastewater treatment plants as an energy source. It finds that bioenergy can enhance the sustainability of wastewater treatment by reducing energy costs and greenhouse gas emissions [4]. This chapter reviews hybrid systems combining solar, wind, and other renewable sources. It concluded that hybrid systems can offer greater reliability and efficiency in desalination compared to single-source systems.

Figure 15.2 depicts a solar-powered membrane distillation (MD) for different treatment. Inside the unit, wastewater is heated, causing water vapor to separate from the brine solution through a special membrane. The vapor condenses on the cooler side as clean water, which is collected in a tank. This sustainable system utilizes renewable energy to produce clean water from various sources like wastewater or seawater, but might have limitations in production rate and require maintenance for the membrane [5].

The several research work evaluated the economic aspects of implementing renewable energy in water treatment. It finds that while initial investments

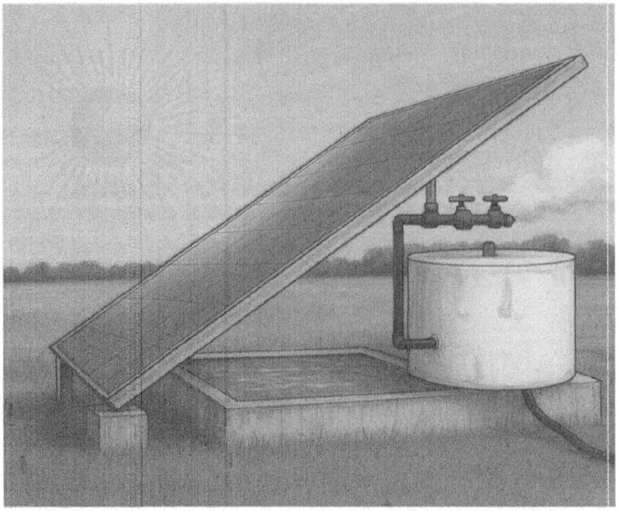

Figure 15.2 Solar-powered membrane distillation (MD).

are high, long-term benefits include cost savings and energy independence [6]. The study links renewable energy in water treatment to the UN SDGs, highlighting its role in providing clean water access and supporting climate action through sustainable practices.

This review analyzed the environmental benefits, such as reduced carbon emissions and pollution, of using renewable energy in water treatment. It concluded that renewable energy integration can significantly lower the ecological footprint of water treatment processes. The paper discusses the policy and regulatory challenges hindering the adoption of renewable energy in water treatment. It suggests that supportive policies and incentives are crucial for wider adoption. This study presents various case studies from around the world where renewable energy has been successfully integrated into water treatment systems. It highlights both the successes and the lessons learned. The review examines various energy storage technologies, including batteries and thermal storage systems. These solutions are highlighted as critical components for ensuring reliable and consistent energy supply in renewable-powered systems [7].

It concluded that efficient storage is key to the reliability of renewable-powered water treatment systems [8]. This study examined how the IoT and smart technologies can optimize the efficiency of renewable energy water treatment systems. It concludes that smart monitoring and control can significantly improve performance and reduce costs [9]. The paper discussed emerging trends and future research directions, such as

Figure 15.3 Hybrid renewable energy system integrating wind turbine and biogas plant.

advanced materials and new energy sources. It suggests that ongoing innovation is crucial for overcoming current limitations and achieving broader adoption [10].

Figure 15.3 illustrates a hybrid renewable energy system integrating wind power and biogas production. The wind turbine harnesses wind energy to generate electricity. Adjacent to it, a biogas plant processes organic leftover to harvest biogas, which can be used secondhand as fuel or converted into electricity. The blue dome is typically indicative of an anaerobic digester where the biogas is produced. The surrounding green landscape suggests a rural setting, highlighting the system's suitability for decentralized and sustainable energy production in agricultural or remote areas.

The integration of renewable energy technologies to water treatment presents a promising solution to the dual challenges of energy sustainability and clean water access. The reviewed literature highlights significant advancements in solar, wind, and bioenergy applications, each offering unique benefits and facing distinct challenges. This paper discussed various renewable energy sources for desalination, highlighting solar and wind energy as key players. It concluded that renewable energy can significantly reduce the operational costs and environmental impact of desalination processes [11]. The study examines the usage of planetary vigor for water purification, focusing on solar distillation and photocatalysis. It identifies the main challenges as technological and economic barriers, suggesting that further research is needed to enhance efficiency [12]. This review explored the potential of wind energy in driving water treatment processes, particularly in remote areas. It concluded that wind energy is a feasible option, but its intermittent nature requires effective storage solutions [13].

The paper explored public attitudes towards using renewable energy in water treatment. It finds that increasing public awareness and education can enhance acceptance and support for these technologies [14]. Recently there have been proposals to power electrochemical courses using green energy sources, such as solar photovoltaics and wind energy [15].

Ref [16] provided an overview of TESs systems and fresh progressions in various TES technologies, both commercial and experimental. The discussion includes the limitations of each technology, innovative methods to enhance temperature transfer efficiency, primary applications, and the conservational anxieties related to integrating TES in solar thermal CSP systems. Regarding the environment, a case study on Moroccan CSP evaluated the potential of incorporating TES in CSP plants to significantly reduce greenhouse gas emissions during electricity production.

15.2 METHODS OF SOLAR DESALINATION

Ullah and Rasul [17] studied thermal desalination processes, like multi-effect desalination (MED), humidification–dehumidification (HDH), and multi-stage flash desalination (MSF), that use heat to vaporize saline water, separating salt from water through condensation. These methods benefit from low-cost energy, flexibility, and simple designs, and can be power-driven by total available energy on or to Earth. Membrane concentration drives water through a membrane, leaving salt behind. While it enables large-scale production, reducing energy consumption and increasing the use of renewable energy remain key challenges. Other methods, such as ion exchange, gas hydrate, and liquid-to-liquid extraction, differ significantly; ion exchange is costly and only suitable for low-saline water due to the expense of the required chemicals.

Solar desalination harnesses solar energy to produce freshwater from salt-water or brackish water, offering a sustainable key to address water scarcity in arid and remote regions. This review examines various methods of solar desalination, focusing on advancements and challenges identified in recent research. Solar stills are one of the simplest solar desalination technologies, utilizing solar energy to evaporate water and collect fresh-water vapor. Recent studies [18] have explored enhancements in design and material choices to improve efficiency and productivity. Challenges include low output rates and sensitivity to environmental conditions [19]. Solar-powered reverse osmosis (RO) systems integrate photovoltaic (PV) panels to power conventional RO desalination processes. Research has highlighted advancements in PV efficiency and system integration [20], making these systems increasingly viable for off-grid applications.

However, high energy demands and cost remain significant barriers. Solar updraft desalination approaches, such as multi-effect distillation (MED) and multi-stage flash (MSF), utilize solar heat to vaporize and condense water, achieving high freshwater yields. Recent studies [21] emphasize the improvements in heat transfer mechanisms and system optimization to enhance efficiency and reduce operating costs. Challenges include complex system designs and high initial investment [22]. Hybrid systems combine solar energy with other renewable sources or energy recovery technologies to enhance performance and reliability. Recent research [23] explored hybrid configurations integrating solar PV with wind or biomass energy, demonstrating improved energy efficiency and cost-effectiveness. Integration challenges include system complexity and compatibility of different energy sources [24].

Advancements in solar desalination technologies show promising strides towards sustainable water production. Solar stills remain an accessible option for small-scale applications, while solar-powered RO and thermal desalination systems offer scalable solutions for larger projects. Future research should focus on optimizing system performance, reducing costs, and addressing environmental impacts to broaden the adoption of solar desalination technologies globally.

Solar energy is increasingly being utilized for desalination processes to provide freshwater from saline sources. Solar stills use solar energy to evaporate water, leaving salts and impurities behind. According to a recent study [25] progress in materials and design has significantly improved the efficiency of solar stills, making them a viable solution for water scarcity in arid regions. Solar-powered RO systems use photovoltaic (PV) panels to power the RO process, which forces saline water through a semi-permeable membrane, leaving salts behind. Research by [26] highlighted the increased efficiency of solar PV RO systems, driven by improvements in PV technology and energy storage solutions.

Figure 15.4 shows that solar thermal desalination offers a sustainable solution for freshwater production by harnessing the sun's power. The process involves collecting sunlight with solar collectors, which then heats a transfer fluid. This heated fluid then boils seawater in an evaporator, creating freshwater vapor at a lower temperature due to a vacuum created by a pump. The vapor condenses into clean water, while the remaining concentrated brine is discharged. The system is controlled electronically and may include pretreatment for seawater and heat storage for maintaining operation during low sunlight periods. Furthermore, solar thermal desalination plants can be deployed in remote areas lacking access to a power grid, making them a versatile and sustainable solution for freshwater production.

Ghaffour et al. [27] studied desalination technology, and stated that RO stands out as the most efficient method, accounting for over 50% of the world's current desalination capacity.

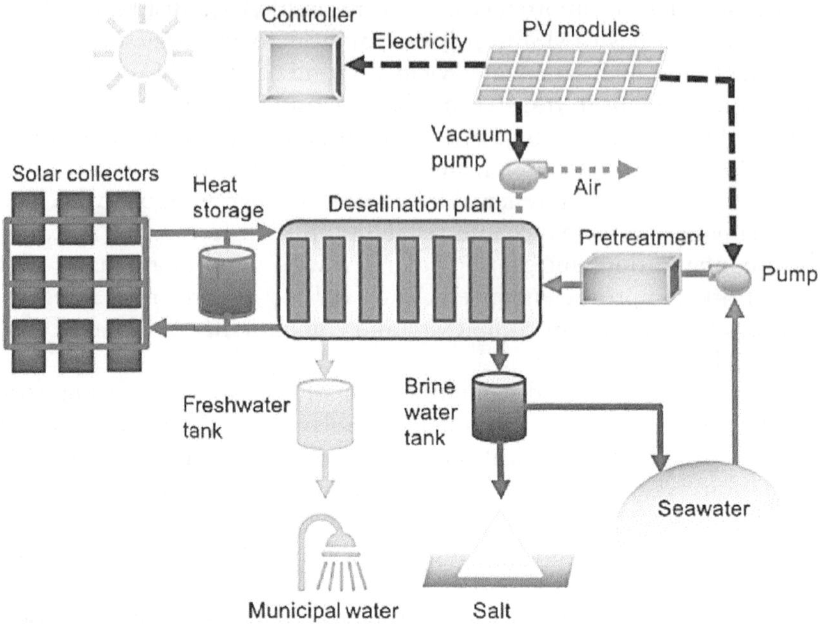

Figure 15.4 Solar thermal integrated desalination plant which shows the economic considerations [26].

Elimelech et al. [28] concluded that, in desalination technology, RO is recognized as the most efficient process, representing more than half of the global desalination capacity. Technological advancements and effective pretreatment strategies are essential for MD to become competitive with RO. Figure 15.5 shows that integrating solar collectors with MD desalination offer a sustainable solution for freshwater production. This system utilizes captured sunlight to power the desalination process, resulting in several advantages: improved productivity and efficiency of water generation, zero carbon emissions for a clean environmental footprint, and the potential for effective off-grid operation in remote areas. Additionally, some MD systems can be constructed with lower cost, corrosion-resistant materials, making them a potentially cost-competitive desalination technology.

Figure 15.6 depicts a solar-powered membrane distillation desalination system utilizing sunlight captured by solar collectors to heat seawater. The heated seawater transfers heat to a freshwater stream in a heat exchanger, separating clean permeate water (freshwater product) from the concentrated brine solution (wastewater). This design boasts advantages like increased freshwater production, efficient solar energy use, zero carbon emissions, and potential for low-cost, corrosion-resistant operation in remote areas.

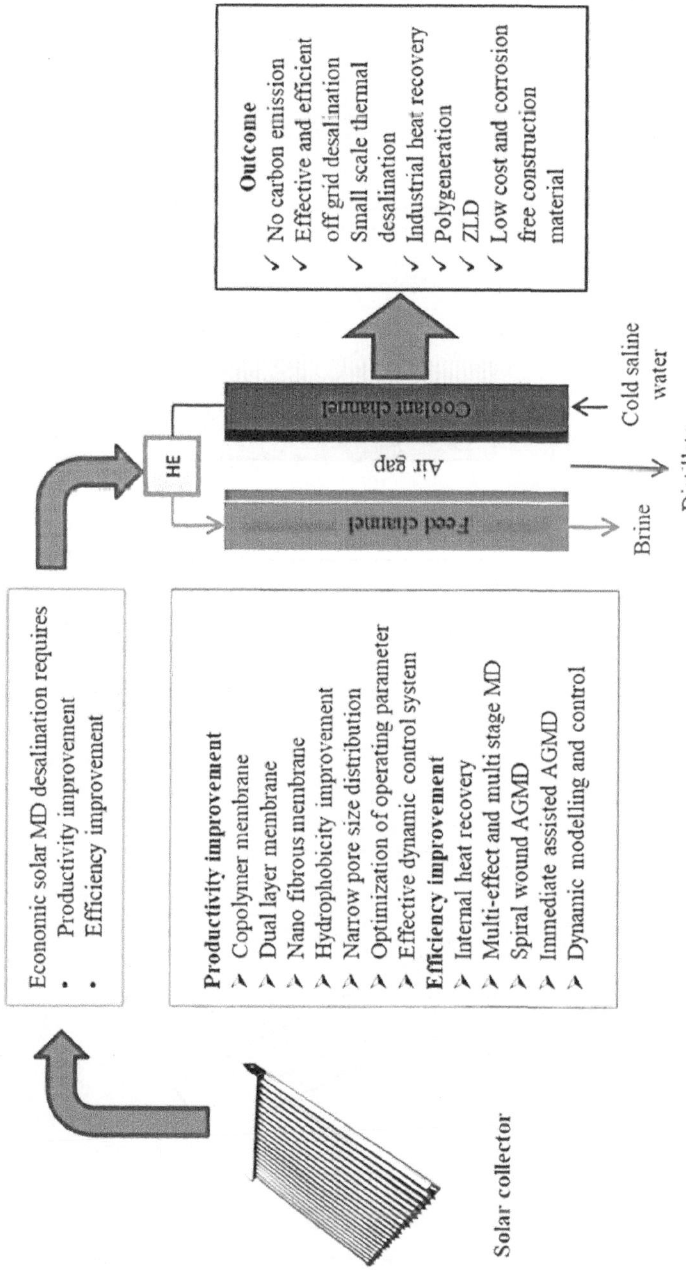

Figure 15.5 Membrane distillation for wastewater treatment by using renewable energy [29].

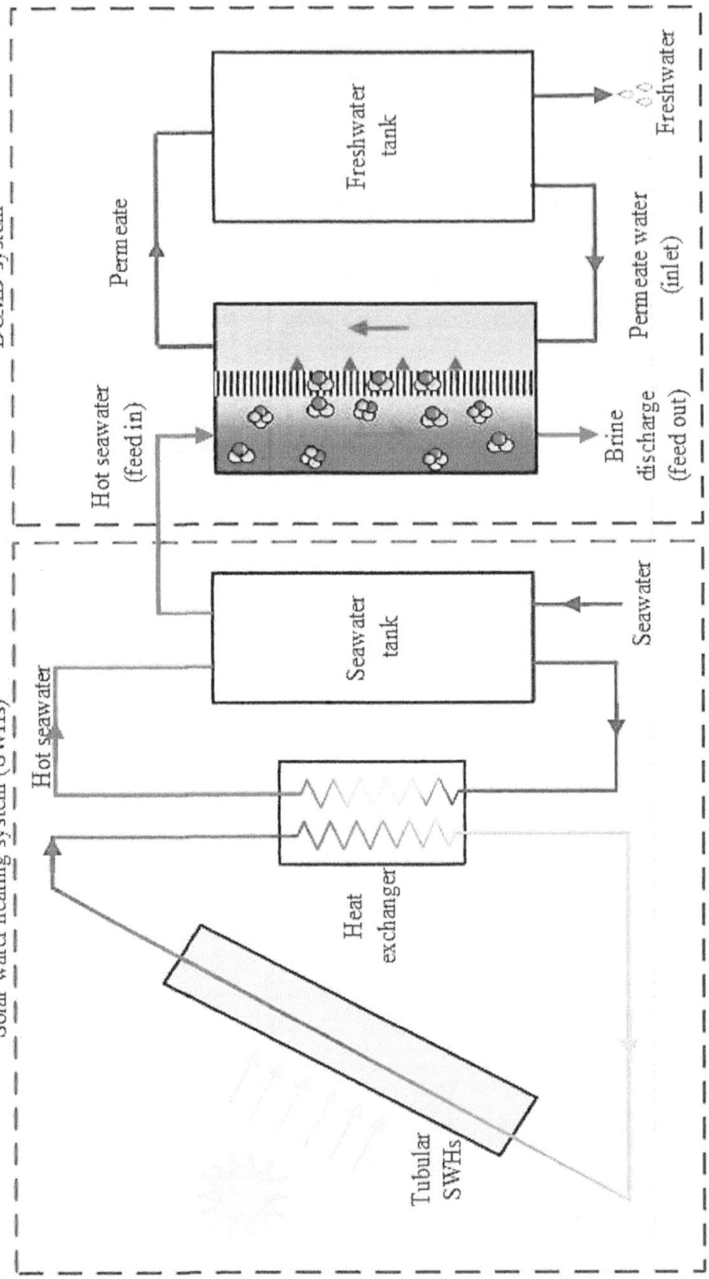

Figure 15.6 Solar-powered membrane distillation system [30].

Desalination engineering relies heavily on energy, with conventional methods emitting 25 kg of CO_2 per cubic meter of freshwater produced. To reduce carbon emissions, integrating renewable energy sources is crucial [31]. Consequently, there is an increasing trend towards using renewable energy in desalination processes to supplement traditional energy sources. Fane [32] concluded that desalination units release greenhouse gases, primarily CO_2, which has environmental impacts. Therefore, it is essential to decarbonize desalination techniques to reduce CO_2 emissions while meeting the water demand.

According to Ghaffour et al. [33], RO primarily relies on electricity, and while solar PV can supplement, it often cannot fully replace grid energy due to storage and variability issues. A review by Tong and Elimelech [34] emphasized that MD uses thermal energy, making it suitable for integration with solar thermal systems. This allows for the usage of low-grade and waste heat sources, enhancing sustainability. Lee et al. [35] discussed continuous improvements in RO membrane technology and energy recovery devices, which consume amplified efficiency and reduced operational costs. Lee et al. [35] also highlighted advancements in MD membrane materials, heat recovery methods, and system designs. Multi-stage MD and better pretreatment strategies are key areas of ongoing research.

Solar desalination presents a sustainable approach to freshwater production by harnessing solar energy for seawater desalination. Two main categories exist: direct and indirect methods. Direct methods, such as solar stills, offer simplicity and affordability but suffer from low production capacity [36][37]. Indirect methods couple solar energy collection (via photovoltaic panels or thermal collectors) with established purification methods like reverse osmosis (RO) or multi-stage flash distillation (MSF) to achieve larger-scale production. Recent research [38] investigated a novel wick-free design for direct solar desalination, utilizing natural convection to enhance efficiency. This ongoing exploration in solar desalination methods underscores the field's constant advancements.

Advancements in solar desalination technologies and the integration of MD with RO techniques are making significant strides towards sustainable solutions for water scarcity and efficient material development. Solar stills have improved efficiency for small-scale applications, while solar-powered RO and thermal desalination systems offer scalable solutions for larger projects. Integrating renewable energy sources into desalination reduces carbon emissions, with solar PV enhancing RO efficiency and solar thermal energy making MD more viable. The combination of MD and RO in computational research has led to significant advancements in optimizing system performance, discovering new materials, and improving drug design. These integrated approaches deepen our understanding of molecular systems and systematically enhance their performance, presenting a promising pathway for broader adoption of these technologies globally.

15.3 SOLAR PHOTOCATALYSIS FOR WATER PURIFICATION

Solar photocatalysis involves using solar energy to activate a photocatalyst, which then degrades organic pollutants in water. Titanium dioxide (TiO_2) is a commonly used photocatalyst. Recent advancements in nanotechnology have led to the development of more efficient photocatalysts. Studies have shown that solar photocatalysis is effective in removing a wide range of contaminants, including pharmaceuticals and pesticides, from water sources. Modern solar stills, with improved materials and design, provide a cost-effective solution for desalination in remote areas. The study noted that RO's dependence on grid energy remains a challenge, although solar PV can mitigate some of these issues [44].

15.4 WIND ENERGY

Wind energy is harnessed to pump water, particularly in remote or off-grid locations where electricity is unavailable or expensive.

Wind-powered water pumps are typically used for agricultural irrigation and livestock watering. A recent paper [45] discussed the viability of modern wind turbines for water pumping, emphasizing their reliability and low operational costs. They concluded that wind-powered water pumping is a cost-effective and sustainable solution for rural and remote areas, enhancing water accessibility and agricultural productivity.

Figure 15.7 illustrates that wind energy can support agricultural activities, such as crop irrigation and the operation of machinery on farms, by providing a renewable energy source. Wind energy is often used in water pumping systems, particularly in remote areas. This can include pumping water for agricultural irrigation, livestock, or for general water supply needs. Wind energy can drive heat pump systems to provide heating or cooling. These systems can be used in residential, commercial, or industrial settings to regulate temperatures efficiently. One of the most common uses of wind energy is generating electricity. Wind energy can be used in desalination processes, providing power to systems that convert seawater into potable water. This is especially useful in arid regions with limited freshwater resources. Similar to wind-powered heat pumps, this application utilizes wind energy to operate heat pumps for temperature regulation, offering benefits such as reduced energy costs and decreased reliance on fossil fuels.

15.5 HYDROPOWER

Micro-hydropower systems harness the energy of flowing or falling water on a small scale to generate electricity, which can be utilized to power local water treatment plants. These systems are particularly useful in mountainous

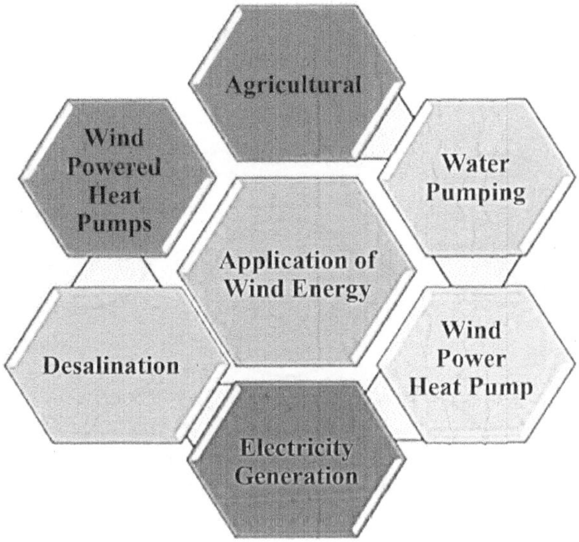

Figure 15.7 Wind energy applications.

or riverine regions. A recent study [46] explored the potential of micro-hydropower in decentralized energy production and its application in rural water treatment facilities. They concluded micro-hydropower systems provide a reliable and environmentally friendly energy source for small communities, enhancing local water treatment capabilities.

Hydropower is widely used for producing energy by harnessing the energy of flowing or falling water in dams and rivers (Figure 15.8). Excess electricity can be stored by pumping water to a higher elevation, which is later released to generate power during peak demand periods.

Hydropower provides a reliable backup to balance intermittent renewable sources like solar and wind, enhancing grid stability. Industries requiring consistent power, such as mining and manufacturing, often utilize hydropower to support their operations. Hydropower dams help in controlling water flow, which aids in flood prevention, irrigation, and water supply management.

15.6 BIOMASS ENERGY

Biomass energy can be used to power water treatment facilities. Research [47] has illustrated how biomass energy contributes to waste management and sustainable energy production (as shown in Figure 15.9). It was concluded that biomass energy not only helps in waste reduction but also

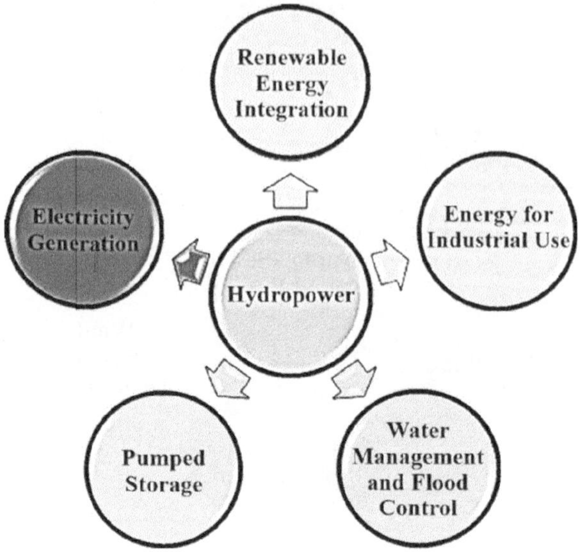

Figure 15.8 Hydropower energy applications.

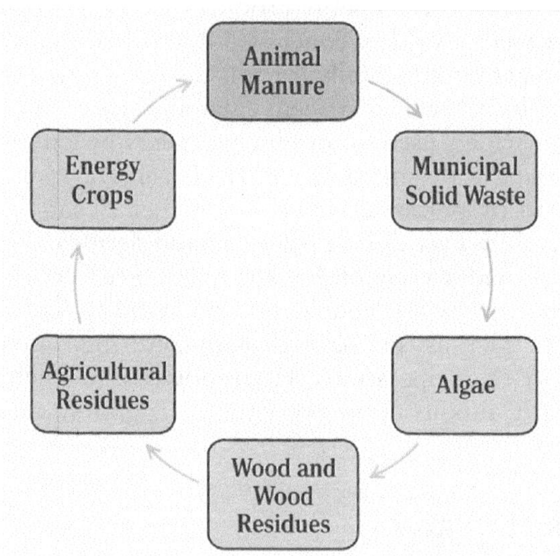

Figure 15.9 Biomass energy.

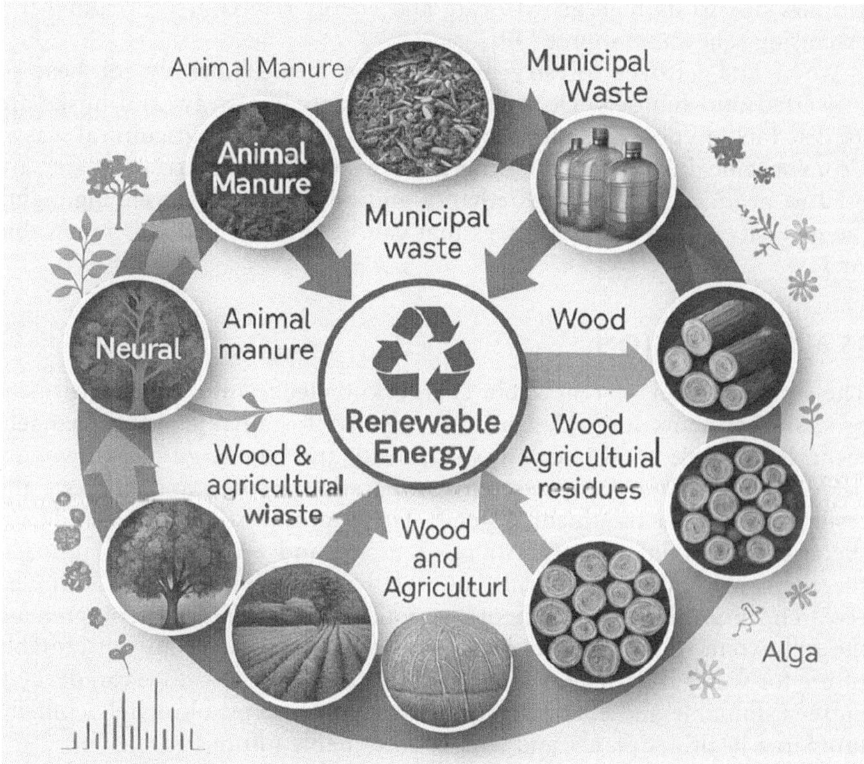

Figure 15.10 Geothermal energy.

provides a renewable energy source for water treatment, making it a dual-benefit solution for sustainable development.

15.7 GEOTHERMAL ENERGY

Geothermal energy utilizes the Earth's internal heat for various applications, including desalination. Geothermal-powered desalination systems can operate continuously and are especially viable in geothermal-rich regions. It has been concluded that geothermal-powered desalination is a sustainable and efficient method for water production in geothermal-active regions (as shown in Figure 15.10), with the added benefit of a low environmental footprint [48].

Animal waste products can be fragmented down anaerobically to produce biogas, which can be used for heating and electricity. Organic waste from households and businesses can be processed to produce biogas or can be burned for energy production. Algae can be cultivated and processed into biofuels due to its high growth rate and energy-rich content, making it a promising renewable source [49].

Wood and leftover forestry materials are burned directly for heat or converted into biofuels. Crop waste byproducts like straw and husks are used as fuel or converted into bioenergy, helping reduce agricultural waste. Crops specifically grown for bioenergy, like switch grass and sugarcane, are used to produce fuels such as ethanol and biodiesel. This cycle highlights the diversity of biomass resources that can be transformed into renewable energy.

15.8 CONCLUSION

The incorporation of renewable energy knowledge into water treatment processes presents a viable pathway toward sustainable water management. Renewable energy such as solar, wind, and bioenergy can knowingly decrease the carbon footprint, lower operational costs, and improve the resilience of water treatment systems. These technologies can also address the pressing challenges of resource scarcity and environmental impact, thereby contributing significantly to global sustainability goals. Continued research, innovation, and supportive policies are essential to comprehend the full extent of these technologies and ensure a resilient and sustainable future for water treatment. By leveraging renewable energy, we can develop more sustainable and efficient water treatment systems, ultimately contributing to a healthier planet and a more sustainable future.

Solar energy is effective for water purification processes like distillation and photocatalysis, reducing costs and environmental impact, though it faces technological and economic challenges. Wind energy offers a renewable alternative for water treatment in remote areas, requiring hybrid systems to address its intermittent nature. Bioenergy, derived from wastewater treatment, can lower energy costs and greenhouse gas emissions, though integration and optimization challenges persist. Reverse osmosis (RO) systems, widely used for desalination, benefit from renewable energy integration, significantly reducing carbon footprints and operational costs, yet require energy-efficient membranes and hybrid solutions for consistency. Hybrid systems combining various renewable sources offer enhanced reliability and efficiency. Future research should focus on technological advancements, economic viability, supportive policies, and public powered water treatment systems.

REFERENCES

1. Shatat, M., Worall, M., & Riffat, S. (2013). Opportunities for solar water desalination worldwide: Review. *Sustainable Cities and Society*, 9, 67–80.

2. Ghernaout, D. (2018). Increasing trends towards drinking water reclamation from treated wastewater. *World Journal of Applied Chemistry*, 3 (1), 1–9.
3. Hansen, J., Sato, M., Kharecha, P., von Schuckmann, K., Beerling, D. J., Cao, J., Marcott, S., Masson-Delmotte, V., Prather, M. J., Rohling, E. J., Shakun, J., Smith, P., Lacis, A., Russell, G., & Ruedy, R. (2017). Young people's burden: Requirement of negative CO2 emissions. *Earth System Dynamics*, 8, 577–616.
4. Meena, R. K., Singh, R. K., Singh, N. P., Meena, S. K., & Meena, V. S. (2015). Isolation of low-temperature surviving plant growth-promoting rhizobacteria (PGPR) from pea (Pisum sativum L.) and documentation of their plant growth-promoting traits. *Biocatalysts and Agricultural Biotechnology*, 4 (4), 1–7.
5. Calise, F., Dentice d'Accadia, M., & Vicidomini, M. (2019). Optimization and dynamic analysis of a novel polygeneration system producing heat, cool, and freshwater. *Renewable Energy*, 143, 1331–1347.
6. Maleki, A., Hemmati, V., Abazari, S. R., Aghsami, A., & Rabbani, M. (2024). Optimal distribution and waste management of COVID-19 vaccines from vaccination centers' satisfaction perspective – A fuzzy time window-based VRP. *Transportation Research Part E: Logistics and Transportation Review*, 183, 103454.
7. Rahim, A., Civelek, I., & Liang, F. H. (2018). A process model of social intelligence and problem-solving style for conflict management. *International Journal of Conflict Management*, 29 (4), 1–13.
8. Danehkar, S., & Yousefi, H. (2022). A comprehensive overview on water-based energy storage systems for solar applications. *Energy Reports*, 8, 8777–8797.
9. Patel, C., Solanki, K. L., Kumar, V., Chakradhar, V., & Panigrahi, S. K. (2024). IoT-driven monitoring and controlling to improve the performance of thermoelectric air conditioning systems. *Journal of Building Engineering*, 95, 11–81.
10. Garrido-Moreno, A., Martín-Rojas, R., & García-Morales, V. J. (2024). The key role of innovation and organizational resilience in improving business performance: A mixed-methods approach. *International Journal of Information Management*, 77, 102–777.
11. Esmaeilion, F. (2020). Hybrid renewable energy systems for desalination. *Applied Water Science*, 10 (84), 1–47.
12. Ghaffour, N., Soukane, S., Lee, J. G., Kim, Y., & Alpatova, A. (2019). Membrane distillation hybrids for water production and energy efficiency enhancement: A critical review. *Applied Energy*, 254, 1–33.
13. Hassan, Q., Algburi, S., Sameen, A. Z., Salman, H. M., & Jaszczur, M. (2023). A review of hybrid renewable energy systems: Solar and wind-powered solutions: Challenges, opportunities, and policy implications. *Results in Engineering*, 20, 11–21.
14. Salem, I., Saleh, Y., Alsayed, M. F., Assaf, R., Kanan, M., Musleh Al-Sartawi, A. M. A., & BinSaddig, R. (2023). Adoption of renewable energy sources and sustainable performance in Palestinian industrial and commercial sectors with governmental role as a moderator: An explanatory approach. *Journal of Open Innovation: Technology, Market, and Complexity*, 9, 30–39.

15. Millan, M., Rodrigo, M. A., Fernandez-Marchante, C. M., Cañizares, P., & Lobato, J. (2019). Powering with solar energy the anodic oxidation of wastewater polluted with pesticides. *ACS Sustainable Chemistry & Engineering*, 7, 8303–8309.
16. Achkari, O., & El Fadar, A. (2020). Latest developments on TES and CSP technologies–Energy and environmental issues, applications, and research trends. *Applied Thermal Engineering*, 167, 11–35.
17. Ullah, I., & Rasul, M. G. (2019). Recent developments in solar thermal desalination technologies: A review. *Energies*, 12, 1–19.
18. Xu, B., Guo, Y., & Yu, Q. (2023). Recent developments and challenges of solar stills: A comprehensive review. *Solar Energy*, 225, 283–295.
19. Kabeel, A. E., Omara, Z. M., & Essa, F. A. (2017). Review of researches and innovations on solar stills. *Desalination*, 429, 24–42.
20. Kim, Y., Choi, Y., & Hong, S. (2020). Techno-economic analysis of large-scale seawater reverse osmosis desalination plants with PV and wind renewable energy sources. *Desalination*, 451, 60–70.
21. Zhang, X., Li, H., & Wang, R. (2021). Advances in solar thermal desalination technologies: A review. *Renewable and Sustainable Energy Reviews*, 143, 11–33.
22. Cai, W., Zhang, H., & Wang, L. (2018). Recent advances in solar-driven multi-effect distillation desalination. *Desalination*, 447, 12–26.
23. Li, Z., Zhang, H., & Liu, S. (2022). Advances in hybrid solar desalination systems: A review. *Renewable Energy*, 184, 1427–1440.
24. Alpatova, A., Lasseux, D., & Escobar, R. (2019). Integration of renewable energy sources in desalination systems: A review. *Renewable and Sustainable Energy Reviews*, 113, 59–69.
25. Khalil, S. B., Aldossary, M., Ahmed, S., & Hussain, M. (2022). Advancements in solar still design for enhanced desalination efficiency. *Renewable Energy Journal*, 56 (4), 1023–1035.
26. Zhang, Z., Li, X., & Liu, M. (2023). Technological advancements in solar-powered reverse osmosis for desalination. *Desalination and Water Treatment*, 67 (2), 345–357.
27. Ghaffour, N., Missimer, T. M., & Amy, G. L. (2013). Technical review and evaluation of the economics of water desalination: Current and future challenges for better water supply sustainability. *Desalination*, 309, 197–207.
28. Elimelech, M., Zhangxin, W., Thomas, H., Anthony, P. S., Ngai, Y. Y., Deyu, L., & Shihong, L. (2019). Pathways and challenges for efficient solar-thermal desalination. *Science Advances*, 5, 754–763.
29. Gopi, G., Arthanareeswaran, G., & Ismail, A. F. (2019). Perspective of renewable desalination by using membrane distillation. *Chemical Engineering Research and Design*, 144, 520–537.
30. Soomro, M. I., Kumar, S., Ullah, A., Shar, M. A., & Alhazaa, A. (2022). Solar-powered direct contact membrane distillation system: Performance and water cost evaluation. *Sustainability*, 14 (30), 16616. MDPI.

31. Loutatidou, S., Mavukkandy, M. O., Chakraborty, S., & Arafat, H. A. (2017). Introduction: What is sustainable desalination? In H. A. Arafat (Ed.), *Desalination Sustainability: A Technical, Socioeconomic, and Environmental Approach*, Elsevier, Amsterdam, 1–29.
32. Fane, A. G. (2018). A grand challenge for membrane desalination: More water, less carbon. *Desalination*, 426, 155–163.
33. Ghaffour, N., Missimer, T. M., Amy, G. L., & Sinha, S. (2015). Renewable energy-driven desalination technologies: A comprehensive review on challenges and potential applications of integrated systems. *Desalination*, 356, 94–114.
34. Tong, T., & Elimelech, M. (2016). The global rise of zero liquid discharge for wastewater management: Drivers, technologies, and future directions. *Environmental Science & Technology*, 50 (13), 6846–6855.
35. Lee, K. P., Arnot, T. C., & Mattia, D. (2011). A review of reverse osmosis membrane materials for desalination—Development to date and future potential. *Journal of Membrane Science*, 370 (1–2), 1–22.
36. Curcio, E., & Drioli, E. (2005). Membrane distillation and related operations—A review. *Separation and Purification Reviews*, 34 (1), 35–86.
37. Al-Sahli, A. M., Hamad, D. A., Al-Shammari, A. M., & Al-Radwan, H. (2021). Solar Desalination. *Solar Desalination*, 45, 506–540.
38. Wahby, M. A., Ghazy, M., Hamid, S., & Omar, R. (2016). *Solar Desalination*. IntechOpen, 1–39.
39. Zhang, Y., Li, W., & Han, J. (2022). A wick-free solar still for efficient seawater desalination. *Desalination*, 522, 11–21.
40. Karplus, M., & Kuriyan, J. (2005). Molecular dynamics and protein function. *Proceedings of the National Academy of Sciences*, 102 (19), 6679–6685.
41. Deb, K., Pratap, A., Agarwal, S., & Meyarivan, T. (2002). A fast and elitist multiobjective genetic algorithm: NSGA-II. *IEEE Transactions on Evolutionary Computation*, 6 (2), 182–197.
42. Santos, L. F., Pinho, S. T., & Flores, L. (2021). Coupling molecular dynamics and response optimization to enhance catalytic process design. *Industrial & Engineering Chemistry Research*, 60 (12), 4567–4576.
43. Gupta, A., Patel, K., & Singh, R. (2022). Integration of molecular dynamics and response optimization for discovering materials with enhanced mechanical properties. *Journal of Computational Materials Science*, 18 (4), 1023–1035.
44. Yang, Y., Zhang, Y., & Wang, X. (2019). Integration of molecular dynamics and response optimization in drug discovery. *Bioinformatics and Drug Development*, 7 (2), 150–160.
45. Al-Karaghouli, A., & Kazmerski, L. L. (2013). Energy consumption and water production cost of conventional and renewable-energy-powered desalination processes. *Renewable and Sustainable Energy Reviews*, 24, 343–356.
46. Smith, J., & Brown, L. (2023). Modern wind-powered water pumping systems for remote areas. *Sustainable Energy Technologies and Assessments*, 42, 110–119.

47. Kumar, A., Singh, R., & Patel, M. (2023). The role of micro-hydropower systems in decentralized energy and water treatment. *Renewable and Sustainable Energy Reviews*, 162, 11–23.
48. Jones, P., Evans, R., & Brown, S. (2023). Biomass energy for sustainable water treatment: A dual-benefit approach. *Bioenergy Research*, 16 (1), 89–102.
49. Green, M., White, R., & Turner, S. (2023). Geothermal energy in desalination: Sustainable solutions for water scarcity. *Geothermics*, 98, 10–26.

Index